Agriculture Issues and Policies Series

AGRICULTURAL WASTES

Agriculture Issues and Policies Series

Agriculture Issues & Policies, Volume I
Alexander Berk (Editor)
2001. ISBN 1-56072-947-3

Agricultural Conservation
Anthony G. Hargis (Editor)
2009. ISBN 978-1-60692-273-6

Hired Farmworkers: Profile and Labor Issues
Rea S. Berube (Editor)
2009. ISBN 978-1-60741-232-8

Environmental Services and Agriculture
Karl T. Poston (Editor)
2009 ISBN: 978-1-60741-053-9

Weeds: Management, Economic Impacts and Biology
Rudolph V. Kingely (Editor)
2009 ISBN 978-1-60741-010-2

Effects of Liberalizing World Agricultural Trade
Henrik J. Ehrstrom (Editor)
2009 ISBN: 978-1-60741-198-7

Economic Impacts of Foreign-Source Animal Disease
Jace R. Corder (Editor)
2009. ISBN: 978-1-60741-601-2

Agricultural Wastes
Geoffrey S. Ashworth and Pablo Azevedo (Editors)
2009. ISBN: 978-1-60741-305-9

Agriculture Issues and Policies Series

AGRICULTURAL WASTES

GEOFFREY S. ASHWORTH
AND
PABLO AZEVEDO
EDITORS

Nova Science Publishers, Inc.
New York

Copyright © 2009 by Nova Science Publishers, Inc.

All rights reserved. No part of this book may be reproduced, stored in a retrieval system or transmitted in any form or by any means: electronic, electrostatic, magnetic, tape, mechanical photocopying, recording or otherwise without the written permission of the Publisher.

For permission to use material from this book please contact us:
Telephone 631-231-7269; Fax 631-231-8175
Web Site: http://www.novapublishers.com

NOTICE TO THE READER

The Publisher has taken reasonable care in the preparation of this book, but makes no expressed or implied warranty of any kind and assumes no responsibility for any errors or omissions. No liability is assumed for incidental or consequential damages in connection with or arising out of information contained in this book. The Publisher shall not be liable for any special, consequential, or exemplary damages resulting, in whole or in part, from the readers' use of, or reliance upon, this material. Any parts of this book based on government reports are so indicated and copyright is claimed for those parts to the extent applicable to compilations of such works.

Independent verification should be sought for any data, advice or recommendations contained in this book. In addition, no responsibility is assumed by the publisher for any injury and/or damage to persons or property arising from any methods, products, instructions, ideas or otherwise contained in this publication.

This publication is designed to provide accurate and authoritative information with regard to the subject matter covered herein. It is sold with the clear understanding that the Publisher is not engaged in rendering legal or any other professional services. If legal or any other expert assistance is required, the services of a competent person should be sought. FROM A DECLARATION OF PARTICIPANTS JOINTLY ADOPTED BY A COMMITTEE OF THE AMERICAN BAR ASSOCIATION AND A COMMITTEE OF PUBLISHERS.

LIBRARY OF CONGRESS CATALOGING-IN-PUBLICATION DATA

Agricultural wastes / edited by Geoffrey S. Ashworth and Pablo Azevedo.
   p. cm.
  Includes index.
  ISBN 978-1-60741-305-9 (hardcover)
  1. Agricultural wastes--Recycling. I. Ashworth, Geoffrey S. II. Azevedo, Pablo.
  TD930.A3925   2009
  628'.746--dc22
                                                                           2009015645

Published by Nova Science Publishers, Inc. ✦ New York

CONTENTS

Preface vii

Chapter 1 Potential Risk and Environmental Benefits of Waste Derived from Animal Agriculture 1
Ajit K Sarmah

Chapter 2 Animal Waste and Hazardous Substances: Current Laws and Legislative Issues 19
Claudia Copeland

Chapter 3 Reprocessing and Protein Enrichment of Agricultural Wastes by Thermophilic Aerobic Digestion (TAD) 29
J. Obeta Ugwuanyi

Chapter 4 Fly Ash Use in Agriculture: A Perspective 77
Wasim Aktar

Chapter 5 Pesticides: Use, Impact and Regulations for Management 93
Vandita Sinha, Vartika Rai and P.K.Tandon

Chapter 6 Carbonization of Rice Husk to Remove Offensive Odor from Livestock Waste and Compost 109
Seiji Kumagai, Koichi Sasaki and and Yukio Enda

Chapter 7 Bio (Single Cell) Protein: Issues of Production, Toxins and Commercialisation Status 129
Ravinder Rudravaram, Anuj Kumar Chandel, Linga Venkateswar Rao, Yim Zhi Hui and Pogaku Ravindra

Chapter 8 Coffee Processing Solid Wastes: Current Uses and Future Perspectives 155
Adriana S. Franca and Leandro S. Oliveira

Chapter 9 Vermi-Conversion of Industrial Sludge in Conjunction with Agricultural Farm Wastes: A Viable Option to Minimize Landfill Disposal? 191
Deepanjan Majumdar

Chapter 10	Agricultural Wastes as Building Materials: Properties, Performance and Applications *José A. Rabi, Sérgio F. Santos, Gustavo H. D. Tonoli and Holmer Savastano Jr.*	**219**
Chapter 11	From Solid Biowastes to Liquid Biofuels *Leandro S. Oliveira and Adriana S. Franca*	**265**
Index		**291**

PREFACE

Agricultural waste, which includes both natural (organic) and non-natural wastes, is a general term used to describe waste produced on a farm through various farming activities. These activities can include but are not limited to dairy farming, horticulture, seed growing, livestock breeding, grazing land, market gardens, nursery plots, and even woodlands. Agricultural and food industry residues, refuse and wastes constitute a significant proportion of world wide agricultural productivity. It has variously been estimated that these wastes can account for over 30% of world wide agricultural productivity. The boundaries to accommodate agricultural waste derived from animal agriculture and farming activities are identified in this book. Examples will be provided of how animal agriculture and various practices adopted at farm-scale impact on the environment. When discharged to the environment, agricultural wastes can be both beneficial and detrimental to living matter and the book will therefore also address the pros and cons of waste derived from animal agriculture in today's environment. Given agricultural wastes are not restricted to a particular location, but rather are distributed widely, their effect on natural resources such as surface and ground waters, soil and crops, as well as human health, will also be addressed.

Chapter 1 - Agricultural waste, which includes both natural (organic) and non-natural wastes, is a general term used to describe waste produced on a farm through various farming activities. These activities can include but are not limited to dairy farming, horticulture, seed growing, livestock breeding, grazing land, market gardens, nursery plots, and even woodlands that are used as ancillary to the use of the land for other agricultural purposes. Given 'agricultural waste' encompasses such a broad class of biodegradable and non-biodegradable components, the focus of this chapter is first to identify and narrow down the boundaries to accommodate agricultural waste derived from animal agriculture and farming activities. Examples are provided of how animal agriculture and various practices adopted at farm-scale impact on the environment. When discharged to the environment, agricultural wastes can be both beneficial and detrimental to living matter, and the review therefore also addresses the pros and cons of waste derived from animal agriculture in today's environment. Given agricultural wastes are not restricted to a particular location, but rather are distributed widely, their effect on natural resources such as surface and ground waters, soil and crops, as well as human health, are addressed.

Chapter 2 - The animal sector of agriculture has undergone major changes in the last several decades: organizational changes within the industry to enhance economic efficiency have resulted in larger confined production facilities that often are geographically concentrated. These changes, in turn, have given rise to concerns over the management of animal wastes and potential impacts on environmental quality.

Federal environmental law does not regulate all agricultural activities, but certain large animal feeding operations (AFOs) where animals are housed and raised in confinement are subject to regulation. The issue of applicability of these laws to livestock and poultry operations — especially the Comprehensive Environmental Response, Compensation, and Liability Act (CERCLA, the Superfund law) and the Emergency Planning and Community Right-to-Know Act (EPCRA) — has been controversial and recently has drawn congressional attention.

Both Superfund and EPCRA have reporting requirements that are triggered when specified quantities of certain substances are released to the environment. In addition, Superfund authorizes federal cleanup of releases of hazardous substances, pollutants, or contaminants and imposes strict liability for cleanup and injuries to natural resources from releases of hazardous substances.

Superfund and EPCRA include citizen suit provisions that have been used to sue poultry producers and swine operations for violations of those laws. In two cases, environmental advocates claimed that AFO operators had failed to report ammonia emissions, in violation of Superfund and EPCRA. In both cases, federal courts supported broad interpretation of key terms defining applicability of the laws' reporting requirements. Three other cases in federal courts, while not specifically dealing with reporting violations, also have attracted attention, in part because they have raised the question of whether animal wastes that contain phosphorus are hazardous substances that can create cleanup and natural resource damage liability under Superfund. Two of these latter cases were settled; the third, brought by the Oklahoma Attorney General against poultry operations in Arkansas, is pending.

These lawsuits testing the applicability of Superfund and EPCRA to poultry and livestock operations have led to congressional interest in these issues. In the 109th Congress, legislation was introduced that would have amended CERCLA to clarify that manure is not a hazardous substance, pollutant, or contaminant under that act and that the laws' notification requirements would not apply to releases of manure. Proponents argued that Congress did not intend that either of these laws apply to agriculture and that enforcement and regulatory mechanisms under other laws are adequate to address environmental releases from animal agriculture. Opponents responded that enacting an exemption would severely hamper the ability of government and citizens to know about and respond to releases of hazardous substances caused by an animal agriculture operation. Congress did not act on the legislation, but similar bills have been introduced in the 110th Congress (H.R. 1398 and S. 807).

Agricultural and food industry residues, refuse and wastes constitute a significant proportion of worldwide agricultural productivity. It has variously been estimated that these wastes can account for over 30% of worldwide agricultural productivity. These wastes include lignocellulosic materials, fruit, vegetables, root and tuber wastes, sugar industry wastes as well as animal/livestock and fisheries operations wastes. They represent valuable biomass and potential solutions to problems of animal nutrition and worldwide supply of protein and calories if appropriate technologies can be deployed for their valorization. Moreover, reutilization of these vast wastes should help to address growing global demands for environmentally sustainable methods of production and pollution control.

Various technologies are potentially available for the valorization of these wastes. In addition to conventional waste management processes, other processes that may be used for the reprocessing of wastes include solid substrate fermentation, ensiling and high solid or slurry processes. In particular, the use of slurry processes in the form of (Autothermal) Thermophilic Aerobic Digestion (ATAD or TAD) or liquid composting is gaining prominence in the reprocessing of a variety of agricultural wastes because of its potential advantages over conventional waste reprocessing technologies. These advantages include the capacity to achieve rapid, cost-effective waste stabilization and pasteurization and protein enrichment of wastes for animal feed use.

TAD is a low technology capable of self heating and is particularly suited for use with wastes being considered for upgrading and recycling as animal feed supplement, as is currently the case with a variety of agricultural wastes. It is particularly suited for wastes generated as slurries, at high temperature or other high COD wastes. Reprocessing of a variety of agricultural wastes by TAD has been shown to result in very significant protein accretion and effective conversion of mineral nitrogen supplement to high-value feed grade microbial/single-cell protein for use in animal nutrition. The application of this technology in reprocessing of wastes will need to take account of the peculiarities of individual wastes and the environment in which they are generated, reprocessed and used. The use of thermopiles in the process has significant safety benefits and may be optimized to enhance user confidence and acceptability.

Chapter 4 - Fly ash has a potential in agriculture and related applications. Physically, fly ash occurs as very fine particles, having an average diameter of <10 mm, low- to medium-bulk density, high surface area and very light texture. Chemically, the composition of fly ash varies depending on the quality of coal used and the operating conditions of the thermal power stations. On average, approximately 95 to 99% of fly ash consists of oxides of Si, Al, Fe and Ca, and about 0.5 to 3.5% consists of Na, P, K and S. The remainder of the ash is composed of trace elements. In fact, fly ash consists of practically all of the elements present in soil except organic carbon and nitrogen (Table 1). Thus, it was discovered that this material could be used as an additive or amendment material in agricultural applications.

In view of the above, some agencies, individuals, and institutes at various locations conducted some preliminary studies on the effect and feasibility of fly ash as an input material in agricultural applications. Some amount of experience was gained in the country and abroad regarding the effect of fly ash utilisation in agriculture and related applications.

Chapter 5 - Modern farming employs many chemicals to produce and preserve large quantities of high-quality food. Fertilizers, pesticides, cleaners and crop preservatives are the major categories that are now abundantly used in agriculture for increasing production. But each of these chemicals poses a hazard—most of the pesticides are degraded very slowly by atmospheric and biological factors, leading to the development of resistant strains of pests, contamination of the environment and food chain, thereby causing serious ecological imbalance. However, in many countries, a range of pesticides has been banned or withdrawn for health or environmental reasons, and their residues are still detected in various substances such as food grains, fodder, milk, etc. The majority of chemical insecticides consist of an active ingredient (the actual poison) and a variety of additives that improve efficacy of their application and action. All of these formulations degrade over time. The chemical by-products that form as the pesticide deteriorates can be even more toxic than the original product.

Often stockpiles of pesticides are poorly stored and toxic chemicals leak into the environment, turning potentially fertile soil into hazardous waste. Once a pesticide enters soil, it spreads at a rate that depends on the type of soil and pesticide, moisture and organic matter content of the soil and other factors. A relatively small amount of spilled pesticide can, therefore, create a much larger volume of contaminated soil. The International Code of Conduct on the Distribution and Use of Pesticides states that packaging or repackaging of pesticides should be done only on licensed premises where staff is adequately protected against toxic hazards. Now, many agencies have come forward to prevent the contamination and accumulation of pesticides in the environment—for example, the issuing of the International Code of Conduct on the Distribution and Use of Pesticides by the United Nations Food and Agriculture Organization (FAO). In addition, the organization works to improve pesticide regulation and management in developing countries. In order to prevent accumulation of pesticides, the WHO works to raise awareness among regulatory authorities and helps to ensure that good regulatory and management systems for the health sector are in place. The United Nations Industrial Development Organization (UNIDO) is supporting cleaner and safer pesticide production with moves toward less hazardous products based on botanical or biological agents. Wider use of these products will result in reductions in the imported chemicals that contribute to obsolete pesticide stockpiles. The World Bank has established a binding safeguard policy on pest management that stipulates that its financed projects involving pest management follow an Integrated Pest Management (IPM) approach.

Chapter 6 - An attempt was made to convert the agricultural waste of rice husk (RH) into an adsorbent to remove the offensive odor released from livestock waste and compost. The ammonia gas adsorption of the RH carbonized at 400°C was much faster than those of several commercial deodorants as well as those of carbonized wood wastes. Acidic functional groups remaining at 400°C were useful to promote adsorption of basic ammonia gas. The actual compost was covered with or mixed with the RH carbonized at 400°C. The covering method reduced the concentration of ammonia gas emitted from the compost much faster than the mixing method, which was connected to volatilization of ammonia gas lighter than ambient air. Wetting the carbonized RH was also effective in reducing the ammonia gas concentration. An assorted feed to which was added the RH carbonized at 400°C at the level of 2 mass% was given to growing pigs. The addition of the carbonized RH reduced about 80% of the concentrations of hydrogen sulfide and mercaptans emitted from the pig dung. The removal of acidic gases of hydrogen sulfide and mercaptans was suggested to result from basic inorganic matter of K, Ca and P, which were intrinsically composed in RH. The testing results showed that the RH carbonized at 400°C was a promising material for removing the offensive odor produced by the livestock industry.

Chapter 7 - The alarming rate of population growth and a regular depletion in food production and food resources are important factors in the present dire need to find new viable options for food and feed sources. Based on scientific developments, particularly in industrial microbiology, one feasible solution could be the consumption of microorganisms as human food and animal feed supplements. Humans have used microbial-based products—like alcoholic beverages, curd, cheese, yogurt, and soya—even before the beginning of civilization. Due to research developments in the scientific arena in the last two decades, (Bio) single cell protein (SCP) has drawn new attention towards its use as supplement in human food, animal feed or staple diet. There are several benefits to using SCP as food or feed, viz. its rapid growth rate and high protein content. The microorganisms involved as SCP have the ability to utilize cheap and plentiful available feedstock for their growth and energy, making them an attractive option. However, in spite of laboratory-based success stories, only a limited number of commercial SCP production plants have been seen worldwide. This review analyzes the possibility of SCP production, various raw materials for its production, available microorganisms with cultivation methods, toxicity assessment and their removal. Also, new developments and risk assessment using SCP along with worldwide industrial SCP production are discussed.

Chapter 8 - The term "coffee" is applied to a wide range of coffee processing products, starting from the freshly harvested fruit (coffee cherries), to the separated green beans, to the product of consumption (ground roasted coffee or soluble coffee). Coffee processing can be divided into two major stages: primary processing, in which the coffee fruits are de-hulled and submitted to drying, the resulting product being the green coffee beans. This is the main product of international coffee trade, and Brazil is the largest producer in the world with production values ranging from 2 to 3 million tons in the years from 2003 to 2007. During this primary processing stage solid wastes are generated, which include coffee husks and pulps, and low-quality or defective coffee beans. Secondary processing includes the stages that comprise the production of roasted coffee and soluble coffee. The major solid residue generated in this stage corresponds to spent coffee grounds from soluble coffee production. These solid residues (coffee husks, defective coffee beans and spent coffee grounds) pose several problems in terms of adequate disposal, given the high amounts generated, environmental concerns and specific problems associated with each type of residue. Coffee husks, comprised of dry outer skin, pulp and parchment, are probably the major residues from the handling and processing of coffee, since for every kg of coffee beans produced, approximately 1 kg of husks are generated during dry processing. Defective beans correspond to over 50% of the coffee consumed in Brazil, being used by the roasting industries in blends with good-quality coffee. Unfortunately, since to coffee producers they represent an investment in growing, harvesting, and handling, they will continue to be dumped in the internal market in Brazil, unless alternative uses are sought and implemented. Spent coffee grounds are produced at a proportion of 1.5kg (25% moisture) for each kg of soluble coffee. This solid residue presents an additional disposal problem, given that it can be used for adulteration of roasted and ground coffee, being practically impossible to detect. In view of the aforementioned, the objective of the present study was to present a review of the works of research that have been developed in order to find alternative uses for coffee processing solid residues. Applications include direct use as fuel in farms, animal feed, fermentation studies, adsorption studies, biodiesel production and others. In conclusion, a discussion on the advantages and disadvantages of each proposed application is presented, together with suggestions for future studies and applications.

Chapter 9 - There are environmental concerns associated with industrial sludge disposal, apart from other issues like logistics of disposal, treatment options, cost of disposal, etc. A customary disposal option for many industries is secure landfilling, but more and more industries are now looking at the possibility of recycling and bioconversion of the solid wastes to value-added products. Agro-based industries have often resorted to composting, vermicomposting or biogas generation from their wastes due to their biological substrate value and negligible toxicity. However, this has not been the case with other types of industries like pharmaceuticals, chemicals/petrochemicals, power plants, iron and steel and many others, where the sludge is may be unsuitable due to the presence of harmful chemicals, volatiles, persistent organic pollutants (POPs), antibiotics, etc. Sludge generated from water treatment plants in the industrial sector forms a major portion of solid waste requiring disposal, and has been used in some reported cases for culturing earthworms and vermicomposting and could be explored for vermicomposting on case-by-case basis. An acceptable approach would be an initial evaluation of the sludge for screening of known harmful agents and factors to earthworms and then conducting proxy vermicomposting trials on these sludges with prior addition of known substrates of earthworms, such as cured animal manures or crop residues or a combination of both. The quality of the final product—or vermicompost—holds great importance, as the end product may not qualify as good manure. But, it is still not clear as to how one could solve the entire sludge disposal problem only by vermicomposting, as it is time-consuming and industries generate sizeable quantities of sludge every day. It appears that vermicomposting could only supplement the normal disposal practices of an industry. This chapter attempts to shortlist the suitable industrial and agricultural wastes for vermiconversion, explores the feasibility of their vermiconversion, and looks at various factors influencing the possible implementation of such a practice in industry.

Chapter 10 - While recycling of low added-value residual materials constitutes a present day challenge in many engineering branches, attention has been given to low-cost building materials with similar constructive features as those presented by materials traditionally employed in civil engineering. Bearing in mind their properties and performance, this chapter addresses prospective applications of some elected agroindustrial residues or by-products as non-conventional building materials as means to reduce dwelling costs.

Such is the case concerning blast furnace slag (BFS), a glassy granulated material regarded as a by-product from pig-iron manufacturing. Besides some form of activation, BFS requires grinding to fineness similar to commercial ordinary Portland cement (OPC) in order to be utilized as hydraulic binder. BFS hydration occurs very slowly at ambient temperatures while chemical or thermal activation (singly or in tandem) is required to promote acceptable dissolution rates. Fibrous wastes originated from sisal-banana agroindustry as well from eucalyptus cellulose pulp mills have been evaluated as raw materials for reinforcement of alternative cementitious matrices, based on ground BFS.

Production and appropriation of cellulose pulps from collected residues can considerably increase the reinforcement capacity by means of vegetable fibers. Composite preparation follows a conventional dough mixing method, ordinary vibration, and cure under saturated-air condition. Exposition of such components under ambient conditions leads to a significant long-term decay of mechanical properties while micro-structural analysis has identified degradation mechanisms of fibers as well as their petrifaction. Nevertheless, these materials can be used indoors and their physical and mechanical properties are discussed aiming at preparing panel products suitable for housing construction whereas results obtained thus far have pointed to their potential as low-cost construction materials.

On its turn, phosphogypsum rejected from phosphate fertilizer industries is another by-product with little economic value up to now. Phosphogypsum may replace ordinary gypsum provided that radiological concerns about its handling have been properly overcome as it exhales ^{222}Rn (a gaseous radionuclide whose indoor concentration should be limited and monitored). Some phosphogypsum properties of interest (e.g., bulk density, consistency, setting time, free and crystallization water content, and modulus of rupture) have suggested its large-scale exploitation as surrogate building material.

Chapter 11 - In a time when a foreseeable complete transmutation from a petroleum-based economy to a bio-based global economy finds itself in its early infancy, agricultural wastes, in the majority currently seen as low-valued materials, are already beginning their own transformation from high-volume waste disposal environmental problems to constituting natural resources for the production of a variety of eco-friendly and sustainable products, with second generation liquid biofuels being the leading ones. Agricultural wastes contain high levels of cellulose, hemicellulose, starch, proteins, and, some of them, also lipids, and as such constitute inexpensive candidates for the biotechnological production of liquid biofuels (e.g., bioethanol, biodiesel, dimethyl ether and dimethyl furan) without competing directly with the ever-growing need for world food supply. Since agricultural wastes are generated in large scales (in the range of billions of kilograms per year), thus being largely available and rather inexpensive, these materials have been considered potential sources for the production of biofuels for quite some time and have been thoroughly studied as such. In the last decades, a significant amount of information has been published on the potentiality of agricultural wastes to be suitably processed into biofuels, with bioethanol as the main research subject. Thus, it is the aim of this chapter to critically analyze the current situation and future needs for technological developments in the area of producing liquid biofuels from solid biowastes. The state-of-the-art in producing bioethanol, bio-oil and biodiesel from agricultural wastes will be discussed together with the new trends in the area. The emerging biowaste-based liquid biofuels (e.g., biogasoline, dimethyl ether and dimethyl furan) currently being studied will also be discussed.

In: Agricultural Wastes
Eds: Geoffrey S. Ashworth and Pablo Azevedo

ISBN 978-1-60741-305-9
© 2009 Nova Science Publishers, Inc.

Chapter 1

POTENTIAL RISK AND ENVIRONMENTAL BENEFITS OF WASTE DERIVED FROM ANIMAL AGRICULTURE

Ajit K Sarmah[*]

Soil Chemical and Biological Interactions, Soils and Landscape Team,
Landcare Research New Zealand, Private Bag 3127, Hamilton 3240, New Zealand

ABSTRACT

Agricultural waste, which includes both natural (organic) and non-natural wastes, is a general term used to describe waste produced on a farm through various farming activities. These activities can include but are not limited to dairy farming, horticulture, seed growing, livestock breeding, grazing land, market gardens, nursery plots, and even woodlands that are used as ancillary to the use of the land for other agricultural purposes. Given 'agricultural waste' encompasses such a broad class of biodegradable and non-biodegradable components, the focus of this chapter is first to identify and narrow down the boundaries to accommodate agricultural waste derived from animal agriculture and farming activities. Examples are provided of how animal agriculture and various practices adopted at farm-scale impact on the environment. When discharged to the environment, agricultural wastes can be both beneficial and detrimental to living matter, and the review therefore also addresses the pros and cons of waste derived from animal agriculture in today's environment. Given agricultural wastes are not restricted to a particular location, but rather are distributed widely, their effect on natural resources such as surface and ground waters, soil and crops, as well as human health, are addressed.

INTRODUCTION

Agricultural waste can be defined as the residues from the growing and first processing of raw agricultural products such as fruits, vegetables, meat, poultry, dairy products, and crops. Although agricultural waste is a general term used to describe waste that is produced on a farm through various farming activities, these activities can include other activities such as seed

---
[*] E-mail: sarmahA@LandcareResearch.co.nz.

growing, nursery plots, and even woodlands that are used as ancillary to the use of the land for other agricultural purposes. Agricultural wastes can be in the form of solid, liquid or slurries, depending on the nature of agricultural activities in a farm or agricultural field, and can be both natural (organic) and non-natural wastes.

Although the quantity of wastes produced by the agricultural sector is significantly low compared with wastes generated by various industries, the pollution potential of agricultural wastes is high on a long-term basis. For example, the land spreading of manures and slurries can cause nutrient and organic pollution of soils and waters. Given animal excreta also contains a plethora of organic chemicals, and pathogens, the risk of surface and groundwater contamination as a result of waste being applied onto the soil can be high. Agricultural waste encompasses a broad class of biodegradable and non-biodegradable components, and the major components of agricultural solid wastes (Table 1) are biodegradable organics. These are unlikely to result in hazardous conditions except when there are inadequate oxygen resources to assimilate the wastes. When this occurs in streams, inadequate dissolved oxygen and high concentrations of ammonia can cause fish kills.

Throughout many developed and developing countries, large quantities of food and crop processing, forestry, and animal solid and liquid wastes are generated each year. Although the majority of these wastes are readily biodegradable, they also contain significant quantities of nutrients (e.g., nitrogen, phosphorous), human and animal pathogens, and various medicinal products and feed additives used in livestock operation. Excretion by grazing livestock and application of effluent onto land as a supplement to fertiliser can cause damage to the receiving environment through surface run-off and/or leaching of many of these pollutants associated with agricultural waste.

Given the broad connotations associated with the term 'agricultural waste', the focus in this chapter is placed mainly on those agricultural wastes derived through animal farming practices and on the environmental implications of applying the waste onto land. In addition, utilization of agricultural waste to obtain environmental and economic benefits is also briefly discussed.

Definition of Agricultural Waste

A general definition of 'agricultural waste' is not available in the literature. According to the United States Environmental Protection Agency (USEPA), agricultural waste is the by-products generated by the rearing of animals and the production and harvest of crops or trees. Animal waste, a large component of agricultural waste, includes waste (e.g., feed waste, bedding and litter, and feedlot and paddock runoff) from livestock, dairy, and other animal-related agricultural and farming practices. Organisation for Economic Cooperation and Development (OECD) defines 'agricultural waste' as waste produced as a result of various agricultural operations including manure and other wastes from farms, poultry houses and slaughterhouses; harvest waste; fertilizer run-off from fields; pesticides that enter into water, air or soils; and salt and silt drained from fields. In the context of this chapter, agricultural waste is defined as waste in the form of the crop residues in the farm, manure from livestock operations, including dairy and piggery effluent, and poultry litter.

Table 1. Characteristics of agricultural solid waste (Source: Loehr 1978)

Agricultural activity	Types of waste	Method of disposal
Crop production and harvest	Straw, stover	Land-application, burning, plowing under
Fruit and vegetable processing	Biological sludges, trimmings, peels, leaves and stems, soil, seeds and pits	Landfill, animal feed, land application, burning
Sugar processing	Biological sludges, bagasse, pulp, lime mud, filter mud	Landfill, burning, composting animal feed
Animal production	Blood, bones, feather, litter, manures, liquid effluent	Land-application, fertiliser
Dairy product processing	Biological sludges	Landfill, land spreading
Leather tanning	Fleshings, hair, raw and tanned hid trimmings, lime and chrome sludge, grease	Rendering, by-product recovery, landfill, land-spreading
Rice production	Bran, straw, hull	Feeds, mulch/soil conditioner, packaging material for glass and ceramics
Coconut production	Stover, cobs, husk trunk, leaves, coco meal	Feeds, vinegar, activated carbon, coir products

POTENTIAL HAZARDS ASSOCIATED WITH AGRICULTURAL WASTE

Given the sources of agricultural wastes are diverse, agricultural solid wastes can often be potentially hazardous and detrimental to the terrestrial and aquatic eco-systems. Uncontrolled and improper handling can often lead to many situations where agricultural waste can become an environmental issue. The following sections discuss some of these issues and provide a perspective on the agricultural waste as potential hazards.

Over-application of agricultural waste in the form manure to crop land and pasture can result in a decrease in crop production due to inhibitory amounts of ammonia of nitrite nitrogen (NO_2-N) or salts in the soil. Application of dairy effluent or feedlot manure to permeable loam and clay loams soils can also reduce the permeability of these soils and thus adversely affecting the crop growth. Excess loadings of nitrogen and phosphorus from agricultural waste applied to land that cause eutrophication of water bodies or contamination of drinking water have been well documented in the literature (Sharpley et al. 1984; Sharpley and Halvorson 1994; Anderson et al. 2002). Apart from excessive nitrogen, phosphorus and other nutrients, salts, pathogens, livestock waste also contain significant amounts of steroid hormones (naturally released by animals of all species in urine and faeces) and their metabolites. Veterinary antibiotics that are fed to the animals during their life period are also excreted in the faeces and urine, which eventually end up in the oxidation effluent pond and finally onto the land as supplement to fertiliser or as disposal option (Boxall et al. 2004). The newly emerging group of chemicals such as steroid hormones and a range of veterinary antibiotics have been detected in various environmental media (e.g., soil, water, manure, sludge) across the globe and concerns are growing because of their potential impact on the terrestrial and aquatic eco-systems (Sarmah et al. 2006). Steroid hormones such as 17β-estradiol (E2) and its metabolite estrone (E1) are known endocrine disrupting chemicals (EDCs) capable of causing adverse effects on terrestrial and aquatic organisms (Jobling et al. 1998).

Natural Steroid Hormones

Livsetsock wastes can be termed as natural agricultural waste and can act as potential sources of EDCs (Figure 1) to the environment. Compounds such as E2, its metabolites E1 and estriol (E3), and male androgen testosterone and its derivative androstenedione are excreted by animals of all species. These hormones are present in faeces and/or urine and reach the environment through the release of animal wastes to receiving waters, animal waste application onto land, or by direct excretal input while grazing (Sarmah et al. 2006). While all species of farm animals excrete these hormones, different species excrete different types and proportions of estrogens. For example, in cattle (Bos Taurus) ≥ 90% of estrogens are excreted as 17α-estradiol, 17β-estradiol, and estriol as free and conjugated metabolites, while pigs (Sus scrofa) and poultry (Gallus domesticus) excrete estradiol, estriol and estrone, and their respective conjugates (Erb et al. 1977). The amounts of hormones excreted by various animals depend on their reproductive stage and the route of excretion, fecal or urinary (Hanselman et al., 2003). Table 2 summarizes the estimated amount of EDCs excreted by a dairy cow of typical weight of 640 kg at different stages in its reproductive cycle.

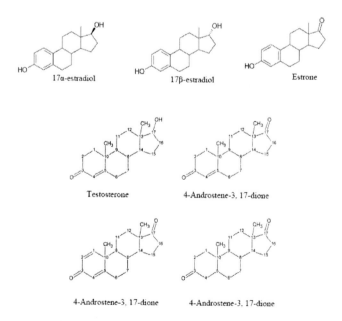

Figure 1. Molecular structures of steroid hormones and their degradation products. The numbers represent carbon position.

Dairy cattle and other animals excrete steroid hormones as free and conjugated compounds. Conjugation of steroid hormones to their respective glucuronides (addition of glucose) and/or sulphate form (by addition of sulphate) neutralises their activity and enhances their removal from the bloodstream via the kidneys. To date, very limited data are available on the proportion of steroid hormones excreted by dairy cattle as free and conjugated metabolites, and how this may be affected by their oestrous cycle, pregnancy, and lactation. It is generally accepted that conjugated forms of steroid hormones are rapidly converted to their respective free and active forms on excretion and release into receiving environments. Many micro-organisms have the ability to rapidly de-conjugate a wide range of organic compounds, and it is reasonable to assume this occurs for steroid hormones. However, studies investigating these processes are limited due to the significant analytical challenges inherent in the analysis of conjugated steroid hormones.

The concentration of 17β-estradiol in various dairy wastes ranges from below detectable limits (BDL) to hundreds of μg/kg, and is comparable to levels measured in other animal wastes (Table 2). Typical levels of steroid hormones in waste from dairy cows are reported as 39 μg/kg of estrone and about 18 μg/kg of 17β-estradiol. Typical slurry pit concentrations of 1.5 μg/L were obtained for 17β-estradiol, while the concentration of estrone was 3-fold higher (Raman et al. 2004).

Table 2. Estimated rates of fecal and urinary estrogen excretion from dairy cattle (Source: cited by Hanselman et al. 2003)

Reproductive stage	No of animals	Excretion rate/ 1000 kg LAM[a] (μg/d)	Measured estrogens
Non-pregnant	21	600 ± 200	E2α
Non-pregnant	7	400 ± 10	E1, E2α, E2β
0–80 d pregnant	10	300 ± nd	E1, E2α, E2β
0–84 d pregnant	7	400 ± 20	E1, E2α, E2β
80–210 d pregnant	10	1500 ± nd	E1, E2α, E2β
140–200 d pregnant	7	11400 ± 1200	E1, E2α, E2β
210–240 d pregnant	10	5400 ± nd	E1, E2α, E2β
Non-pregnant	7	500 ± 40	E1, E2α, E2β
55–81 d pregnant	5	700 ± 60	E1, E2α, E2β
101–123 d pregnant	13	14400 ± nd	E1, E2α, E2β, E3
111 d pregnant	3	34300 ± nd	E1, E2α, E2β, E3
107–145 d pregnant	4	3400 ± 1200	E1, E2α, E2β
165–175 d pregnant	5	28800 ± nd	E1, E2α, E2β, E3
205–209 d pregnant	4	22300 ± 2500	E1, E2α, E2β
250–254 d pregnant	5	86800 ± 28000	E1, E2α, E2β, E3
271–285 d pregnant	13	163000 ± 20000	E1, E2α, E2β, E3

a = live animal mass; nd = no data; E1 = estrone; E2 = 17β-estradiol; E3 = Estriol.

Table 3. Examples of levels of steroid hormones (μg/kg) reported in dairy wastes (Source: Hanselman et al. 2003)

Waste type	17α-estradiol	17β-estradiol	Estrone	References
Press-cake solids,	139 (± 7)	BDL	426 (± 78)	Raman et al. (2001)
Dry-stack (semi-solid),	603 (± 358)	236 (± 216)	349 (± 339)	Williams (2002)
Dry-stack (solid),	289 (± 207)	113 (± 67)	203 (± 176)	
Holding ponds	370 (± 59)	239 (± 30)	543 (± 269)	

BDL = below detection limit.

The concentration of steroid hormones in dairy waste and sludge is dependant on the level of waste effluent treatment and storage and this is reflected in the concentration ranges shown in Table 2. Furthermore, analysis of total estrogens within a pile of dairy cow manure demonstrated estrogen levels as high as 1000 μg/kg at the surface of the heap, but only about 3% of that within the heap (Möstl et al. 1997).

The variability in concentrations of steroid hormones measured in dairy waste effluents suggests the composition of the waste entering treatment systems varies considerably over time. The relatively high concentrations of steroid hormones measured in diary waste effluents and lagoons can be several orders of magnitude higher than that typically encountered in municipal wastewater treatment plants. Furthermore, the measured levels are significantly higher than the nanograms per litre (ppt) levels at which feminisation of fish has been observed to occur (Jobling et al. 1998).

Steroid Hormones Associated with Animal Waste in New Zealand

New Zealand is a small country with numerous lakes, rivers and streams, a rapidly expanding dairy industry, and established beef, sheep, pig and poultry production. Pasture grass and farm animals dominate more than half the country's land surface, and affect nearly all catchments. For many years, wastes from New Zealand dairy farms were largely treated in oxidation ponds before discharge to nearby waterways. However, in recent years, application of animal wastes to the land has become a permitted activity allowing farmers to apply effluent directly to the land without resource consent as long as they follow prescribed conditions set out by their respective regional councils (Sarmah and Northcott 2004).

It has been reported that the New Zealand livestock population excretes about 40 times more waste than the human population (Sarmah 2003; Sarmah et al. 2008). Within the Waikato region, animal waste is now applied to pasture by 70% of dairy farms and this proportion is steadily growing (Ministry for the Environment 1997). This proportion is higher in some other regions of New Zealand where land application of dairy effluents predominates as a means of disposal, for instance in Canterbury. In addition to the application of the effluent onto land by irrigation, a significant quantity of waste is excreted directly onto land by grazing animals. The potential therefore exists for contamination of surface and groundwater by EDCs sourced from diary operations due to run-off and/or leaching processes within receiving soils.

Steroid hormone levels measured in 5 dairy farms randomly selected within the Waikato region showed that significant amounts of steroid hormones were present in the dairy effluent samples collected from the point of discharge from oxidation ponds. The combined load for these natural excreted estrogens varied from 60 ng/l to > 4000 ng/l in the analysed effluents, and estriol was not detected in any of the collected samples (Table 4). The α epimer of estradiol was more prevalent in dairy effluent compared with the β epimer.

Table 4. Concentration of estrogens (μg/l) in dairy effluent across selected dairy farms within the Waikato region, NZ

	Farm 1	Farm 2	Farm 3	Farm 4	Farm 5
17α-estradiol	40.0	592	458	1028	18.8
17β-estradiol	28.8	289	147	331	ND
Estriol	ND	ND	ND	ND	ND
Estrone	66.9	3123	696	3057	40.9
Total	135.7	4004	1301	4416	59.7

ND = not detected.

A comparison of levels of estrogens concentrations in animal waste effluents reported in overseas literature and in New Zealand (Table 2 and 3) shows that the concentration of these compounds varies significantly depending on the number of animals and their reproductive stages. For example, a studies conducted in the USA demonstrated the highest average concentrations of estradiol and its metabolites were observed in effluents from dairy holding ponds (Williams 2002; Raman et al. 2001).

In the New Zealand study, dairy effluent samples were collected from either 1- or 2-stage oxidation pond systems. The number of cattle on each farm may have had a significant influence on the concentration of estrogens found in these samples. For instance, some samples showed much higher values than others, such as the sample in Farm 4, where the total load of estradiol (α and β epimers) was a few orders of magnitude higher than the sample collected from Farm 5. Farms 2 and 4 had greater numbers of animals (250 and 262 dairy cows, respectively) than other farms, and this was reflected in the total estrogens present in the effluent. The ratio of steroid hormones excreted in faeces and urine can dramatically change during pregnancy and lactation periods (cited by Hanselman et al. 2003). The 17α-estradiol epimer is reportedly more prevalent in dairy cattle than its β-epimer (Hanselman et al. 2003) and it was therefore not surprising to find greater values of 17α-estradiol in dairy farm effluents. The α and β stereochemical distinction of estradiol might be useful for identifying the livestock species contributing to waterway contamination (cattle vs pig or poultry); however, this possibility has yet to be investigated.

Bacterial Pathogens from Animal Waste

Animal wastes applied to soil may be a potential source of pathogens and parasites. Besides, heavy loads of faecal pollution are common from outdoor feedlots where storm runoff may be equivalent to the discharge of raw sewage from a city (Geldreich 1990). Once drainage or runoff from animal production unit reach a watercourse, a potential chain for the spread of disease is initiated (Loehr, 1978). According to reports, the severity of diseases may be increased following manure application (Osunlaja 1990). Land application of animal effluent is associated with bacteria such as *salmonella* and *Escherichia coli*, while other organisms such as *Bacillus anthracis*, *Mycobacterium tuberculosis*, *Clostridia spp.* and *Lptospira spp.* can survive and be spread in effluent (Kelly and Collins 1978; Thurston-Enriquez, 2005; Unc and Goss, 2003).

It has been reported that there are an estimated 376,000 livestock operations in the United States, generating about 58 million tons of manure each year (Guber et al. 2007). According to the 1998 National Water Quality Inventory produced by the United States Environmental Protection Agency (USEPA), approximately 60% of the pollution in rivers and 45% in lakes come from agricultural sources (EPA, 2001). Sixty-five outbreaks of human infections linked to water have been reported in the United Kingdom between 1991 and 2000 (Hunter, 2003), while 230 outbreaks have been reported in the United States between 1991 and 1998 (Croun et al. 2002).

Many bacteria in solid and liquid wastes from farm animals have been screened for resistance to antibiotics commonly used as growth promoters and/or therapeutically to treat diseases (Sarmah et al. 2006). In one study, 80% terramycin resistance was exhibited by solid waste isolates, with 30% terramycin and 100% sulfamethazine resistance shown by liquid waste isolates (Bromel et al. 1971). However, the risk from animal to human being is minimal, with the exception of those personnel of agriculture or slaughterhouses who are in close contact with the animals and may develop resistant strains of bacteria of animal origin. It must be noted, however, that clear evidence of development of antibiotic resistance transfer from animal to humans has not yet been reported.

AGRICULTURAL WASTE UTILIZATION

Agricultural waste utilization is of comparatively recent origin. Agricultural solid wastes are widely recognized as potential sources of nutrients for direct or indirect use in animal agricultural production. Based on the nutrient status of the wastes, some portion of animal wastes is used as a feed supplement for livestock; however, caution is warranted for some unknown compounds that are likely to be present in the wastes in trace amounts, such as drugs fed to animals through feed additives or administered to the body of the animals during production to treat them during diseases. Agricultural waste derived from animal production facilities around the world can act as various sources for renewable energy, carbon sequestration, and reduction of emission of greenhouse gases. In the following section various benefits of agriculturally derived waste are briefly discussed.

Production of Bioethanol

Ethanol is an alternative fuel derived from biologically renewable resources and can be employed to replace octane enhancers such as methylcyclopentadienyl manganese tricarbonyl (MMT) and aromatic hydrocarbons such as benzene or oxygenates such as methyl tertiary butyl ether (MTBE) (Champagne, 2008). A potential source for low-cost ethanol production is the utilization of ligonocellulosic materials (e.g., crop residues, grass, livestock manure); however, the high cost of biethanol production using current technologies does not make this a feasible option. There has been considerable effort from various research groups to produce bioethanol from lignocellulosic waste material using crop residues (Kim and Dale, 2004) and animal manure (Chen et al. 2003, 2004; Wen et al. 2004). For instance, a research group at Washington State University developed a process for hydrolyzing lignocellulosic materials from cattle manure into fermentable sugars (Chen et al. 2003, 2004). According to the authors, when raw dairy manure was pretreated with 3% H_2SO_4 (at 110 $^{\circ}$C) for an hour, hemicellulose was degraded to form arabinose, galactose and xylose, which were then treated with celluloytic enzymes to hydrolysze the cellulose. While these recent preliminary studies using agricultural wastes as lignocellulosic feedstock for ethanol production seem promising, more work is required to develop the technology on a larger scale. Nevertheless, this laboratory-based research will help develop an innovative waste management approach that uses agriculturally derived wastes as a renewable resource both for the extraction of value-added product, cellulose, and for its conversion to bioethanol (Champagne 2008).

Energy Production from Agricultural Waste

Although it is clear from above discussion that ethanol production through fermentative methods from crops and other renewable biomass sources has received much attention recently, crop-based feed-stocks are subject to seasonal fluctuations in supply, ultimately limiting ethanol generation (Kasper et al. 2001). The energy cost in harvesting these feed-stocks (e.g., corn stubble) as well as their lost value as soil amendments can make ethanol production costly for farmers (Pimental, 1992). Animal manures avoid many of these problems because they are a truly renewable feedstock. Production of ethanol from animal waste through the process of gasification is another new technology that has been trialed (Kaspers et al. 2001).

Gasification of biomass has received much attention as a means of converting waste materials to a variety of energy forms such as electricity, combustible gases, synfuels, various fuel alcohols, etc. Gasification is a two-step, endothermic process in which solid fuel is thermo-chemically converted into a low or medium Btu gas. In the first step, pyrolysis of the biomass takes place; in the second step either direct or indirect oxygen-deprived combustion takes place during the gasification process. This process converts raw biomass into a combustible gas, retaining 60–70% of the feedstock's original energy content. A recent cost and performance analysis of biomass (i.e. wood) gasification systems for combined power generation indicated that such a steam system (Battelle Columbus Laboratory) had the lowest capital cost and product electricity cost (Craig and Mann 1997).

Huge amounts of swine waste are produced annually in many parts of the world. For instance, the quantity of swine manure produced in the USA is estimated to be 5 billion kg dry matter per year, sufficient to contribute substantially to ethanol supplies. Assuming a conversion efficiency of 40%, this is a theoretical ethanol yield of 500 million gallons per year (Kaspers et al. 2001).

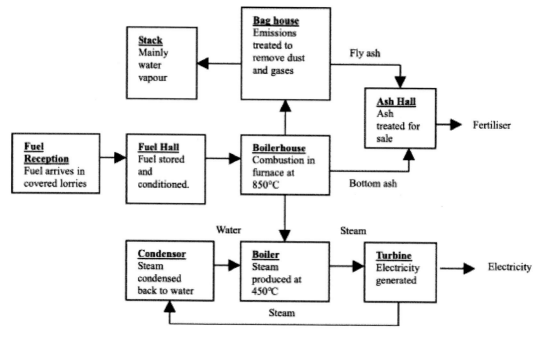

Figure 2. Flow diagram of poultry litter fuelled power plant (Source: Staff report, Modern Power System, 2000).

In the UK, poultry litter has been used for large-scale off-site electricity generation and on-farm space heating of broiler houses using two separate stages-gasification and combustion (Dagnell 1992). Figure 2 shows a flow diagram of poultry litter fuelled power plant. The size of poultry litter production in many countries worldwide indicates a sustained and increasing trend. For example, in the UK, the poultry farming industry produces nearly 1.5 million tons of litter annually per year (Fibrowatt). If land-application of litter is not a viable option due to the potential contamination of aquatic bodies through the run-off of nutrients associated with litter, an alternative mean of disposal is the production of energy, as has been demonstrated in the USA. Many promising projects are under way, both in the USA and Europe, researching the environmental effects and economic benefits of poultry litter biomass combustible. One such example of power generation, the Fibrowatt has built three power plants in the UK, consuming 800000 t of litter annually to generate approximately 64 MW of electricity (Fibrowatt; Morisson 1997). In one of the three plants, nearly 400000 t of poultry litter per year is used to produce enough electricity to power about 93,000 homes in Thetford, UK. The facility at Thetford has a maximum feed rate of 55 tonnes per hour. Feedstock is delivered from surrounding operations by covered trucks from nearby farms. The poultry litter is fed to boilers using spiral screw augers blown into the combustion chamber and burnt at 850°C. Water in the boiler is heated to 450°C and steam from the process turns the turbine connected to an electrical generator. This process is continued, steam is cooled, and water is recycled for boiler use, while ash from combustion is conditioned through the precise addition of nutrients to create a fertiliser product (Fibrophos), marketed as a concentrated organic fertiliser (ECW, 2006). Dávalos et al. (2002) recently evaluated the usefulness of poultry litter as a feasible fuel and found that the calorific values or the massic energy combustion of dry-poultry litter was 14 447 kJ/kg. The authors also reported that if the water content is < 9%, it can burn without extra fuel.

Biogas Production

Biogas is generated when bacteria degrade biological material in the absence of oxygen, in a process known as anaerobic digestion. Biogas consists of about 50–75% methane and about 25–50% carbon dioxide. According to the United Nations' Food and Agricultural Organization (FAO), there are about 1.3 billion cattle worldwide (one for every five people), slightly more than 1 billion sheep, around 1 billion pigs, 800 million goats and 17 billion chickens. The total fecal matter produced by these animals is around 13 billion tons per year according to various estimates. It has been estimated that one cubic foot of biogas can be produced from one pound of cow manure (heated at around 28° C, or 82.4° F), which is enough to cook one day's worth of meals for four to six people in India (Ecofriend 2005). One cow in one year can produce enough manure, which when converted into methane can meet the fuel needs equivalent to 200 liters-plus (about 53 US gallons) of gasoline. According to a report from the University of Alberta, Canada, around 7,500 cattle can produce 1 megawatt (MW) of electricity (1 MW can power the average home in the developed world). The manure of 6 million cows can fulfill the needs of 1 million homes or about six cows per home. Table 5 shows the annual biogas energy potential in Alberta, Canada, with all feed material contributing to > 50% of methane production.

Table 5. Inventory of livestock materials and biogas energy potential in Alberta, Canada (Source: Alberta Agriculture and Rural Development)

Feed material	Total Solids (%)	Volatile solids (%) of total solids	Biogas yield (m³/tonne)	Annual biomass production (tonnes)	Annual energy potential (PJ)	Methane content (%)
Beef cattle manure	8–12	80–85	19–46	51890736	20–48	53
Hog manure-(grower-finisher)	9–11	80–85	28–46	2452800	1.4–2.3	58
Dairy manure	12	80–85	25–32	3994195	2.0–2.6	54
Poultry manure	25–27	70–80	69–96	1728987	3.3	60

New Zealand has a relatively large primary agricultural industry with many opportunities for producing biogas from animal manure. Potential methane resource generated from piggery waste alone at farm scale could potentially provide up to 0.05PJ (14G Wh) a year, equivalent to the amount of electricity used by around 1,700 houses (EECA 2005). There is also further potential for biogas from the poultry and dairy industry in New Zealand. The potential production of biogas and energy output from various fresh feedstocks in New Zealand has not yet been evaluated. The yield of biogas, however, varies significantly from one feedstock to another due to the percentage of dry matter (or solids) and, in turn, to the percentage of volatile solids within the feedstock. However, given New Zealand has an estimated cattle population of > 5 million cattle, the total amount of waste that can be converted to produce biogas energy is enormous. Therefore it would be prudent to explore opportunities for biogas production and its utilization for energy supply especially within the rural sectors in New Zealand.

The production and use of biogas as an alternative fuel in many developing countries is fast growing, with India currently leading the way. Biogas digesters to produce methane for cooking, lighting and heating have become widespread in China and India following progressive national policies to aid their adoption (Ministry of Agriculture 2000, 2001; Somashekhar et al. 2000; Wenhua 2001). Animal waste and plant material are added to the digester to produce methane, and the remaining sediment, which is high in organic matter, is returned to the soil. There are 8.48 million biogas digesters in China, up from 4.5 million in 1990, with projections for an additional million per year for 2010 (Shuhong 1998; Ministry of Agriculture 2000).

In India, there are some 2.5 million family-sized biogas plants, constructed by the National Project on Biogas Development, plus another 500 larger community biogas systems (Ravindranath and Ramakrishna 1997; Shukla 1998; Somashekhar et al. 2000). Shukla (1998) estimated that nationwide there are 60% are functioning well (1.5 million units). Depending on the size of these domestic digesters, and their efficiency and productive period during the year, each can save the annual combustion of 1.5–4.0 tonnes of fuelwood, equivalent to an avoided emission of 0.75–2.0 tC per digester (Shuhong 1998; Shukla 1998; Wenhua 2001).

Production of Biochar

The term 'Biochar', or more appropriately 'biocarbon', refers to all products made from the process of pyrolysis that decomposes organic materials at temperatures generally between 350 and 500°C in the absence of oxygen or with limited oxygen. During pyrolysis, an average of 50% of feedstock carbon content is converted to char; however, this percentage varies by feedstock and pyrolysis conditions. Due to the environmental and agronomic values of biochar, production of biochar from agricultural wastes (e.g., green plant material, feedlot manure, poultry litter, wood waste, bagasse, etc.) has recently been trialled in many countries (Figure 3). Although various types of feedstock can be used to produce biochar (Figure 4), not all chars produced from different feedstock have similar characteristics.

Benefits from Biochar

A range of environmental benefits can be obtained from biochar: reduction in greenhouse gas emissions; reduction in nutrient leaching; improvement in soil structure; water retention; and higher crop productivity. In NZ, most soils have > 3% carbon; however, pH can range from around 4.5 to 6.5, so liming is often required. The application of biochar could reduce lime use as well as provide other benefits. NZ soils also have a finite ability to store nitrogen, and nitrogen-saturated soils create the risk of nitrogen leaching into waterways. Soils with high carbon:nitrogen ratios usually have a greater capacity to store nitrogen and thereby reduce nitrous oxide emissions and nitrate leaching. Locking carbon in soil through the application of biochar (70–80% carbon level) seems a novel idea to reduce both greenhouse gas emissions and nitrate leaching. Preliminary results indicate that biochar amendments to soil appear to decrease emissions of nitrous oxide as well as methane, which is a greenhouse gas 23 times more potent than CO_2. In greenhouse and field experiments in Colombia, nitrous oxide emissions were reduced by 80% and methane emissions were completely suppressed with biochar additions to a forage grass stand (Lehmann and Rondon 2006).

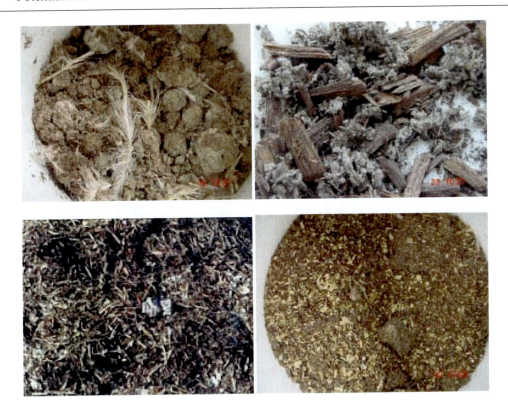

Figure 3. Range of biomass feedstocks derived from agricultural activities used in the production of biochar (Source: Best Energies Australia).

Figure 4. Biochar produced from agricultural waste through pyrolysis process (Photo: Best Energies Australia).

Biochar can influence the global carbon cycle in two main ways. If biochar is produced from material that would otherwise have oxidized in the short to medium term, and the resultant carbon-rich char can be placed in an environment in which it is protected from oxidation, it may provide a means to sequester carbon that would otherwise have entered the atmosphere as a greenhouse gas (Woolf 2008). In addition, the gaseous and liquid products of pyrolysis may be used as a fuel that can offset the use of fossil fuels (Lehman 2007). Biochar can potentially be used as a soil amendment for improving the quality of agricultural soils (Glaser et al. 2002a, 2002b; Lehmann et al. 2003). For example, Chan et al. (2007) observed that while there were significant changes in soil quality, including increases in pH, organic carbon, and exchangeable cations as well as reduction in tensile strength at higher rates of biochar application (>50 t/ha), long-term field experiments are required to confirm and quantify the eventual long-term benefits from biochar use.

Beneficial effects of biochar in terms of increased crop yield and improved soil quality have been reported (e.g., Iswaran et al. 1980; Glaser et al. 2002a, 2002b). However, review of previous research showed a huge range of biochar application rates (0.5–135 t/ha of biochar) as well as a huge range of plant responses (29–324%) (Glaser et al. 2002a). Recently, the use of poultry litter biochar applied at 10 t/ha as soil amendments on an Australian hard-setting soil showed significant increases in the dry matter yield of radishes. These yield increases were largely due to the ability of this biochar to increase nutrient availability, particularly nitrogen.

It must be noted that in much of the research undertaken so far, the properties of the biochar used in the investigations were not reported. Biochar can be produced from a range of organic materials and under different conditions resulting in products of varying properties (Baldock and Smernik 2002; Nguyen et al. 2004; Guerrero et al. 2005). However, little research has been published elucidating the mechanisms responsible for the reported benefits of the biochar on crop growth, production, and soil quality, and such understanding is essential for the development both of agricultural markets for biochar and of technology for the production of biochar products with improved quality and value. Although research on biochar is still in its infancy, and the various benefits that it can offer to the environment are yet to be fully demonstrated through long term field trials, the effectiveness of the use of 'agricultural waste' as a potential source for alternative energy needs greater appreciation by regulatory authorities for managing agricultural wastes.

REFERENCES

Alberta Agriculture and Rural Development. 2008. Biogas energy potential in Alberta. http://www1.agric.gov.ab.ca/$department/deptdocs.nsf/all/agdex11397

Anderson DM, Gilbert PM, Bukholder JM (2002). Harmful algal blooms and eutrophication: nutrient sources, composition, and consequences, *Estuaries* 25, 704–726.

Baldock JA, Smernik RJ (2002). Chemical composition and bioavailability of thermally altered *Pinus resinosa* (Red pine) wood. *Organic Geochemistry* 33, 1093–1109. doi: 10.1016/S0146-6380(02)00062-1

Boxall ABA, Fogg LA, Blackwell P, Kay P, Pemberton EJ, Croxford a. 2004. (2004). Veterinary medicines in the environment. *Rev. Environ. Contam. Toxicol.* 180, 1–91.

Bromel M, Lee, YN, Baldwin, B. (1971). Antibiotic resistance and resistance transfer between bacterial isolates in a waste lagoon. In: Livestock Waste Management and Pollution Abatement. American Society of Agricultural Engineers, 1971. 283 p.

Champagne, P (2008). Bioethanol from agricultural waste residues. *Environmental Progress* 27(1), 51–57.

Chan KY, van Zwieten L, Meszaros I, Downie A, Joseph S (2007). Agronomic values of green waste biochar as a soil amendment. *Australian Journal of Soil Research* 45, 629–634. doi: 10.1071/SR07109

Chen S, Liao W, Liu C, Wen Z, Kincaid RL, Harrison JH, Elliot DC. Brown, M.D., Solana AE, Stevens DJ (2004). Value added chemicals animal manures 135 pp. Northwest Bioproducts Research Institute Technical Report, US Department of Energy Contract DE-AC06-76RLO1830.

Chen S, Liao W, Liu C, Wen Z, Kincaid RL, Harrison JH (2003). Use of animal manure as feedstock for bio-products. In Proceedings of Ninth International Animal, Agricultural and Food Processing Wastes Symposium (pp. 50–57). Research Triangle Park, NC.

Craun GF, Calderon RL, Nwachuku N (2002). Causes of waterborne outbreaks reported in the United States, 1991–98. In: PR Hunter, M Waite, E Ronchi (eds), Drinking water and infectious disease: establishing the link (pp. 105–117). CRC Press, Boca Raton, FL.

Craig KR and Mann MK Cost and Performance Analysis of Three Integrated Biomass Gasification Combined Cycle Power Systems. DOE BioPower Program Technical Reports, Aug. 1997.

Dávalos J, Roux MV, Jiménez P (2002). Evaluation of poultry litter as a feasible fuel. *Thermochimica Acta* 394, 261–266.

EECA (Energy Efficiency and Conservation Authority) (2005). Fact sheet 9: Biogas and Landfill gas. www.eeca.govt.nz

Energy Centre of Wisconsin (ECW) (2006). Wisconsins biobased industry: Opportunities and advantages study, volume 2: Technical Analysis Report. ECW Report No: 238-1. 248 p.

Ecofriend. (2005). http://www.ecofriend.org/

Fibrowatt. Industrial "Fibrowatt Group". http://www.fibrowatt.com.

Geldreich EE (1990). Microbiological quality of source waters for water supply. In: McFeters GA (ed.), Drinking water micro-biology (pp. 293–367). Springer-Verlag, New York.

Glaser B, Lehmann J, Zech W (2002a). Ameliorating physical and chemical properties of highly weathered soils in the tropics with charcoal – a review. *Biology and Fertility of Soils* 35, 219–230.

Glaser B, Lehannes J, Steiner C, Nehls T, Yousaf M, Zech W (2002b). Potential of pyrolyzed organic matter in soil amelioration. 12th ISCO Conference, Beijing 2002. Pp. 421–427.

Guber AK, Pachepsky YA, Shelton DR, Yu O (2007). Effect of Bovine Manure on Fecal Coliform Attachment to Soil and Soil Particles of Different Sizes. *Appl Environ Microbiol.* 73(10): 3363–3370.

Guerrero M, Ruiz MP, Alzueta MU, Bilbao R, Millera A (2005) Pyrolysis of eucalyptus at different heating rates: studies of biochar characterization and oxidative reactivity. *Journal of Analytical and Applied Pyrolysis* 74, 307–314. doi: 10.1016/j.jaap.2004.12.008

Hanselman TA, Graetz DA, Wilkie AC (2003). Manure borne estrogens as potential environmental contaminants: a review. *Environ Sci Technol* 37: 5471–5478.

Hunter PR (2003). Climate change and waterborne and vector-borne disease. J. Appl. Microbiol. 94, 37S-46S.

Iswaran V, Jauhri KS, Sen A (1980). Effect of charcoal, coal and peat on the yield of moong, soybean and pea. *Soil Biology and Biochemistry* 12, 191–192. doi: 10.1016/0038-0717(80)90057-7

Jobling S, Nolan M, Tyler CR, Brighty G, Sumpter JP (1998). Widespread sexual disruption in wild fish. *Environ Sci Technol* 32: 2498–2506.

Kaspers B, Koger J, Gould R, Wossink A, Edens R, van Kempen T (2001). Fformation of fuel-grade ethanol from swine waste via gasification.

Kelly WR, Collins JD (1978). The health significance of some infectious agents present in animal effluents. *Vet Sci Comm* 2, 95-103.

Kim S, Dale BE. (2004). Global potential bioethanol production from wasted crops and crop residues, *Biomass and Bioenergy*. 2004, 26, 361-375.

Lehmann J, de Silva JP Jr, Steiner C, Nehls T, Zech W, Glaser B (2003). Nutrient availability and leaching in an archaeological Anthrosol and a Ferralsol of the Central Amazon basin: fertilizer, manure and charcoal amendments. *Plant and Soil* 249, 343–357. doi: 10.1023/A:1022833116184

Lehmann J, Gaunt J, Rondon M (2006). Bio-char sequestration in terrestrial ecosystems – a review. *Mitigation and Adaptation Strategies for Global Change* 11, 403–427. doi: 10.1007/s11027-005-9006-5

Lehmann J (2007). Bio-energy in the black. Frontiers in Ecology and the Environment 5:381–387

Loehr R (1978). Hazardous solid waste from agriculture. *Environmental Health Perspectives* 27, 261-273.

Ministry of Agriculture (2000). China biogas. Report. Department of Science, Education and Rural Environment, Beijing.

Ministry of Agriculture (2001). Statistical data on renewable energy resources in China 1996{2000. Report. Department of Science, Education and Rural Environment, Beijing, China.

Ministry for the Environment (1997). The state of our waters: State of New Zealand's environment. Wellington, New Zealand: Ministry for the Environment and GP Publications. 1997.

Morrison, EM. In: AG Innovation News, Vol. 8, No. 4, MN, USA, October 1999. http://www.auri.org/news/ainoct99/08poult.htm.

Möstl E, Dobretsberger A, Palme R. (1997). Östrogenkonzentration im stallmist trächtiger rinder. *Wien Tieärztl Mschr*. 84, 140–143 (In German)

Nguyen TH, Brown RA, Ball WP (2004) An evaluation of thermal resistance as a measure of black carbon content in diesel soot, wood char, and sediment. *Organic Geochemistry* 35, 217–234. doi: 10.1016/j.orggeochem.2003.09.005

Osunlaja SO (1990). Effect of organic soil amendments on the incidence of stalk rot of maize. *Plant and Soil* 127, 237-241.

Pimental, D (1992). Energy inputs in production agriculture. Energy in World Agriculture. ed. R.C. Fluck. Amsterdam; Elsevier. Pgs. 13-29.

Raman DR, Williams EL, Layton AC, Burns RT, Easter JP, Mullen MD, Sayler GS (2004). Estrogen content of dairy and swine wastes. *Environ. Sci. Technol* 38: 3567–3573.

Ravindranath NH, Ramakrishna J (1997). Energy options for cooking in India. *Energy Policy* 25, 63-75.

Sarmah AK, Northcott GL (2004). Animal waste as a potential source for steroid hormones: a New Zealand perspective. Programme and abstract book, SETAC Europe 14th Annual Meeting, Prague, Czech Republic, 18–22 April 2004.

Sarmah AK, Northcott GL, Leusch FDL, Tremblay LA (2006). A survey of endocrine disrupting chemicals (EDCs) in municipal sewage and animal waste effluents in the Waikato region of New Zealand. *Sci. Total. Environ.* 355: 135–144.

Sarmah AK, Northcott GL, Scherr FF (2008). Retention of estrogenic steroid hormones by selected New Zealand soils. *Environment International 34*: 749–755

Sharpley AN, Smith SJ, Stewart BA, Mathers AC (1984). Forms of phosphorous in soil receiving cattle feedlot waste, *J. Environ. Qual.* 13, 211–216.

Sharpley AN, Halvorson AD (1994). The management of soil phosphorous availability and its transport in agricultural runoff. In Soil Processes and Water Quality, R. Lal, Ed., Advances in Soil Science, Lewis Publishers, Boca Raton, Fl, 1–84.

Shuhong C (1998). Biomass energy for rural development in China. In *Biomass energy: data, analysis and trends*. OECD, Paris.

Shukla P (1998). Implications of global and local environment policies on biomass energy demand: a long-term analysis for India. In *Biomass energy: data, analysis and trends*. OECD, Paris.

Somashekhar HI, Dasappa S, Ravindranath NH (2000). Rural bioenergy centres based on biomass gasiers for decentralised power generation: case study of two villages in southern India. *Energy Sustainable Development IV*(3), 55–63.

Staff Report, Modern Power Systems (2000). Staff Report 2000. Chicken runs provide the fuel for Fibrowatt's global ambitions. *Modern Power Systems 20*, 41–45.

Thurston-Enriquez, JA, Giley JE, Eghball B. (2005). Microbial quality of runoff following land application of cattle manure and swine slurry. *Journal of Water Health* 3, 157–171.

Unc A, Goss MJ. (2003). Movement of faecal bacteria through the vadose zone. *Water Air Soil Pollution* 149, 327–337.

Williams EL. (2002). MS Thesis, University of Tennessee, Knoxville, USA.

Wen Z, Liao W, Chen S. (2004). Hydrolysis of animal manure lignocelluloseics for reducing sugar production, *Bioresource Technology* 91, 31–39.

Wenhua L (2001). Agro-ecological engineering in China. Man and the Biosphere (MAB) Series, vol. 26. UNESCO, Paris.

Woolf D (2008). Biochar as soil amendment: A review of the environmental implications. http://orgprints.org/13268/01/Biochar_as_a_soil_amendment_-_a_review.pdf.

In: Agricultural Wastes
Eds: Geoffrey S. Ashworth and Pablo Azevedo

ISBN 978-1-60741-305-9
© 2009 Nova Science Publishers, Inc.

Chapter 2

ANIMAL WASTE AND HAZARDOUS SUBSTANCES: CURRENT LAWS AND LEGISLATIVE ISSUES[*]

Claudia Copeland

ABSTRACT

The animal sector of agriculture has undergone major changes in the last several decades: organizational changes within the industry to enhance economic efficiency have resulted in larger confined production facilities that often are geographically concentrated. These changes, in turn, have given rise to concerns over the management of animal wastes and potential impacts on environmental quality.

Federal environmental law does not regulate all agricultural activities, but certain large animal feeding operations (AFOs) where animals are housed and raised in confinement are subject to regulation. The issue of applicability of these laws to livestock and poultry operations — especially the Comprehensive Environmental Response, Compensation, and Liability Act (CERCLA, the Superfund law) and the Emergency Planning and Community Right-to-Know Act (EPCRA) — has been controversial and recently has drawn congressional attention.

Both Superfund and EPCRA have reporting requirements that are triggered when specified quantities of certain substances are released to the environment. In addition, Superfund authorizes federal cleanup of releases of hazardous substances, pollutants, or contaminants and imposes strict liability for cleanup and injuries to natural resources from releases of hazardous substances.

Superfund and EPCRA include citizen suit provisions that have been used to sue poultry producers and swine operations for violations of those laws. In two cases, environmental advocates claimed that AFO operators had failed to report ammonia emissions, in violation of Superfund and EPCRA. In both cases, federal courts supported broad interpretation of key terms defining applicability of the laws' reporting requirements. Three other cases in federal courts, while not specifically dealing with reporting violations, also have attracted attention, in part because they have raised the question of whether animal wastes that contain phosphorus are hazardous substances that can create cleanup and natural resource damage liability under Superfund. Two of these

---

[*] This is an edited, reformatted and augmented version of a Congressional Research Service publication, Report RL33691, dated April 11, 2007.

latter cases were settled; the third, brought by the Oklahoma Attorney General against poultry operations in Arkansas, is pending.

These lawsuits testing the applicability of Superfund and EPCRA to poultry and livestock operations have led to congressional interest in these issues. In the 109th Congress, legislation was introduced that would have amended CERCLA to clarify that manure is not a hazardous substance, pollutant, or contaminant under that act and that the laws' notification requirements would not apply to releases of manure. Proponents argued that Congress did not intend that either of these laws apply to agriculture and that enforcement and regulatory mechanisms under other laws are adequate to address environmental releases from animal agriculture. Opponents responded that enacting an exemption would severely hamper the ability of government and citizens to know about and respond to releases of hazardous substances caused by an animal agriculture operation. Congress did not act on the legislation, but similar bills have been introduced in the 110th Congress (H.R. 1398 and S. 807).

INTRODUCTION

The animal sector of agriculture has undergone major changes in the last several decades, a fact that has drawn the attention of policymakers and the public. In particular, organizational changes within the industry to enhance economic efficiency have resulted in larger confined production facilities that often are geographically concentrated.[1] Increased facility size and regional concentration of livestock and poultry operations have, in turn, given rise to concerns over the management of animal wastes from these facilities and potential impacts on environmental quality, public health and welfare.

Animal manure can be and frequently is used beneficially on farms to fertilize crops and add or restore nutrients to soil. However, animal waste, if not properly managed, can adversely impact water quality through surface runoff and erosion, direct discharges to surface waters, spills and other dry-weather discharges, and leaching into soil and ground. It can also result in emission to the air of particles and gases such as ammonia, hydrogen sulfide, and volatile organic chemicals. According to the U.S. Department of Agriculture (USDA), in 1997, 66,000 operations had farm-level excess nitrogen (an imbalance between the quantity of manure nutrients produced on the farm and assimilative capacity of the soil on that farm), and 89,000 had farm-level excess phosphorus.[2] USDA believes that where manure nutrients exceed the assimilative capacity of a region, the potential is high for runoff, leaching of nutrients, and other environmental problems. Geographically, areas with excess farm-level nutrients correspond to areas with increasing numbers of confined animals.

Federal environmental law does not regulate all agricultural activities. Some laws specifically exempt agriculture from regulatory provisions, and others are structured so that farms escape most, if not all, of the regulatory impact. Still, certain large animal feeding operations (AFOs) where animals are kept and raised in confinement are subject to environmental regulation. The primary regulatory focus on environmental impacts has been on protecting water resources and has occurred under the Clean Water Act. In addition, facilities that emit large quantities of air pollutants may be regulated under the Clean Air Act. Some livestock operations also may be subject to requirements of the Comprehensive Environmental Response, Compensation, and Liability Act (CERCLA, the Superfund law) and the Emergency Planning and Community Right-to-Know Act (EPCRA).[3] The issue of applicability of these laws to livestock and poultry operations — especially CERCLA and EPCRA — has been controversial and has drawn congressional attention.

This chapter describes the provisions of Superfund and EPCRA, and enforcement actions under these laws that have increasingly been receiving attention. Congressional scrutiny in the form of legislative proposals and a House hearing in the 109th Congress are discussed. Bills intended to exempt animal manure from the requirements of Superfund and EPCRA were introduced in the 109th Congress, but no legislation was enacted. Similar bills have been introduced in the 110th Congress (H.R. 1398 and S. 807). Issues raised by the legislation are analyzed.

CERCLA AND EPCRA

Both the Comprehensive Environmental Response, Compensation, and Liability Act (the Superfund law, 42 U.S.C. §§9601-9675) and the Emergency Planning and Community Right-to-Know Act (42 U.S.C. §§11001-11050) have reporting requirements that are triggered when specified quantities of certain substances are released to the environment.[4] Both laws, which are administered by the Environmental Protection Agency (EPA), utilize information disclosure in order to increase the information available to government and citizens about the sources and magnitude of chemical releases to the environment. In addition to reporting requirements, CERCLA includes provisions authorizing federal cleanup of releases of hazardous substances, pollutants, or contaminants that may present an imminent and substantial danger to the public health or welfare (Section 104), and imposing strict liability for cleanup and damages for injury to, destruction of, or loss of natural resources resulting from releases of hazardous substances (Section 107). At issue today is how the reporting requirements and other provisions of these laws apply to poultry and livestock operations.

Superfund authorizes programs to remediate uncontrolled or abandoned hazardous waste sites and assigns liability for the associated costs of cleanup. Section 103(a) of CERCLA requires that the person in charge of a facility (as defined in Section 101(9)) that releases a "reportable quantity" of certain hazardous substances must provide notification of the release to the National Response Center.

EPCRA establishes requirements for emergency planning and notification for storage and release of hazardous and toxic chemicals. Section 304(a)(1) of EPCRA requires the owner or operator of a facility (as defined in Section 329(4)) to report to state and local authorities any releases greater than the reportable quantity of substances deemed hazardous under Superfund or extremely hazardous under EPCRA. Under Superfund, the term "release" (Section 101(22)) includes discharges of substances to water and land and emissions to the air from "spilling, leaking, pumping, pouring, emitting, emptying, discharging, injection, escaping, leaching, dumping, or disposing into the environment." Under EPCRA, the term "release" (Section 329(8)) includes emitting any hazardous chemical or extremely hazardous substance into the environment. Superfund excludes the "normal application of fertilizer" from the definition of release, and EPCRA excludes from the definition of hazardous chemicals any substance that is "used in routine agricultural operations or is a fertilizer held for sale by a retailer to the ultimate customer."

The CERCLA definition of "hazardous substance" (Section 101(14)) triggers reporting under both laws. Among the reportable substances that may be released by livestock facilities are hydrogen sulfide, ammonia, and phosphorus. The reportable quantity (RQ) for both hydrogen sulfide and ammonia is 100 pounds per day, or 18.3 tons per year; the RQ for phosphorus is 1 pound per day. Section 109 of Superfund and Section 325 of EPCRA authorize EPA to assess civil penalties for failure to report releases of hazardous substances that equal or exceed their reportable quantities (up to $27,500 per day under CERCLA and $27,500 per violation under EPCRA).

ENFORCEMENT AGAINST AFOs

EPA has enforced the Superfund and EPCRA reporting requirements against AFO release of hazardous pollutants in two separate cases. The first involved the nation's second largest pork producer, Premium Standard Farms (PSF) and Continental Grain Company. In November 2001, EPA and the Department of Justice announced an agreement resolving numerous claims against PSF concerning principally the Clean Water Act, but also the Clean Air Act, Superfund, and EPCRA. More recently, in September 2006, the Department announced settlement of claims against Seaboard Foods — a large pork producer with more than 200 farms in Oklahoma, Kansas, Texas, and Colorado — and PIC USA, the former owner and operator of several Oklahoma farms now operated by Seaboard. Like the earlier PSF case, the government had brought complaints for violations of several environmental laws, including failure to comply with the release reporting requirements of CERCLA and EPCRA.

Both Superfund and EPCRA include citizen suit provisions that have been used to sue poultry producers and swine operations for violations of the laws (Section 310 of CERCLA and Section 326 of EPCRA). In two cases, environmental advocates claimed that AFO operators had failed to report ammonia emissions, putting them in violation of Superfund and EPCRA. In both cases, federal courts supported broad interpretation of key terms defining applicability of the laws' reporting requirements.

In the first of these cases, a federal district court in Oklahoma initially ruled in 2002 that a farm's individual barns, lagoons, and land application areas are separate "facilities" for purposes of CERCLA reporting requirements, rather than aggregating multiple emissions of pollutants across the entire site. This court held that Superfund's reporting requirements would only apply if emissions for *each individual facility* exceed 100 pounds per day. However, the district court's ruling was reversed on appeal (*Sierra Club v. Seaboard Farms Inc.*, 387 F.3d 1167 (10th Cir. 2004)). The court of appeals ruled that the whole farm site is the proper entity to be assessed for purposes of CERCLA reporting and determining if emissions of covered hazardous substances meet minimum thresholds.

In the second case, a federal district court in Kentucky similarly ruled in 2003 that the term "facility" should be interpreted broadly to include facilities operated together for a single purpose at one site, and that the whole farm site is the proper entity to be assessed for purposes of the Superfund and EPCRA reporting requirements (*Sierra Club v. Tyson Foods, Inc.*, 299 F. Supp. 2d 693 (W.D. Ky. 2003)). While Superfund provides that a continuous release is subject to reduced reporting requirements, and EPCRA provides an exemption for reporting releases when the covered substance is used in routine agricultural operations or is used on other farms for fertilizer, the court found that these exemptions did not apply to the facts of this case. The ruling was not appealed.

EPA was not a party in either of these lawsuits. The U.S. Court of Appeals for the 10th Circuit invited EPA to file an amicus brief in the *Seaboard Farms* case in order to clarify the government's position on the issues, but EPA declined to do so within the time frame specified by the court.

Three other cases in federal courts, while they do not include reporting violations, also have drawn attention, in part because they raised the question of whether animal wastes that contain phosphorus are hazardous substances that can create cleanup and natural resource injury liability under Superfund.[5] Animal wastes typically contain low levels of phosphorus, and animal wastes are beneficially used as fertilizer on farms. Over the long term, however, the application of animal waste fertilizers may result in phosphorus buildup in soils which may be released to watersheds through surface runoff. In 2003, a federal court in Oklahoma held that phosphorus contained in poultry litter in the form of phosphate is a hazardous substance under Superfund and thus could subject poultry litter releases to provisions of that law (*City of Tulsa v. Tyson Foods, Inc.*, 258 F. Supp. 2d 1263, (N.D. Okla. 2003)). This ruling was later vacated as part of a settlement agreement, but some observers believe that the court's reasoning may still be persuasive with other courts. The second case, *City of Waco v. Schouten* (W.D. Tex., No. W-04-CA-118, filed April 29, 2004), was a suit by the city against 14 dairies alleging various causes of action based on disposal of wastes from those operations. It was resolved by a settlement agreement early in 2006.

The third case, *State of Oklahoma v. Tyson Foods, Inc.* (N.D. Okla, No. 4:05-cv00329, filed June 13, 2005), is still pending. This suit, brought by the Oklahoma Attorney General, asserts various claims based on the disposal of waste from 14 poultry operations in the Illinois River Watershed. The state principally seeks its past and present response costs and natural resource injuries under CERCLA due to release of wastes from these facilities.

The net result of these lawsuits is growing concern by the agriculture community that other legal actions will be brought and that the courts will continue to hold that the Superfund and EPCRA reporting requirements and other provisions apply to whole farm sites, thus potentially exposing more of these operations to enforcement under federal law.

In 2005, a group of poultry producers petitioned EPA for an exemption from EPCRA and CERCLA emergency release reporting requirements, arguing that releases from poultry growing operations pose little or no risk to public health, while reporting imposes an undue burden on the regulated community and government responders.[6] While the agency has not formally responded to this petition, early in 2007 EPA formed an internal workgroup to review information on animal waste as it relates to CERCLA and to possible exemptions from emissions reporting. Further, EPA Administrator Stephen Johnson told congressional committees that the agency will propose a rule to exempt routine animal waste air releases from emergency notification requirements. He did not provide details on how broad a waiver might be proposed. While such a regulatory exemption might satisfy many agriculture industry groups, environmental advocates and others oppose the exemption. Some argue that an exemption is premature, since EPA is moving forward with research on emissions levels (see CRS Report RL32947, *Air Quality Issues and Animal Agriculture: EPA's Air Compliance Agreement*). State air quality officials have said that they oppose blanket regulatory or legislative exemptions, and they have recommended that if the agency considers any action, it should only be a narrow exemption, such as one based on a size threshold for farms.[7]

CONGRESSIONAL INTEREST

The court cases testing the applicability of Superfund and EPCRA to poultry and livestock operations have led to congressional interest in these issues. In March 2004, a number of senators wrote to the EPA Administrator to ask the agency to clarify the reporting requirements of the two laws so as to limit their impact on poultry operations. The senators' letter said that because of unclear regulations and a lack of scientific information about emissions, poultry and livestock producers are uncertain about the laws' requirements and are vulnerable to enforcement actions.[8] In report language accompanying EPA's FY2006 appropriations, the House Appropriations Committee urged EPA to address the issues.

> The Committee continues to be concerned that unclear regulations, conflicting court decisions, and inadequate scientific information are creating confusion about the extent to which reporting requirements in [CERCLA] and [EPCRA] cover emissions from poultry, dairy, or livestock operations. Producers want to meet their environmental obligations but need clarification from the Environmental Protection Agency on whether these laws apply to their operations. The committee believes that an expeditious resolution of this matter is warranted.[9]

Also in 2004, some in Congress considered proposing legislation that would amend the definition of "release" in Superfund (Section 101(22); 42 USC §9601(22)) to clarify that the reporting requirements do not apply to releases from biological processes in agricultural operations and to amend EPCRA to exclude releases of hazardous chemicals produced through biological processes in routine agricultural operations. For some time, there were indications that an amendment containing these statutory changes would be offered during debate on FY2005 consolidated appropriations legislation, but this did not occur.[10]

Some Members sought to amend the FY2006 Agriculture appropriations bill, H.R. 2744, with a provision exempting releases of livestock manure from CERCLA and EPCRA. The proposal was promoted by Senate conferees on the bill, but it was not accepted by House conferees. Proponents, including Senator Larry Craig, contended that the proposed language was consistent with current law, because in their view CERCLA and EPCRA were never intended to apply to agriculture. Environmentalists objected to the language, arguing that it could prevent public health authorities from responding to hazardous substance releases from AFOs, would block citizen suits against agriculture companies for violations of reporting requirements, and would create an exemption from Superfund liability for natural resource injuries that might result from a large manure spill. EPA's congressional affairs office released an unofficial analysis criticizing the bill. It argued that, by eliminating federal liability for manure releases under Superfund and EPCRA, the provision could interfere with EPA's Air Compliance Agreement, because companies would have much less incentive to participate in the agreement. The agreement is a plan that EPA announced in January 2005 to collect air quality monitoring data on animal agriculture emissions.[11] The House and Senate gave final approval to H.R. 2744 in November 2005 (P.L. 109-97), without the language that Senate conferees had proposed.

Also in November 2005, legislation was introduced that would amend CERCLA to clarify that manure is not a hazardous substance, pollutant, or contaminant under that act and that CERCLA's notification requirements would not apply to releases of manure (H.R. 4341). The bill was similar to the legislative language that Senator Craig had proposed to conferees as a provision of the FY2006 Agriculture appropriations bill with a broad definition of "manure" that includes, for example, bedding commingled with animal waste.

H.R. 4341 was introduced the same day that a House Energy and Commerce subcommittee held a hearing on animal agriculture and Superfund. The Subcommittee on Environment and Hazardous Materials heard from agriculture industry witnesses who urged Congress to provide policy direction on the issue that has developed as a result of recent and potential litigation. Other witnesses testified that the reporting and notification requirements of Superfund and EPCRA provide a safety net for making information on releases available to government and citizens, and that other environmental laws, such as the Clean Air Act, cannot function in that manner. An EPA witness said that the agency is considering ways to reduce the paperwork burdens for large AFOs to report their emissions, but has not yet formalized a proposal. Similar legislation was introduced in the Senate (S. 3681). No further action occurred on either bill. Similar legislation has been introduced in the 110th Congress (H.R. 1398 and S. 807).

POLICY ISSUES

Supporters and opponents of the 109th Congress and 110th Congress legislation have raised a number of arguments for and against the proposals. For example, proponents of the exemption proposed in these bills, representing the agriculture industry, especially livestock and poultry producers, say that animal manure has been safely used as a fertilizer and soil amendment by many cultures all over the world for centuries and thus should not be considered a hazardous substance. Opponents — including environmental activists, public health advocates, and state and local governments — agree that when properly managed, manure has beneficial uses. Superfund's reporting and cost recovery requirements do not threaten responsible operators who manage manure as a valuable fertilizer, they say. However, these groups say that when improperly managed and in the massive amounts produced at today's large feedlot operations, animal waste can release a number of polluting substances to the environment. Releases to surface water, groundwater, and the atmosphere may include nutrients, organic matter, solids, pathogens, volatile compounds, particulate matter, antibiotics, pesticides, hormones, gases that are associated with climate change (carbon dioxide and methane), and odor.

Proponents of the legislation argue that neither Superfund nor EPCRA was intended by Congress to apply to agriculture and that the pending legislation would simply clarify congressional intent. CERCLA exempts "normal application of fertilizer" from the definition of "release" and also exempts releases of "naturally occurring organic substances." Animal waste arguably was intended to be covered by these existing exemptions, they say. Opponents respond that there is little firm evidence either way on this point, as there is limited legislative history concerning this language. The exemption for "normal application of fertilizer," enacted in CERCLA in 1980, applies to application of fertilizer on crops or cropland for beneficial use, but does not mean dumping or disposal of larger amounts or concentrations than are beneficial to crops.[12]

EPA has not issued guidance to interpret what constitutes "normal application of fertilizer," and the only court decision so far addressing this issue (the vacated 2003 *City of*

Tulsa case discussed above) held that neither plaintiffs nor defendants in that case had presented evidence sufficient for a fact-based determination of what constitutes "normal application." Opponents of the legislation also argue that animal manure consists of a number of substances that are nutritional and pharmaceutical elements of the feed provided to animals (trace elements, antibiotics, nutrients), and releases are the result of inadequate waste disposal, not "naturally occurring" substances and activities.

Proponents argue that enforcement and regulatory mechanisms exist under the Clean Water Act (CWA) and other media-specific statutes, such as the Clean Air Act (CAA), making it unnecessary to rely on Superfund or EPCRA for enforcement or remediation. In particular, both the Clean Water Act and Clean Air Act require that regulated facilities obtain permits that authorize discharges or emissions of pollutants. Enforcement of permit requirements has been an important tool for government and citizens to address environmental concerns of animal agriculture activities.

Opponents respond that enforcement under Superfund fills critical gaps in these other environmental laws, because not all pollutants are covered by other laws. For example, releases of ammonia and hydrogen sulfide are listed under CERCLA but are not currently regulated as hazardous pollutants under the CAA. Clean Water Act AFO permits primarily address discharges of nutrients, but not other components of manure waste (e.g., trace elements, metals, pesticides, pathogens). Moreover, neither of these laws provides for recovery of costs for responding to or remediating releases, nor for natural resource injuries. Opponents also argue that, while "federally permitted releases" are exempt from CERCLA's reporting requirements, CWA and CAA permit requirements apply only to facilities that meet specified regulatory thresholds (for example, CWA permit rules apply to about 14,000 large AFOs, less than 6% of all AFOs in the United States).[13]

Finally, proponents of the legislation argue that if animal manure is considered to be a hazardous substance under Superfund, farm operations both large and small potentially could be exposed to costly liabilities and penalties. Opponents note that the purpose of release reporting is to keep federal, state, and local entities informed and to alert appropriate first responders of emergencies that might necessitate response, such as release of hazardous chemicals that could endanger public health in a community. The exemption proposed in pending legislation, they point out, U.S. Environmental Protection Agency, "National Pollutant Discharge Elimination System Permit Regulation and Effluent Limitation Guidelines and Standards for Concentrated Animal Feeding Operations (CAFOs); Final Rule," 68 *Federal Register* 7179, February 12, 2003. would apply not only to CERCLA and EPCRA reporting requirements but also to other provisions (such as Superfund's authority for federal cleanup of releases, cleanup liability, and liability for natural resource injuries).

According to states and some other interest groups, liability, which arises when manure is applied in amounts that exceed what is beneficial to support crops, is necessary to bring about improvements in waste handling practices of large AFOs. Enacting an exemption would severely hamper the ability of government to appropriately respond to releases of hazardous substances and pollution caused by an animal agriculture operation, they argue. On the issue of penalties, opponents note that penalties are not available under Superfund for removal or remedial actions (except for failure to comply with information gathering and access related to a response action), regardless of whether initiated by government or a private party. CERCLA does authorize civil penalties for violation of the Section 103 reporting requirements (up to $27,500 per day), but neither of the two key citizen suit cases decided thus far (*Sierra Club v.*

Tyson Foods, Inc., and *Sierra Club v. Seaboard Farms Inc.*) involved penalties for failure to report releases.

CONCLUSION

Issues concerning the applicability of Superfund and EPCRA to animal agriculture activities have been controversial and have drawn considerable attention. Bills in the 109th Congress gained much support (when the 109th Congress adjourned in December 2006, H.R. 4341 had 191 co-sponsors, and S. 3681 had 35 co-sponsors), but were not enacted. They also drew opposition from environmental advocacy groups and state and local governments. The Bush Administration did not present an official position on the legislation. Continuing interest in the issue is evident from the fact that similar legislation has been introduced in the 110th Congress.

ENDNOTES

[1] For additional information, see CRS Report RL33325, *Livestock Marketing and Competition Issues*, by Geoffrey S. Becker.

[2] U.S. Department of Agriculture, Natural Resources Conservation Service, "Manure Nutrients Relative to the Capacity of Cropland and Pastureland to Assimilate Nutrients: Spatial and Temporal Trends for the United States," Publication no. nps00-579, December 2000, p. 85.

[3] For additional information, see CRS Report RL32948, *Air Quality Issues and Animal Agriculture: A Primer,* by Claudia Copeland.

[4] For additional information on Superfund and EPCRA, see CRS Report RL30798, *Environmental Laws: Summaries of Statutes Administered by the Environmental Protection Agency*, coordinated by Susan Fletcher, and CRS Report RL33426, *Superfund: Overview and Selected Issues*, by Jonathan Ramseur and Mark Reisch.

[5] Unlike the citizen suit cases discussed above, these lawsuits do not address what is a "facility," for purposes of determining whether a release has occurred. EPA also is not a party in any of these cases.

[6] In 1998, EPA granted an administrative exemption from release reporting requirements for certain radionuclide releases. EPA cited authority in CERCLA sections 102(a), 103, and 115 for granting administrative reporting exemptions where "releases of hazardous substances that pose little or no risk or to which a Federal response is infeasible or inappropriate." See 63 *Federal Register* 13461 (March 19, 1998).

[7] National Association of Clean Air Agencies, letter to the Honorable Barbara Boxer, chairman, Senate Environment and Public Works Committee, March 20, 2007.

[8] Senator Blanche L. Lincoln et al., letter to Michael Leavitt, EPA Administrator, March 12, 2004.

[9] U.S. Congress, House Committee on Appropriations, *Report accompanying H.R. 2361, Department of the Interior, Environment, and Related Agencies Appropriation Bill, 2006*, H.Rept. 109-80, 109th Cong., 1st sess., p. 87.

[10] "Spending Bill Excludes Proposal for Farms; Craig Plans Separate Legislation Next Year," *Daily Environment Report*, Nov. 23, 2004, p. A-10.

[11] For information, see CRS Report RL32947, *Air Quality Issues and Animal Agriculture: EPA's Air Compliance Agreement*, by Claudia Copeland.

[12] U.S. Senate, Committee on Environment and Public Works, *Environmental Emergency Response Act, Report to Accompany S. 1480*, 96th Cong., 2nd sess., S.Rept. 96-848, p. 46.

In: Agricultural Wastes
Eds: Geoffrey S. Ashworth and Pablo Azevedo

ISBN 978-1-60741-305-9
© 2009 Nova Science Publishers, Inc.

Chapter 3

REPROCESSING AND PROTEIN ENRICHMENT OF AGRICULTURAL WASTES BY THERMOPHILIC AEROBIC DIGESTION (TAD)

J. Obeta Ugwuanyi[*]

Department of Microbiology, University of Nigeria, Nsukka, Nigeria

ABSTRACT

Agricultural and food industry residues, refuse and wastes constitute a significant proportion of worldwide agricultural productivity. It has variously been estimated that these wastes can account for over 30% of worldwide agricultural productivity. These wastes include lignocellulosic materials, fruit, vegetables, root and tuber wastes, sugar industry wastes as well as animal/livestock and fisheries operations wastes. They represent valuable biomass and potential solutions to problems of animal nutrition and worldwide supply of protein and calories if appropriate technologies can be deployed for their valorization. Moreover, reutilization of these vast wastes should help to address growing global demands for environmentally sustainable methods of production and pollution control.

Various technologies are potentially available for the valorization of these wastes. In addition to conventional waste management processes, other processes that may be used for the reprocessing of wastes include solid substrate fermentation, ensiling and high solid or slurry processes. In particular, the use of slurry processes in the form of (Autothermal) Thermophilic Aerobic Digestion (ATAD or TAD) or liquid composting is gaining prominence in the reprocessing of a variety of agricultural wastes because of its potential advantages over conventional waste reprocessing technologies. These advantages include the capacity to achieve rapid, cost-effective waste stabilization and pasteurization and protein enrichment of wastes for animal feed use.

TAD is a low technology capable of self heating and is particularly suited for use with wastes being considered for upgrading and recycling as animal feed supplement, as is currently the case with a variety of agricultural wastes. It is particularly suited for wastes generated as slurries, at high temperature or other high COD wastes. Reprocessing of a variety of agricultural wastes by TAD has been shown to result in very significant protein

---

[*] Phone +234 (0)802 621 2756; Fax +234 (0)42 256 831. E.mail: jerryugwuanyi@yahoo.com

accretion and effective conversion of mineral nitrogen supplement to high-value feed grade microbial/single-cell protein for use in animal nutrition. The application of this technology in reprocessing of wastes will need to take account of the peculiarities of individual wastes and the environment in which they are generated, reprocessed and used. The use of thermopiles in the process has significant safety benefits and may be optimized to enhance user confidence and acceptability.

INTRODUCTION

Waste treatment by microbiological methods is probably as old as the generation of waste itself. In spite of this, scientific knowledge of the processes involved in biological waste treatment is relatively recent. Traditionally, wastes have been treated to remove them from areas in which they are not wanted (Bewely et al., 1991). Hence, early methods of waste treatment were essentially waste relocation or disposal processes, inspired by man's desire to protect his immediate environment. The processes of disposing of wastes applied to both human and agricultural wastes, and therefore did not discriminate between wastes that were potentially reusable with only minor reprocessing and, those that needed to be disposed of on account of their limited reuse value. An increase in the scale of the problems of environmental pollution, as well as changes in social attitude, have led to multidisciplinary approaches to the problems of waste management and pollution control (Grainger, 1987a). Increasing pressures on resources also mean that the vast quantities of organic materials that remain from human productive activities can no longer be seen strictly as wastes that need disposal.

The ever-increasing world population, with the attendant food supply problems, as well as the lessons of the energy crisis of the 1970s, combined to force a change in the global attitude towards waste. Wastes, particularly organic and agricultural wastes, are progressively being seen as resources in the wrong location and form that should be recycled, rather than as refuse that must be disposed of. Thus, biomass, reprocessing, recycling and reuse are progressively gaining increased importance in discussions on waste management, at the expense of refuse, disposal and waste treatment, which can no longer be seen solely as exercises in pollution control.

Conventional treatment processes are rarely linked to any form of reprocessing, recycling or reuse. Where reuse occurs, it is almost solely as organic fertiliser, as a means of disposal rather than the target. The need to recycle waste is most relevant in the food, agricultural and related industries (Grainger, 1987a,b), where biomass estimated to be in excess of 10^{17} tonnes are generated annually, on a worldwide basis (Bath, 1991). Refuse amounting to up to 40% of world food production is generated following diverse agricultural and food industry processes. These are very important energy-rich resources, the reprocessing and recycling of which offer the possibility of returning them to more beneficial use than is currently possible. The key to successful processes of this nature is an economical "no loss" process, in which the cost of processing should be offset, at least in part, by the possibility of producing a valuable product, with the added benefit of stabilising the otherwise environmentally hazardous waste. Such a process should be satisfactory if it causes the overall cost of waste management to be significantly less than the classical approach (Loehr, 1977). Today, agricultural and food industry residues, refuse and wastes constitute a significant proportion of worldwide agricultural productivity. The continued disposal of these "wastes" constitutes significant loss of agricultural productivity and calories which, if properly harnessed, can impact positively on worldwide food supply, animal production and global food security.

Traditional subsistence agriculture led to the production of only limited agricultural refuse, which was generally disposed of untreated to land at zero cost, or actually at a credit because of its use as farmyard manure and soil conditioner. Introduction of cheap inorganic fertiliser obviated much of the need for farmyard manure. Increasing world population and industrialisation led to the introduction of intensive agricultural practices, the latter resulting in accumulation of large quantities of agricultural refuse in small land areas. Recent advances in the field of food technology, and the increasing demand for factory processed foods have led to a considerable increase in the quantity of food industry / processing waste. Additionally, the tendency to locate food processing factories close to food producing areas has led to a concentration of agricultural refuse disposal problems in limited land areas.

An associated problem of waste disposal on land is concern over the survival of pathogens on pasture land, and pollution of underground waters at areas of high water tables. The introduction of various noxious/man-made chemicals into agriculture has also introduced new requirements and pressures that conventional treatments are not able to handle. This is in addition to new needs to utilise agricultural products that are not conventionally used due to their natural content of toxins. Thus, waste management has progressed through the stages of disposal and treatment, beginning with the time when the only wastes that attracted attention were municipal solid wastes and sewage, to one of reprocessing and recycling.

OVERVIEW OF CONVENTIONAL WASTE DISPOSAL PROCESSES

Incineration, Overland Disposal and Sea Disposal

Traditional methods of waste disposal include disposal to sea (Rees, 1985), incineration and burning (Gandolla and Aragno, 1992; Bath, 1991) and overland disposal. These methods have little future due to increasingly stringent legislation for the protection of the environment. Consequently, they will not be discussed further in this review.

Landfilling

Archaeological records indicate that landfills have been around since the stone-ages (White-Hunt, 1980; 1981a,b). It is unlikely though, that the stone-age communities dug pits specifically for the disposal of wastes. Such landfills may have developed because a natural pit was available, hence traditional processes of landfilling simply involved dumping of refuse. It is not known whether such landfills were used specifically for the disposal of agricultural wastes or other types of wastes. Currently, a variety of agro-industrial wastes and sludge are disposed of to landfills, in addition to its established use for disposal of municipal solid waste and wastewater sludge, a variety of hazardous and other industrial and domestic wastes (Nguyen et al., 2007; Monte et al., 2008). In 1993, landfills were used to dispose 62% of all municipal solid wastes generated in the United States (Barlaz, 1997). Landfilling has the advantage of reclaiming devastated and ruined lands, and also has viable recycling potential, because of its capacity to generate usable off gas as methane. In addition to conventional landfilling which is an anaerobic process, aerobic landfilling has been reported to offer rapid waste stabilization and to have potential for long term use in the management of municipal solid waste, particularly where the principal objective is waste stabilization and reuse as soil amendment and manure rather than for biogas generation (Erses et al., 2008).

Limited availability of sites and the risk of contamination of underground waters are the major impediments to the wide applicability of landfills. To contain the latter problem, and also prevent the biogas from escaping into the atmosphere, has necessitated intensive research into the engineering of landfill for environmentally sound procedures. The fact that some wastes may not be recycled using other available technologies and are therefore technically non reusable at the moment, in addition to its comparatively low cost, assures landfill a future that can only be limited by the availability of land. However, this potential long term value is constrained by increasing stringent concerns for the safety of environment and possible contamination of underground waters. In addition, the very long life span of a landfill requires that the process or site is continually monitored. It is on account of these that various legislations exist that seek to constrain the use of landfills (Council of European Union 1999; Erses et al., 2008). Sound landfill processes also involve complex and often expensive engineering procedures that may constrain the use of the process in less developed countries.

Biochemical and Microbiological Changes in Landfill

Waste decomposition in a landfill is typically an anaerobic process involving the co-ordinated activity of several groups of micro-organisms. Based on laboratory scale experiments, the microbiological processes that occur in a landfill have been delineated into four phases viz.: (1) aerobic phase; during which both oxygen and nitrates are consumed using sugars and organics. (2) anaerobic acid phase; (3) accelerated methanogenic phase; and (4) decelerated methanogenic phase. In the anaerobic acid phase, carboxylic acids accumulate, the pH drops and degradation of complex polymers sets in. However, the level of methane remains low. In phase three, methane production is extensive, and rapid consumption of acids leads to a rise in pH, while degradation of polymers, like cellulose, also increases. In the deceleration phase, methane production remains at levels similar to, or less than in phase three, while degradation of cellulose and hemicellulose increases again (Erses et al., 2008; Kulikowska and Klimiuk 2008; O'Sullivan et al., 2008).

It is not known whether this is a cyclic process in which phases 2–4 repeat until waste mineralisation is complete, as none has been monitored scientifically, long enough for waste to be completely stabilised (Barlaz, 1997), and the boundaries between the phases are also not clearly defined. Biochemical changes associated with the different phasesmay take place simultaneously. The processes involved in phases 1–4 take several years to occur, and complete stabilisation of the waste, even longer and may only be slightly improved by the inoculation of the landfill with select microbial innoculant (O'Sullivan et al., 2008).

OVERVIEW OF CONVENTIONAL WASTE TREATMENT PROCESSES

Composting

Composting is the controlled aerobic process carried out by successive microbial populations, combining both mesophilic and thermophilic activities, leading to the production of carbon-dioxide, water, minerals and stabilised organic matter (Subba-Rao, 1987; Chang et al., 2006; Huang et al., 2006). It is an age-old and well established process, which has been applied to agricultural waste management and utilisation, particularly for production of manure. It probably evolved following the heaping of agricultural refuse prior to overland dispersal. Wastes amenable to composting vary from highly heterogeneous organic mixtures, including polymers in urban refuse, to the more homogeneous agricultural wastes and residues, and sewage sludge (Maso and Blasi 2008; Sellami et al., 2008; Li et al., 2008; Lu et al., 2008; Chang et al., 2006; Huang et al., 2006; Yun et al., 2000; Bhamidimarri and Pandey, 1996; Mergaert and Swings, 1996; Thambirajah et al., 1995; Shuval et al., 1991).

Development of Compost

The development of compost has been delineated into four stages viz. (a) Latent; (b) Growth; (c) Thermophilic; and (d) Maturation phases (Saludes et al., 2008; Abouelwafa et al., 2008; Sanchez-Arias et al., 2008; Raut et al., 2008; Yu et al., 2008; Cayuela et al., 2008; Alfano et al., 2008). The latent phase is associated with the acclimatisation of the microorganisms in the waste. During the growth phase, microbial population and the temperature of the digesting mass rise rapidly. This stage is mostly associated with the activity of mesophilic microorganisms. In the thermophilic phase the temperature rises to peak level, often in excess of $60°C$ and rapid stabilisation of waste takes place. Pathogen destruction is highest at this stage, depending on the peak temperature and its duration (Turner 2002; Nakasaki et al., 1985a,b). The maturation phase is associated with a drop in temperature to mesophilic and then ambient levels. Nitrification and humification reactions are extensive at this stage (Metcalf and Eddy, 1991).

Microbiology of Compost

Composting is a complex and dynamic biological process, resulting from the activities of a succession of mixed populations of microorganisms (and perhaps also some higher organisms). The activity of each microbial group is often limited by narrow and exacting environmental requirements (mostly temperature and oxygen tension). At the early stages of composting mesophilic bacteria and fungi multiply rapidly, by metabolising easily degradable compounds. Bacterial populations of up to 10^8-10^9 g^{-1} of moist compost have been reported (Abouelwafa et al., 2008; Raut et al., 2008; Saludes et al., 2008; Heerden et al., 2002; Andrews et al., 1994; Davis et al., 1992; Strom, 1985). As metabolism continues the temperature rises to thermophilic range of up to $60°C$ and higher. Mesophilic organisms, including most of the non-spore forming bacteria are rapidly inactivated, (or at least cease to be active at this temperature (Droffner et al., 1995ab)).

During the thermophilic stage, most of the biological activity is due to spore forming thermophiles (Elango et al., 2008; Fujio and Kume, 1991) and actinomycetes (Durak and Ozturk, 1993; Singh et al., 1991). The majority of the microbial activity at this stage is directed towards the breakdown of high molecular weight polymers, including cellulose, hemicelluloses, proteins, lipids and starch. As readily degradable compounds are exhausted, the temperature begins to drop leading to the re-establishment of mesophiles in the final or maturation stage. At this stage, actinomycetes and fungi continue to breakdown the high molecular weight polysaccharides and formation of humic acid takes place.

Environmental Requirements for Composting

The rate of composting is believed to depend on a number of rate limiting steps, which include production and release of hydrolytic enzymes needed for the breakdown of substrates; diffusion of solubilized substrate molecules, and oxygen transport and availability within the composting mass (Huang, 1980). Optimisation of the composting process depends on the management of a number of variables such as; (a) nutrient balance; an important component of which is the carbon/ nitrogen balance. A ratio of 25–30:1 is believed to be optimal, in addition to the presence in adequate amounts of all other macro- and micro- nutrients needed by the vast array of micro-organisms that take part in composting (Jimenez and Perez 1991); (b) particle size; the optimum particle size in compost varies with the aeration rate employed, but sizes of 12–50mm are considered appropriate for most processes (Biddlestone and Gray, 1985); (c) moisture content; levels of 50–70% are considered optimum (Inaba et al., 1996). Moisture content influences oxygen transfer and attainable temperature in compost (Tiquia et al., 1996; Nakasaki et al., 1985a,b). Airflow rates of 0.6–1.8 m^3 air day^{-1} kg^{-1} volatile solids are considered adequate. More recently, composting is being considered as a means of developing inocula for the bioremediation of contaminated soils (Laine and Jorgensen, 1996), in addition to its established use for growing edible mushrooms (Miller et al., 1990).

ANAEROBIC DIGESTION (AD)

This is the treatment of organic wastes of solid, liquid or slurry consistency in the absence of molecular oxygen (Ward et al., 2008). The ability of this process to produce biogas as methane has helped to improve its appeal and economic acceptability, particularly since the energy crises of the 1970s, and increasing concern over greenhouse gas emission and global warming. Mass reduction, methane production and improved dewatering are considered to be some of the principal attractions of AD (Ferrer et al., 2008). On the other hand, slow digestion leading to long retention times is a draw back, particularly at mesophilic temperatures. It has been in use since the middle or late nineteenth century, but its application in wastewater treatment did not grow rapidly, due to a lack of fundamental understanding of the process (Wheatley, 1990). It is currently the principal method for treatment of waste sludge, particularly following the invention and success of the up-flow anaerobic sludge blanket

(UASB) (Lettinga, 1995; Lettinga et al., 1980), and variants thereof. It has consequently been extensively studied (Verstreate et al., 1996; Luostarinen et al., 2009), and is in use in many parts of the world, for the treatment of a variety of wastes and effluents including those from food, fermentation, brewery, beverage and paper pulp industries, in addition to domestic, agricultural and municipal wastes and waste water (Ortega et al., 2008; Marcias-Corral et al., 2008; Forster-Carneiro et al., 2008abc; Fezzani and Cheikh 2008; Alvarado-Lassman et al., 2008; Yilmaz et al., 2008; Zupancic et al., 2007), and farm animal house slurries, effluents and wastes (Tricase and Lombardi 2008; Cantrell et al., 2008) as well as slaughter house and meat process wastes (Luostarinen et al., 2009; Buend et al., 2008; Cuetos et al., 2008;).

Continuing research effort has resulted in improved understanding and optimisation of anaerobic digestion and methanogenesis (Cantrell et al., 2008; Forster-Carneiro et al., 2008abc). A number of AD plants currently operate worldwide as means of (centralised) waste treatment and biogas production (Ahring, 1995). Also, a number of small scale household/ farm digesters are being built for dual purpose of gas generation and waste treatment in many developing, warm climate countries. It is estimated that over 5 million cottage digesters operate producing methane and treating wastes in China and India (Poh and Chong 2009; Lansing et al., 2008; Zhao et al., 2008; Coombs, 1994). The use of AD offers a number of other advantages including low cost, low technology, ease of scaling up/ down; considerable stability and ease of start up. It is not energy intensive (if it is operated at mesophilic temperatures), and since anaerobes conserve less energy than aerobes, AD results in less sludge (Schink, 1997). It also achieves reliable waste stabilisation, if it is recognised that AD cannot be used alone for complete waste treatment (Verstreate et al., 1996).

Fundamentals of Anaerobic Digestion

The operation of AD requires the co-operation of fermentative bacteria, hydrogen producing acetogenic bacteria, hydrogen consuming acetogenic bacteria; carbon-dioxide reducing methanogens and aceticlastic methanogens (Appels et al., 2008; Lee et al., 2008; Montero et al., 2008; Zinder, 1986; Archer and Kirsop, 1990). Organic polymers are first hydrolysed to simple soluble substrates which are then fermented to yield various organic acids including acetate, formate and reduced organics such as propionate and butyrate (Cantrell et al., 2008; Siller and Winter, 1998; Sarada and Joseph, 1993a,b). These acids are then metabolised to methanogenic precursors, hydrogen and carbon-dioxide (Montero et al., 2008; Tatara et al., 2008; Forster-Caneiro et al., 2008abc; Stams, 1994). About 76% of the metabolisable organics are degraded via reduced intermediate. Acetogenic bacteria metabolise long chain fatty acids to acetate, hydrogen and carbon-dioxide, assisted by hydrogen consuming methanogens. The latter helps to keep hydrogen concentration at a low level, thus making acetogenesis a key process in anaerobic digestion (methanogenesis) (Montero et al., 2008; Tatara et al., 2008; McHugh et al., 2003).

The synthrophic interaction of the variety of anaerobes that operate in the process is essential for methanogenesis, and success of the reaction. By this interaction, several anaerobes can share the energy available in the metabolism of a compound to methane and carbon-dioxide, and so drive endergonic reactions that would have been mechanistically/ energetically difficult under standard conditions (McHugh et al., 2003; Schink, 1997; Dolfing, 1992;). This synthrophy, which is the principal strength of AD, is also the major source of its weakness since various organisms with differing growth rates and requirements have to grow in a synchronised fashion. As a result, the pace of AD is determined by the growth rate of the slowest growing organism in this "symbiosis", which may be as low as 0.08–0.15 d (Verstreate et al., 1996). Pace of digestion and methanogenesis are also influenced by the flux of metabolites, and efficiency of transfer between producing and consuming partners.

AD often has retention times of more than seven days, but usually up to 30 days or more, to avoid washout of methanogens. Effective operation requires retention of viable sludge, and the maintenance of sufficient contact between sludge and wastewater. The system is very sensitive to drop in pH, hence acid consumption must be balanced with generation to avoid failure (Cantrell et al., 2008). This need for growth synchronisation sometimes makes AD precarious and unreliable, requiring highly trained personnel for its operation (Zinder, 1986). This also makes the multi-stage digesters preferred option to the single stage batch processes.

A variety of complex engineering designs have been developed to be able to uncouple biomass from hydraulic retention, and so reduce system fragility and stabilisation time, including anaerobic contact processes with sludge recycle. Some prominent variants of this are AD-ultra filtration, anaerobic filters and fluidised beds (Wheatley, 1990; Coombs, 1990; Hobson, 1990; Noone, 1990; Zinder, 1986;). By far the most popular variant is the upflow anaerobic sludge blanket, UASB (Lettinga et al., 1980), which involves slow upward pumping of fluid waste in a reactor under anaerobic conditions. Selection takes place resulting in the microorganisms growing as granules, which then act as the catalyst, converting organic waste to biogas (Verstreate 1996; Lettinga 1995; Lettinga and Hulshoff Poll 1991). Since its introduction, improved modifications of UASB have been developed, including expanded granular sludge blanket (EGSB), internal circulation reactor (ICR), tubular reactors, etc.

Anaerobic digestion requires careful and close monitoring and control, and usually produces effluents that require further aerobic polishing. This is because individual anaerobes have low substrate affinity (high K_s), resulting in high residual volatile fatty acid content in the effluent, even when stable synthrophic associations have developed. This tends to restrict the use of AD to treatment of highly concentrated wastes as a pre-process to be followed by aerobic polishing. AD cannot be relied upon to detoxify noxious chemicals being often poisoned by them, and has poor pathogen kill capacity (Verstreate et al., 1996; Lettinga, 1995), except when they are operated as thermophilic processes (Aitken et al., 2007; Ahn and Forster, 2000). Attempts to reduce the stabilisation time and increase process efficiency have led to the development of thermophilic anaerobic digestion.

Thermophilic Anaerobic Digestion

Thermophilic anaerobic digestion evolved as an attempt to harness the intrinsic advantages of AD on the one hand, and of bioprocessing with thermophiles on the other (Ahn and Forster 2000; Mackie and Bryant, 1995). Thermophilic anaerobic digesters are relatively new (Zinder, 1990; Rimkus et al., 1982). However, a lot of research efforts have been made to understand and optimise the processes, in studies that have employed a variety of wastes, mostly in laboratory and pilot processes (Espinoza-Escalante et al., 2008; Forster-Carneiro et al., 2008ab; Kaparaju et al., 2008; Montero et al., 2008; Linke 2006; Angelidaki et al., 2006; Bouallagui et al., 2004; Ahring et al., 2001; Solera et al., 2001; Ahn and Forster 2000; 2002ab; Lier, 1996; Lier et al., 1993a,b; 1994; 1995;). More recently, attempts are being made to apply it to digestion of real wastes in full scale processes, particularly wastes discharged at elevated temperatures. In general, wastes that have been subjected to this form of treatment include sugar beet waste, coffee, brewery, distillery and other beverage wastes as well as slaughter house and fish process effluents, vegetable / potato wastes and paper pulping wastes as well as organic fraction of municipal wastes and household wastes (Forster-Carneiro et al., 2008abc; Koppar and Pullammanappallil 2008; Lee et al., 2008; Ortega et al., 2008; Yilmaz et al., 2008; Zupancic et al., 2007; Linke 2006; Angelidaki et al., 2006; Hartmann and Ahring 2005; Bouallagui et al., 2004).

Temperatures that have been investigated range from 35° to 65°C (Lee et al., 2008; Zinder, 1986), and various temperature optima have been reported during thermophilic digestions (Perez et al., 2001; Ahring et al., 2001; Lier, 1996;). For instance, whereas 70°C is the highest temperature recorded for anaerobic fermentation of cellulose, anaerobic metabolism of soluble sugars have been reported at 80°C (Maden, 1983; Zeikus et al., 1979). So far, 65°C is the highest temperature at which stable methanogenesis has been reported in waste treatment (Zinder, 1990), even though methanogenesis has been reported at 100°C in microbial communities sampled from super heated deep sea waters (Baross et al., 1982; Baross and Deming, 1983). Ahring (1995), reported methanogenesis during waste treatment at up to 75°C, though most of the processes were adjudged unstable, above 60°C. Recovery of thermophilic process following temperature shift-up induced deterioration often involve slow re-growth of a new populations, with growth optima in the new temperature range (Ahring, 1994). However, following temperature shift-downs the thermophiles retain considerable activity, often enough to drive the process, albeit sub-optimally, until a new population with optima at the new temperature is established (Lettinga, 1995; Waigant et al., 1986). By far the most thermophilic processes have been run in the temperature range 55°- 65°C (Kaparaju et al., 2008; Lee et al., 2008; Ortega et al., 2008). In comparison to mesophilic process, thermophilic AD has the advantage of pathogen destruction and reduced retention time (Aitken et al., 2007). However, it is energy intensive, being incapable of self-heating. Its products also require aerobic polishing prior to disposal (Lier, 1996). The reported instability of the process is aided by poor granulation of thermophiles, due to their often low content of extracellular polysaccharides and low viscosity (Schmidt and Ahring, 1994), leading to dispersed growth. Thermophilic processes are also easy to washout due to high turbulence during active gas formation (Uemera and Harada, 1993, 1995; Soto et al., 1992).

Thermophilic anaerobic waste processing has proved to be sensitive to slight upward shifts in temperature and requires precise temperature and pH control (Cecchi et al., 1993ab). Its temperature sensitivity is such that even slight upward shifts can cause death of active microbial populations. If acid consumers are affected, the process may take many days to recover due to their slow growth rate (Zinder, 1986) or fail completely, due to significant drop in pH. It is also adversely affected by slight increases in the concentration of ammonia, hydrogen sulphide and other toxic reaction by-products, particularly at the upper ranges of temperature (Schink, 1997; Ahring, 1994; Angelidaki and Ahring, 1993, 1994). For these reasons, thermophilic anaerobic digestions have remained slow in gaining acceptability for use in rural communities since successful operation of reactors will require the services of highly trained and often expensive labour.

Overview of Silage Making (Ensiling)

Although ensiling is not a conventional process for the management of waste, it has found significant application in the reprocessing/ treatment and preservation of a variety of agricultural residues for use in animal feeding. Silage making is the (lactic) fermentation/storage of (forage) for use in animal (ruminant) feeding. Ensiling is a multistage process which ultimately results in low pH (< 4.0) products that have extended resistance to spoilage (and often having appealing flavour to ruminants). During ensiling, some bacteria are able to break down some cellulose and hemicellulose to their components sugars which are subsequently metabolized to low molecular weight acids, mostly lactic acid. This can also be encouraged by the use of appropriate mix of enzymes and microbial (lactic acid bacteria) silage inoculants (Okine et al., 2005; Aksu et al., 2004; Zahiroddini et al., 2004; Colombatto et al., 2004abc; Gardner et al., 2001). The lactic acid bacteria are also believed to produce bacteriocins that discourage the growth of and spoilage by unwanted populations. Efficient fermentation ensures a palatable and digestible feed. Production of good quality silage requires that anaerobiosis be achieved quickly to enable the lactic acid bacteria to develop and predominate, and in the process further bring down the pH of the mass. This discourages spoilage of the silage by putrefactive aerobic populations and ensures the retention of the most nutrients in the final product (Arvidsson et al., 2008).

Silage making starts with the impounding of the biomass and is initiated by aerobic populations. During this stage the aerobic organisms scavenge oxygen and bring about anaerobiosis. This phase is undesirable, because the aerobic bacteria consume soluble carbohydrates that should otherwise be available for the beneficial lactic acid bacteria. It also leads to the production of moisture, and heat generation, which if not properly managed are capable of destroying the process. Proteinaceaous materials may also be rapidly broken down during this phase and this can lead to loss of nutrients and the accumulation of ammonia (Slottner and Bertilsson, 2006). This phase may not be completely avoided but must be reduced to the minimum for successful silage making. To encourage rapid acidification during ensiling, fermentable sugars and lactic acid bacteria inoculants are often added to the silage (Okine et al., 2005; Yang et al., 2006). This is common during ensiling of protein rich feeds such as manure, slaughter house and fish wastes as well as many agricultural residues such as wheat straw, tomato or apple pomace and citrus waste (Santana-Delgado et al., 2008; Vazquez et al., 2008; Bampidis and Robinson, 2006; Denek and Can, 2006; Pirmohammadi et al., 2006; Yang et al., 2006; Volanis et al., 2006; Vidotti et al., 2003; Oda et al., 2002; Scerra et al., 2001; Chaudhry et al., 1998; Shaw et al., 1998).

In the anaerobic stage of ensiling, mixed populations of lactic acid bacteria predominate and metabolize fermentable sugars, producing lactic acid and reducing the pH of the mass to acidic levels. As the pH drops, minor acetic fermentation (if present) ends. This process continues until most of the available sugars have been consumed, and the pH has dropped to a level low enough to discourage bacterial activity. The duration of this stage varies with the nature of the biomass being ensiled, particularly, the initial concentration of fermentable sugars and the population of lactic acid bacteria. Ensiling by itself hardly leads to protein enrichment of the biomass except if mineral nitrogen such as urea is included. However, its capacity to achieve conservation of waste protein for use in animal feeding makes it important in schemes for the reuse of agricultural refuse.

THERMOPHILIC AEROBIC DIGESTION

TAD is a process in which the metabolic heat of aerobically growing microbial cells is conserved in an insulated system, leading to the elevation of the temperature of the digesting mass to thermophilic range (Gomez et al., 2007; Coulthard et al., 1981; Surucu et al., 1975, 1976;). It arose as a modification of the conventional activated sludge process (Adav et al., 2008). Given efficient aeration, microbial metabolism leads to production of sufficient heat to cause a rise to, and maintenance of thermophilic temperatures in the digesting mass, provided that heat loss is minimised by good insulation of the digestion vessel and control of evaporative cooling. It is believed that achievable temperature may be manipulated by varying the concentrations of oxygen, and biodegradable organic materials in the waste (Messenger et al., 1990 Hamer and Bryers, 1985).

Like all waste treatment processes, TAD is a mixed culture process. During start-up, a variety of mesophilic and thermotolerant bacteria interact, metabolising readily available waste components to generate heat (Yun et al., 2000). As the temperature increases, microbial succession and selection takes place until, at thermophilic temperatures, only a few species of micro-organisms remain active. The selection process results in the inactivation of sensitive mesophiles (including most vegetative and non spore-forming organisms which include, luckily, all important animal pathogens). And, at thermophilic temperatures, hydrolysis of complex and otherwise recalcitrant molecules (hopefully, including noxious and xenobiotic compounds) is enhanced. Considering the limitation of competing biomass reprocessing technologies, particularly anaerobic digestion in the handling of noxious compounds, the ability of TAD to achieve the degradation of xenobiotics and noxious compounds makes it attractive for the treatment of a variety of wastes that have recycling potentials (Adav et al., 2008). TAD resembles composting in many respects and has been described as liquid (slurry) composting (Yun et al., 2000; Jewell, 1991).

The process is considered to be very flexible and versatile, for which reason it may be adaptable, not just to waste treatment, but also as a means of generating useful by-products (as animal feed supplements and biochemicals) from wastes. The use of TAD in the protein enrichment of a variety of agricultural and food industry refuse for animal feed supplementation has been studied and continues to attract a lot of research interest (Ugwuanyi et al 2009; Ugwuanyi 2008; Ugwuanyi et al., 2006; 2008ab; Couillard and Zhu, 1993). The high temperature of operation of TAD, and metabolic versatility of thermophiles, ensures that the processes take place very rapidly, and under conditions that may require less control than conventional bioprocessing (Lee and Yu, 1997). A striking advantage of TAD is its ability to achieve the pasteurisation of processes waste. This has implication for the use of the technology for the reprocessing of a variety of biomass intended for reuse in animal nutrition and also for land application (Ugwuanyi et al., 1999).

Compared to older and more established biomass handling and processing techniques TAD is still poorly understood as a biotechnological process. However, work has continued into the optimization of various aspects of this process including aeration efficiency of equipment used in TAD, and its use in replacement of conventional waste biomass management methods. Considerable efforts are also being expended to achieve improved understanding of the microbiology and biochemistry of the process to enhance its ultimate applicability (Cibis et al., 2006, 2002; Heinonen-Tanski et al., 2005; Agarwal et al., 2005;

Ugwuanyi et al., 2004ab; 2005ab; Kosseva et al., 2001;). Although thermophilic microbiology has a fairly long history, the application of thermophiles in biotechnology is only just beginning to attract attention, and application of thermophiles in biomass management such as in TAD, has received even less attention. Consequently few large scale operations currently employ this potentially versatile process.

Aeration of TAD Processes

TAD uses the metabolic heat generated from microbial oxidation of organic compounds to raise and maintain the temperature of digesting biomass at thermophilic levels. To achieve this, the metabolism of the microbial population has to be strongly oxidative, and rapid. Heat evolution is enhanced as the reaction temperature increases, particularly if uncoupling of metabolism occurs, leading to reduced cell yield, and dissipation of a greater amount of the energy content of the substrate as heat (Birou et al., 1987; Hamer and Bryers 1985). The process requires efficient transfer of oxygen into solution, to be able to sustain the rapid metabolism of thermophiles (Burt et al., 1990a). Conventional sparged, impeller agitated reactors are considered unsatisfactory for the level of aeration required for auto-thermophilic aerobic digestions (Messenger et al., 1990), particularly as oxygen solubility decreases with increase in temperature.

The problem of oxygen transfer has had profound impact on the economics, applicability, and development of this process. The result is that early development works on TAD were centred on the development of aeration processes, with little attention paid to the potentially versatile application of the process and the associated population in waste biomass handling and processing. Another implication of this was that most of the early studies on the process were directed at its application in the pasteurisation of sewage sludge prior to treatment by anaerobic digestion and disposal to agricultural land. In the early stages of development of TAD, it was believed that thermophilic temperatures could only be achieved if pure oxygen (Gould and Drnervich, 1978) or oxygen enriched air (Bruce 1989) was used. Aeration with pure oxygen was employed in pilot plant studies for thermophilic treatment of sewage sludge at Ponthir Water Works UK, where a temperature of 63°C was easily obtained (Morgan and Gumson, 1981). Although the process was uneconomic, it provided a basis for studies on TAD leading to the development of alternative aeration system to achieve thermophilic temperatures with air (Jewell et al., 1982). Unfortunately, some of the processes were considered uneconomic because significant proportion of the heat was of mechanical origin from agitators.

A variety of aeration devices have been developed for use in TAD. The Venturi Aerator was shown to aerate sewage sludge to thermophilic temperature inexpensively (Morgan et al., 1986). It was based on a nozzle type aerator which has been applied to a variety of biochemical processes (Jackson, 1964), and achieves aeration (and mixing) by having the digesting mass pumped through an external recirculation loop and back into the reactor, through a nozzle. The air/ liquid mixture emerging from the nozzle is then discharged at the base of the reactor vessel, where large quantities of oxygen are forced into solution in a well-mixed mass. These self-aspirating aerators were used to achieve over of 20% oxygenation with air. Oxygenation efficiency equal to or greater than 10% (i.e., capable of solubilising at least 10% of the available oxygen) has been recommended as the minimum needed to maintain thermophilic temperatures (Jewell, 1991). It has also been suggested that the reduced amount of nitrification (if any), taking place at high temperatures, means that TAD may have significantly lower oxygen requirement than aerobic mesophilic digestion for the same amount of solid material destroyed (Hawash et al., 1994; Jewell and Kabrick, 1980).

Additional problems associated with the aeration of TAD are the limited solubility of oxygen at elevated temperatures (50°C and above), and the high solids content often desired in such operations. However, in spite of the low solubility of oxygen at high temperatures, the molecular diffusivity of oxygen increases with temperature (Surucu et al., 1975). This compensates for the low solubility, leading to comparable or better oxygen transfer at the higher temperatures. In fact, oxygen-transfer efficiencies comparable to, or better than that obtainable at mesophilic temperatures have been demonstrated during the fermentation of some caldoactive organisms, including *Pyrococcus furiosus,* and *Sulfolobus shibatae* at 90°C (Krahe et al., 1996). Oxygen transfer at high temperature is also aided by the decrease in viscosity of the reaction medium with increase in temperature.

Various levels of aeration have been applied to TAD with different results. Following pilot trials in Canada, Kelly (1991) and Kelly et al. (1993) recommended that aeration rate in the digester should be maintained at 0.5 to 2.0vvh (volume air per volume medium per hour) depending on the organic load. Hawash et al. (1994) reported increase in efficiency of TAD (destruction of volatile suspended solids (VSS) and reduction of biochemical oxygen demand (BOD)) with increase in aeration rate up to 0.5vvm (the highest employed to minimise evaporation and frothing). Frost et al. (1990) reported oxygen consumption rates between 1–2 kg^{-1} kg^{-1} VSS destroyed in sewage sludge. Vismara (1985) proposed that the aeration rate required for TAD would be midway between that needed in classical aerobic systems, and that of typical anaerobic mesophilic digestions. Sonnleitner and Fiechter (1983ab) reported aeration rates of 0.02 to 0.3vvm (volume air per volume medium per minute), leading to stable final temperatures between 50°C and 67°C, during a two-year study of TAD. In studies that employed CSTR, Ugwuanyi et al. (2004ab) demonstrated that destruction of VSS and soluble COD, as well as development of critical microbial enzymes increased with the aeration rate of TAD process from 0.1vvm to a peak at 0.5vvm, before declining with further increase in the aeration rate applied to the process. Microbial population however, did not decline with the increase in aeration rate to 1.0vvm.

As the main cost of TAD is that of aeration, running it at microaerophilic rates should make it more economical (Ugwuanyi et al., 2004ab; Carlson 1982). However, a highly aerobic system will be desirable if TAD is to find a role in highly aerobic oxidations such as may be required in detoxification of aromatic and haloaromatic chemicals. Since heat generation in microbial systems is a function of oxygen consumption, at least to a point, it follows that highly aerobic systems will be more efficient in heat generation than less aerobic ones (Messenger et al., 1990; Williams et al., 1989; Birou et al., 1987; Cooney et al., 1967). An additional advantage of extensive aeration could be located in increased evaporation of waste liquid, if this is desired to reduce the amount of waste available for disposal, provided of course, that it does not lead to evaporative cooling of the waste. Achievement and maintenance of thermophilic conditions should also benefit from effective insulation, and where possible the use of heat exchangers between effluent and influent waste, particularly in very cold climates. This has led to various modification being made to different digesters that have been employed for TAD (Messenger et al., 1990; Morgan et al., 1986). These modifications include the use of stainless steel tanks with insulation cladding, as well as concrete tanks with plastic or fibre glass interior. Insulation of digesters will enable the operation of TAD with the minimal amount of aeration necessary to achieve thermophilic temperature.

Operational Temperatures Employed in TAD

There is as yet, no consensus on what constitutes or optimal temperatures, for the operation of TAD. The fact of the fluidity of definition of thermophilic temperatures as applied to microbial processes has complicated its application in aerobic thermophilic digestion. Difficulties in setting temperature standards arise from the subjective and imprecise definition of thermophily (Brock, 1986), and also from the heterogeneity of microbial population likely to operate in TAD, which will result in a wide band and overlaps of growth temperature optima. Thermophily applied to waste treatment has also been considered in relation to mesophilic treatment, or the differential between the reactor and feed temperatures. In temperate countries feed temperatures may vary from under 5°C to more than 20°C depending on the season (Vismara, 1985), while approaching 40°C in some tropical countries. The implication of this for the definition of thermophilic digestion as applied to TAD is considerable. Surucu et al. (1976, 1975) consider that TAD would be a process that operates between 50° and 60°C. Vismara (1985) considered a process that operates in a range of 40°–50°C as TAD, while Frost et al. (1990), defined TAD as a digestion which operates above 35°C, and up until temperature becomes the limiting factor (at about 70°C). Matsch and Drnevich (1977) consider TAD as that operating above 45°C, while Jewell (1991), defined it as that which achieves 43°C or above.

Notwithstanding the seeming requirement for temperatures of up to 45°C and above, the relationship, if any, between operational temperature and stabilisation efficiency in TAD is subject of considerable controversy. This has led to the imposition of mesophilic digestion standards on TAD (Koers and Marvinic, 1977; Vismara, 1985). In their simulation studies, Kambhu and Andrew (1969) considered 45°–60°C as acceptable range for TAD, and believed that the highest reaction rate constant would be achieved at 55°C while decreasing to zero at 75°C. Carlson (1982), reported increase in sludge degradation, as the temperature increased to 57°C. Hawash et al. (1994) also reported increase in the rate of waste stabilisation with temperature, with the kinetic parameters of stabilisation nearly doubling with each 10°C increase in temperature within the permissible range. Tyagi et al. (1990) reported an increase in digestion efficiency between 45° and 55°C, followed by a gradual decline thereafter. Ugwuanyi et al. (2004b) reported increase in the digestion and stabilization efficiency during TAD as the temperature increased from 45°C to 55°C, followed by slight decline thereafter. However, above 55°C, waste pasteurisation and consumption of soluble COD increased up to 60°C before declining drastically thereafter. On the other hand the degradation and solubilization of particulate waste matter increased as the temperature decreased to 45°C. The implication of these variations in the response of different waste components to changes in the temperature of the process is significant, and suggests that the choice of operation temperature will vary with the nature of the waste and the principal reason for the operation. It has also been demonstrated that the level of accumulation and quality of protein achieved during the protein enrichment of agro-food waste by TAD vary with the temperature of digestion of waste

(Ugwuanyi et al., 2006; Ugwuanyi, 1999). This is expected to impact significantly on the choice of temperature for digestion, particularly if protein enrichment is intended in the process.

Temperatures so far reported as optima for TAD correspond to the approximate optimum for growth of a range of common aerobic thermopiles (Brock, 1986). By comparison, in various studies of thermophilic anaerobic digestion, different optimum temperatures have been reported in the range of 50°–65°C, and these appear to also vary considerably with the waste type (Verstreate et al., 1996; Lettinga, 1995; Ahring, 1994). Sonnleitner (1983), Sonnleitner and Fiechter (1985, 1983bc), in some of the few studies on the microbiology of TAD to date, consider that extreme thermopiles (caldoactive organisms) are relatively fastidious with respect to their nutritional requirements for growth, and hence are poorly suited for application in TAD. Thus, only thermotolerant and moderately thermophilic organisms are likely to play a role in TAD. These organisms have been classified by Hamer and Bryers (1985) as those growing optimally at 40°–50°C and 50°–65°C respectively, thus giving a wide temperature range of 40°–65°C within which TAD may be operated efficiently. This has been demonstrated to be the case during the digestion of vegetable waste (Ugwuanyi et al., 2007).

Although there are only few full or even pilot scale experiences with TAD to check these projections, Kelly et al. (1993) and Edginton and Clay (1993) successfully operated pilot scale digesters within this range for over one year. Ponti et al. (1995ab) reported that TAD could be efficiently run at 65°C, particularly if the primary motive is to achieve waste pasteurisation. The requirement for pasteurisation is also the basis upon which Messenger et al. (1993ab) and Messenger and Ekama (1993ab) operated dual TAD-AD system at 65°C and above. Since the TAD was designed principally for pasteurisation, the waste was subsequently stabilised in the mesophilic AD system. They however, reported progressive loss of stabilisation efficiency in the TAD phase as the temperature approached and exceeded 65°C. Although no reason was advanced for the decline in efficiency, the high temperature could have robbed the system of its metabolic versatility by restricting the variety of active microbial population. This has been demonstrated to be the case (Ugwuanyi et al., 2008b). The implication of this population selection and restriction are considerable and dependent on the target of the digestion process. For instance, very high temperatures may not be an asset if TAD is to be used as a stand alone waste treatment process. In self heating systems, precise control of temperature is unlikely, and the emphasis will continue to be on a range which gives acceptable performance. The preferred range of temperature within the band at which efficient treatment have been reported will depend on a variety of factors, such as system design and treatment target, type and organic loading of waste, pathogen content and the need for pasteurisation among others.

Effect of Waste Load on Process Heating

The nature of the waste determines the amount of biodegradable chemical oxygen demand (COD), which directly affects heat evolution. As the biodegradable organic load increases, the amount of material removed increases and so does heat evolution, not least because the amount of ballast water present per unit mass of organic matter decreases. Jewell and Kabrick (1980), and Jewell (1991), have related heat evolution to COD removal by the expression: $\Delta F = 3.5 \Delta COD$;

- Where ΔF is the heat released in kilocalories per litre and
- ΔCOD is the measured change in COD in gl^{-1}.
- One kilocalorie per litre is equal to 1°C change in temperature.

The maximum temperature reached depends on the balance between heat loss and heat input. This equilibrium state is facilitated by two prominent factors,

a) as the biodegradable organic load increases the system soon becomes mass transfer limited, a situation accelerated by reduction in oxygen solubility with increase in waste (solid) load, and rapid consumption of oxygen by the proliferating microbial populations, putting more pressure on the available oxygen.

b) as the temperature of the digesting mass increases to 60°C and above, the number of viable thermopiles gradually starts to decline leading to decrease in organic matter removal, and consequently a decrease in heat production (Matsch and Drnevich, 1977; Ugwuanyi et al., 2004b). This self-regulation is very similar to what obtains in classical composting.

Within the permissible range, increase in temperature will be accompanied by an increase in digestion rate, following the classical van Hoff Arrhenius relationship, provided that care is taken to account for transition from growth of mesophiles to that of thermophiles if such temperature transition is involved. Hisset et al. (1982) reported less respiration in piggery slurry at 50°C than at 35–40°C. And Ponti et al. (1995ab) reported that respiration is greater at 60°C than at 50°C. Also, Surucu et al. (1975) reported greater activity at 55° and 58°C than at 50°C. This supports the presence of transition temperature (range) between upper limits of thermotolerant mesophiles and the lower limits of facultative thermophiles at which neither of the two groups of organisms is at a metabolic advantage (Ugwuanyi et al., 2004b). This particular range does not take account of the behaviour of the more obligate thermophiles. As of now however, the choice of temperature seems to be a matter of trial and error, and varies widely with the process and the operator.

Effect of Substrate Load on Process Development

The solids content (biodegradable solids/chemical oxygen demand (COD) and biochemical oxygen demand (BOD)) required to achieve and maintain thermophilic temperatures in TAD is not well defined. In sewage sludge treatment, it has been recommended that a high solids content is essential to achieve thermophilic temperature, hence the slurry nature of TAD (Hamer and Bryers, 1985). This draws from the experience of composting, in which high temperature is aided by high solids content and reduced moisture. However, as sewage sludge consists essentially of partially degradable microbial cells (within the context of the duration of digestion), the solid content of sludge cannot be used as a direct index of biodegradable COD. Theoretically, since the principal source of energy of the process is the enthalpic content of organic matter, the temperature reached in the reactor depends on the concentration of biodegradable material in the waste (Ugwuanyi 1999; Vismara, 1985; Wolinski and Bruce, 1984; Jewell and Kabrick, 1980).

In the absence of heat loss, the minimum theoretical concentration of biodegradable solids needed to produce a 50°C rise in temperature has been reported to be 3 g l^{-1} (Vismara, 1985). Jacob et al. (1989) recommended a minimum solid content of 2.5% for the attainment of thermophilic temperature in sewage sludge, while Jewell and Kabrick (1980) recommended 5%, to be able to sustain thermophilic temperatures for long enough to achieve waste pasteurisation. Successful treatment of wastes whose concentrations range between 2% and 6% have been reported at different thermophilic temperatures (Ugwuanyi et al., 2004b; Hawash et al., 1994; Tyagi et al., 1990; Kelly et al., 1993; Edginton and Clay, 1993; Morgan and Gumson, 1981). Because of the importance of degradable organic load in heat evolution in TAD, Surucu et al. (1975) recommended it for treatment of high strength agro-food wastes with potential for protein recovery, especially in warm climate countries and in particular for the treatment of wastes generated at high temperatures.

Microbiology of Aerobic Thermophilic Digestion

There have been only a few studies on the microbiology of TAD, mainly because the process is relatively new. Besides, application of thermophilic organisms in bioprocesses was essentially unknown before the 1970s (Brock, 1986). And studies with thermophiles have since concentrated on the caldoactive thermophiles (mostly Archaea), with a view to pure culture biotechnological application (Krahe et al., 1996), particularly the extraction of high value biochemicals, rather than biotransformation. This is in addition to their limited use in biogasification of organic wastes. As a result, the microbiology of TAD is poorly understood, and its potentials have remained largely unexploited (Fiechter and Sonnleitner, 1989). As in the case with composting, the micro-organisms responsible for TAD develop from the proliferation of thermophiles and facultative thermophiles indigenous in the waste, whose growth would have been suppressed at the mesophilic or ambient temperature of the influent waste.

The stability, or otherwise, of microbial populations during the operation of TAD is not well understood, neither is the effect of different substrate types and waste load on the (selection of) populations, since the few studies on microbiology of TAD have employed sewage sludge. An unstable thermophilic population, or one with a long doubling time, would require long retention times for waste processing, as in the case with thermophilic anaerobic digestion (Verstrate et al., 1996). On the other hand, a rapidly metabolising population with short doubling time will have the advantage of rapid waste stabilisation, short retention time and greater process stability. Sonnleitner and Bomio (1990), Sonnleitner and Fiechter (1983ab) studied the microbiology of TAD of sewage sludge at temperatures ranging from 50° to 67°C. They characterised at least 95% of the isolates as members of the genus *Bacillus* with maximal growth temperatures in excess of 70°C. The balance of 5% would have been so classified but for their inability to produce endospores in culture. The isolates showed very rapid growth rates (μ_m =0.7–2.2 h^{-1}), but low final biomass yield (0.2 to 0.3g g^{-1}) when grown in carbohydrate medium in shake flask cultures. In pilot studies they exhibited rapid adaptation at various retention times. The authors concluded that the thermophilic population responsible for heat generation in TAD consists entirely of members of the extremely thermophilic *B. stearothermophilus* group. Loll (1976, 1989) also reported that thermophilic and thermotolerant *Bacillus* spp. were responsible for the stabilisation of continuously treated model wastewater.

During TAD of swine waste, Beaudet et al. (1990), counted microbial populations at 55°C, varying from 10^4 to 10^7 ml^{-1} of waste. The organisms were identified as *Bacillus* spp., including *B. licheniformis*. Peak population was shown to vary with the final pH of the digesting waste, which seemed to vary with the COD load. During TAD of sewage sludge, 65° and 55°C thermophiles in excess of 10^6 and 10^8 ml^{-1} respectively were reported (Burt et al. 1990b). These were approximately 10^2 fold greater than the thermophilic population in feed sludge. It was considered that the 65°C population was made up of obligate thermophiles since their population did not rise significantly above that of feed sludge until waste temperature exceeded 54°C, unlike the 55°C population. Malladi and Ingham (1993) reported thermophilic aerobic spore-former population of up to 10^9 ml^{-1} during TAD of potato process waste water. They also identified *Lactobacillus* spp. during digestion at 55°C.

The paucity of information on the microbiology of TAD has left room for speculations on the diversity of microorganisms in the process. Even the limited information that exists has been based on reactions that employed sewage sludge (Sonnleitner and Fiechter, 1983ab). This waste type has limitations as the basis for projection to other more diverse and potentially reusable (particularly in animal nutrition) wastes (Ugwuanyi et al., 2006; Couillard and Zhu 1993). The possibility of upgrading and recycling of wastes by TAD requires that further studies on the microbiology of the process be implemented, particularly using such wastes that have potential for reuse in animal nutrition. In a study on the microbiology of TAD using potato process wastes, Ugwuanyi et al., (2008b) reported that the principal populations that drive the process include facultative thermophiles identified as *B. coagulans* and *B. licheniformis* (which predominated when the temperature was below 55°C) as well as obligate thermophiles identifies as *B. stearothermophilus*. These populations developed rapidly and fluctuated with changes in temperature, with the obligate thermophiles predominating as the temperature increased to beyond 55°C, such that at above 60°C they were present as nearly pure cultures. The predominant populations did not change in response to waste type, load, operational pH or aeration rate but changed in response to temperature. It is safe to state then, that a eurythermal population of thermotolerant and thermophilic organisms carry out TAD, with selection and succession responding to the local environment, particularly temperature.

Physiological Features of Thermophiles Important in TAD

Within each microbial group, thermophiles that have a wide growth temperature range are known as eurythermal, while those with more restricted range are stenothermal. Generally, thermophiles have been considered as those organisms able to grow in the temperature range of 55°C and above. This range is selected for ecological reasons. For instance, while temperatures below 50°C are common on earth surfaces, associated with sun heated habitats, temperatures above 55°C are rare as biological/ natural habitats. Additionally, 60°C is the maximum for most eukaryotic life. However, a temperature continuum exists within and between groups, making sharp delineation impracticable (Wolf and Sharp, 1981). Suutari and Laakso (1994) adopted a liberal delineation of thermophiles, in which thermotolerant organisms were defined as those with optimum growth temperature ≤45°C and T_{max} >45°C, and thermophiles as those with T_{opt} >50°C and T_{max} >60°C. They also defined as extreme thermophiles or caldoactive, those organisms with T_{opt} >65°C and T_{max} >90°C, and hyperthermophiles as the more exotic isolates able to grow at 100°C and above.

Thermophiles are mostly heterotrophic, nutritionally versatile, and capable of utilising a wide variety of organic carbon sources, including simple sugars, alcohols, organic acids, and polysaccharides. Many are able to utilise more recalcitrant and exotic compounds such as phenols, cresols, benzoates and hydrocarbons (Mutzel et al., 1996). Some may grow prototrophically, or exhibit a requirement for growth supplements such as vitamins, amino acids or complex organic mixtures (Sundaram, 1986). Oxygen is an important nutrient for the growth of aerobic thermophiles, but the low solubility of oxygen at high temperatures, makes supply difficult during large scale cultivation of thermophiles, and may impose severe limitations on the attainable cell density (Krahe et al., 1996). Thermophiles grow optimally in either acidic environment of pH 1.5–4.0 (thermoacidophiles), or in neutral to moderately alkaline pH of 5.8 and above (neutrophiles). There are also thermophilic alkalophiles, although these appear to be more restricted in abundance. The ability of thermophiles to play important roles in the process of valorisation of waste biomass (particularly of the genre of TAD) derives from some of their intrinsic biochemical features that make bioprocessing with them attractive. Some of these interesting features of thermophiles that enhance their appeal in bioprocessing are highlighted briefly below.

Growth Rate, Yield and Maintenance of Thermophiles

Thermophiles are remarkable for their ability to grow rapidly at their optimum temperature (Couillard et al., 1989). Several thermophiles with temperature optima between 55° and 70°C have generation times of the order of 11 to 16min as compared to 26min for mesophilic *B. subtilis* (Brock, 1967). However, based on theoretical expectations (of growth rate) from their high growth temperatures, it is believed that thermophiles do not grow efficiently when compared to mesophiles (Sonnleitner and Fiechter, 1983a). Thus biomass accumulation, particularly in batch cultures of thermophiles is low. This may be due to an inability to sustain their growth rate or due to low catalytic efficiency of some key thermophile enzymes, or high substrate affinity (K_s) constant of the thermophiles for their principal carbon sources (Brock, 1967). They exhibit very short exponential growth phases, due perhaps to inadequate oxygen supply at high temperature, and cell density, particularly in batch culture (Kuhn et al., 1980,1979). It is also possible that thermophiles exhibit enhanced susceptibility to toxic metabolites at high temperatures. During growth of thermophilic *Bacillus* spp isolated from TAD of potato process waste in glucose mineral medium Ugwuanyi (2008) reported peak specific growth rate of *B. stearothermophilus* as $2.63\mu h^{-1}$ at 1.0vvm aeration rate and 60°C, declining with decrease in aeration rate and with increase or decrease in growth temperature. *B. coagulans* had maximum specific growth rate of $1.98\mu h^{-1}$ at 55°C and 1.0vvm while *B. licheniformis* had its peak of $2.56\mu h^{-1}$ at 50°C and 1.0vvm. For both organisms also, the specific growth rates declined as the aeration rates decreased.

Growth, yield and maintenance requirements of a variety of thermophiles have been studied in batch and continuous culture. A yield of 65g cells mole^{-1} glucose was reported for *B. stearothermophilus nondiastaticus* growing optimally at 55° to 58°C in batch conditions. This value is lower, than the range of 70 to 95g mole^{-1} glucose reported for a variety of mesophilic aerobes (Payne, 1970). The lower yield is believed to be due to the high maintenance requirement of thermophiles (Sundaram, 1986). Low yield of thermophiles has also been reported during TAD of slaughterhouse effluent (Couillard et al., 1989). In continuous cultures however, thermophiles exhibit yields similar to mesophiles. Thus, *B. caldotenax* yielded 89g mole^{-1} glucose at 65° and 70°C (Kuhn et al., 1980), while *B. acidocaldarius* gave 83.5g mole^{-1} glucose at 51°C pH 4.3 (Farrand et al., 1983). Thermophilic populations isolated from TAD of potato process waste namely *B. licheniformis, B.coagulans* and *B. stearothermophilus* have been studied for their yield under conditions similar to waste digestion (Ugwuanyi, 2008). Peak biomass yield of 72.72gmol^{-1} was obtained at 50°C for batch culture of *B. stearothermophilus* growing in glucose mineral medium.

There is also variability in yield of thermophiles based on oxygen uptake. For instance, *B. caldotenax* has a yield of 50g mole^{-1} at 65° and 70°C compared to 56g mole^{-1} at 30°C, and 28g mole^{-1} at 35°C for *E. coli*, or 52g mole^{-1} at 35°C for *Klebsiella aerogenes*, suggesting that thermophiles may be more efficient than mesophiles within their permissive temperature in continuous culture (with regards to yield on oxygen). Similar variability based on oxygen supply was reported in respect of TAD associated *Bacillus* spp growing in glucose mineral medium (Ugwuanyi, 2008). Some thermophiles, particularly *Bacillus* spp have high maintenance requirement, which has been postulated as the reason, in addition to high decay rate, for their low yield in batch culture (Couillard et al., 1989). For instance, *B. caldotenax* has a maintenance requirement of 4.1 and 20 mmoles g^{-1} h^{-1} of glucose and oxygen respectively at 70°C and 3.8 and 20 mmoles respectively at 65°C. These figures are up to ten times the maintenance need of mesophiles. However, wide variations in maintenance requirements exist among thermophiles and this may be influenced by the environment.

Effect of Temperature on Growth Yield and Maintenance

Reports on the effect of temperature on the yield of thermophiles seem somewhat contradictory. During batch cultivation of *B. stearothermophilus*, Coultate and Sundaram (1975) found that yield on glucose decreases with increase in temperature. They proposed that this was due to inefficient co-ordination of oxidative and non-oxidative phases of metabolism, leading to accumulation of incompletely utilised glucose carbon as organic acids. Acetate accumulation has also been reported in mesophiles (Holms and Bennett, 1971), where yields seem to remain constant over a wide temperature range, but drop sharply above the optimum temperature (Sundaram, 1986). However, incomplete utilisation of glucose does not seem to be significant during continuous culture of *B. acidocaldarius*, particularly if it is glucose limited (Farrand et al., 1983). During culture of thermophilic *Bacillus* spp. isolated from TAD process in glucose mineral medium, decrease in biomass yield with increase or decrease in temperature away from the optimum, was associated with increased accumulation of overflow metabolites as volatile fatty acids, with acetate being the predominant metabolite (Ugwuanyi, 2008).

A *Bacillus* sp. studied by Matsche and Andrew (1973) gave constant yield on glucose between temperature of 45° and 60°C. However, for *B. caldotenax*, yield increased with temperature between 50° and 70°C (Kuhn et al., 1980). In spite of contradictions relating to yield, maintenance coefficient and death rates of thermophiles increase consistently with increase in temperature (Brock, 1986). It is envisaged that high growth rates and maintenance requirements, and low yield will lead to enhanced breakdown of organic materials, thus combining the low sludge production advantage of anaerobic digestion with the efficiency of aerobic processes. This is in addition to the accumulation of useful metabolic products at rates higher than would be expected of their mesophilic counterparts (Sonnleitner and Fiechter, 1983c), and could be the foundation for the exploitation of thermophiles in waste treatment, biotransformation and production of biochemicals.

DETERMINANTS OF MINIMUM AND MAXIMUM GROWTH TEMPERATURE

All bacteria have characteristic minimum (T_{min}) and maximum (T_{max}) temperatures for growth. The determinants and regulators of these limits are not well understood, but extensive research efforts go into determining the mechanism of thermophily, particularly with regards to the nature of thermophile proteins and membranes (Lieph et al., 2006). The observations that heat sensitive mutations lead to a lowering of T_{max} but not T_{min} suggests that both are independent, single genome traits (Sundaram, 1986). Membrane lipids of thermophiles tend to have a greater abundance of high melting point (saturated long chain) fatty acids than their mesophilic counterparts (Suutari and Laakso, 1994). These can be further adjusted, depending on the growth temperature. Bacterial phospholipid bilayer membranes undergo a thermotrophic reversible transition between an ordered rigid gel (solid phase), and a fluid (liquid crystalline) phase. This transition involves the melting of the hydrocarbon chains in the interior of the bilayer. The temperature at which this takes place depends on the nature (melting point) of the membrane fatty acids (Cronan and Gelman, 1975).

The heterogeneity of membrane lipid composition implies that phase changes take place over a range of temperature, rather than at a discrete point, with liquid crystalline and solid gel phases existing simultaneously. Above the upper boundary, liquid crystals obtain while solid gels exist below the lower boundary. Depending on the type of organism, membrane bound proteins may, or may not, have any effect on the phase transition temperature (McElhany and Souza, 1976; Lipowsky, 1991). The quantity and quality (ratio and types) of membrane lipids and proteins of thermophiles have also been shown to vary with the growth temperature (Suutari and Laakso, 1994). In the eubacterium *B. stearothermophilus* this manifests as a decrease in total lipid, and an increase in membrane protein, whereas in the archae *T. aquaticus* the quantity of lipid increases with temperature. In *B. stearothermophilus* and *E. coli*, the existence of membranes in liquid crystalline phase is believed to be essential for performance of membrane functions. This implies a role for the membrane transition point in setting T_{min} and T_{max}. Changes in conformation of essential enzymes, leading to loss or reduction of catalytic activity, or alteration of important regulatory characteristics, which may lead to arrest of growth below certain temperature (T_{min}), may also affect minimum temperature of thermophiles. Additionally, ribosomal assembly in thermophilic bacteria has been shown to be energy intensive and may thus contribute in setting a high minimum temperature for growth (Suutari and Laakso, 1994).

As for T_{min}, determinants of T_{max} are not well understood. Early suggestions were that the membrane lipid phase transition temperature may be important in setting the upper temperature. This position was prompted by observations in some thermopiles that increase in growth temperature led to an increase in the melting point of membrane lipids. It is contradicted however, by observations that the phase transition temperature may be varied by manipulating the lipid composition without affecting the T_{max}, while (in *B. stearothermophilus*) mutations which lowered T_{max} did not affect the lipid transition temperature (Kawada and Nosoh, 1981). In spite of these observations, membrane stability may be a major determinant of T_{max} in so far as cell death may be accelerated by heat damage to the membrane above its optimum temperature (Kuhn et al., 1980). The fluidity of membrane lipid (transition phase) will therefore be an index of stability at a given temperature. Membranes of thermophiles have been shown to be more stable than that of mesophiles, and to increase in stability with temperature (Suutari and Laakso, 1994).

It is also believed that there is a limit to the gap that can exist between the transition temperature and T_{max}, i.e., a limit to the level of fluidity of membranes in growing bacteria. Mutants of *E. coli* defective in lipid synthesis were unable to grow under conditions that led to the incorporation of very low melting point lipids. It is presumed that incorporation of such lipids in the membrane led to production of membranes that were too fluid and therefore unstable at the T_{opt}. A similar observation was made in the growth of heat sensitive *B. stearothermophius* (Sundaram, 1986). This organism could not adjust its membrane phase transition point beyond 41°C, and rapidly lost the ability to grow beyond 58°C, which became the T_{opt} and T_{max}, as well as the upper limit for membrane stability. The wild type organism could adjust its upper phase transition point to 65°C, and could grow at 72°C. The T_{max} of microbes may also be affected by temperature stability of various macromolecules, particularly enzymes (Lindsay, 1995).

TURNOVER OF MACROMOLECULES IN THERMOPHILES

Early theories of thermophily suggested that high temperatures led to a rapid turnover of macromolecules, particularly proteins. This, however, has not been demonstrated in *B. stearothermophilus nondiastaticus*, which does not turn over its proteins and RNA (other than mRNA) faster than do mesophiles (Sundaram, 1986). Low protein turnover has also been reported in *T. aquaticus* (Kenkel and Trela, 1979). It has been reported that thermophile proteins are more stable than their mesophile counterparts, hence the low turnover at high temperatures (Moat and Foster, 1995). However, turnover of macromolecules in thermophiles may be significantly affected by medium composition (Sundaram, 1986). Macromolecule stability may be due to their association with high concentrations of carbohydrates and their ability to bind certain ions more strongly (Bergquist et al., 1987). Increased stability of proteins has also been attributed to strategic substitutions in their amino acid sequence. Enhanced DNA stability has been attributed to a higher G+C content in thermophiles than in mesophiles (Lindsay, 1995). Enhancement of tRNA stability with growth temperature has been reported in *T. thermophilus* in which the ribothymidylate, a normal component of tRNA is replaced with 5-methyl.2-thiouridylate, and the level of substitution increases with temperature. The higher growth and decay rate reported in thermophiles from the TAD of slaughterhouse waste, than mesophiles indicate that there are differences in the stability and turnover of macromolecules between thermophiles and mesophiles (Couillard et al., 1989).

Development, Adaptation and Stability: Thermophiles in TAD

Thermophiles responsible for TAD of sewage sludge develop from the proliferation of populations naturally present in the sewage (Ugwuanyi et al., 2008b; Sonnleitner and Fiechter, 1983a). As with compost (Abouelwafa et al., 2008; Raut et al., 2008; Yu et al., 2008; Cayuela et al., 2008; Adams and Frostic 2008; Li et al., 2008; Cunha-Queda et al., 2007; Mari et al., 2003; Heerden 2002; Hassen et al., 2001; Thambirajah et al., 1995), these remain dormant during startup while mesophiles, and then facultative thermophiles, build up the temperature until a thermophilic range is reached, when they begin to grow (Yun et al., 2000; Burt et al., 1990b). Although Sonnleitner and Fiechter (1983a) observed up to 5-log orders in the fluctuations of populations of thermophiles, washout was never observed. Over a two-year observation of a pilot scale continuous TAD, the viable thermophilic population remained at or greater than $10^5 g^{-1}$ of sludge even when hydraulic retention time (HRT) was reduced to as low as 10h, at aeration rates ranging between 0.02 and 0.3vvm. Extreme thermophiles (growth at $\geq$ 60°C) were particularly well maintained. Rapid and comparable development of different thermophiles was observed during the TAD of potato process waste at aeration rates ranging from 0.25vvm to 1.0vvm and different temperatures (Ugwuanyi et al., 2008b). Similar stability was reported in the thermophilic biofilm responsible for the stabilisation of swine waste (Beaudet et al., 1990). This has led to suggestions that TAD should be operated at the highest possible temperature, since that could result in substantial reduction in HRT without compromising efficiency of waste stabilisation and pasteurisation.

Although high thermophilic populations are maintained during TAD, Beaudet et al. (1990) however, reported a drop in treatment efficiency as the temperature of the waste exceeded 60°C due to a restriction in the diversity (and population) of the digesting thermophiles. The persistence and stability of thermophiles in TAD has been attributed to the selection of populations with very rapid growth rate (Ugwuanyi et al., 2008b; Sonnleitner and Fiechter, 1983a,b). Unlike (thermophilic) anaerobic digestion where the overall pace of the process is determined by the growth rate of the slowest growing populations in a synthrophic interaction, the fastest growing populations in TAD determines the stability and pace of digestions.

Development of Hydrolytic Enzymes During TAD

The development of enzyme activities during TAD has not been actively investigated, even though it is believed that this may aid the development of control parameters for the process (Ugwuanyi 1999; Bomio et al., 1989). This is unlike the case with composting in which hydrolytic enzyme development has been studied in considerable detail and has been used variously to characterise and monitor process development (Abouelwafa et al., 2008; Raut et al., 2008; Yu et al., 2008; Cayuela et al., 2008; Adams and Frostic 2008; Li et al., 2008; Mari et al., 2003; Heerden et al., 2002; Hassen et al., 2001; Thambirajah et al., 1995). Protease was been reported as the major activity present in sewage sludge during TAD (Ponti et al., 1995a,b; Bomio et al., 1989). Bomio et al. (1989) also detected low activity of lysozyme, and concluded that other macromolecules whose corresponding enzymes were not detected were either not present (hence did not induce activities) or passed out of the system with only minor modifications.

However, Burt et al. (1990b) recorded considerable degradation of lipids in sludge during TAD, and suggested that lipases are probably produced during the process but may be easily degraded by proteases in the digesting waste (and so not readily measured). Grueninger et al. (1984) also suggested that amylases produced by the *Bacillus* spp. in TAD may be degraded by protease. During TAD of potato process waste, Ugwuanyi et al. (2004ab) demonstrated rapid development of hydrolytic enzymes including amylases, cellulose, proteases and xylanases. These enzymes were constitutive but their activities also responded to the dynamics of substrate concentration and population development in the digesting waste.

Microorganisms associated with TAD have shown considerable capacity to degrade various organics in the treated wastes, and have been considered as a possible reservoir of organisms capable of producing thermostable, industrially useful enzymes (Ugwuanyi et al., 2004ab; Malladi and Ingham, 1993). The elaboration of a variety of the hydrolytic enzymes is probably repressed in the presence of easily metabolisable waste components (Ugwuanyi 1999; Sonnleitner and Fiechter, 1983b). It is reasonable to expect that enzymatic activities associated with TAD will vary with the composition of waste, and knowledge of these may be employed to improve the digestion efficiency in the process (Kim et al., 2002).

Efficiency and Performance of TAD

Standards for comparing the efficiency of TAD processes are non-existent or at best discordant, apparently due to the variety in the type of wastes to which TAD may be applied as well as the potential products of the digestion processes. Performance of processes designed to achieve protein enrichment of waste for reuse in animal nutrition will need to be assessed differently from those intended to pasteurise waste prior to treatment by another method. These may also be assessed differently from those in which the process is employed to achieve waste stabilization, prior to disposal. Where TAD is used for waste treatment, prior to disposal, the desired performance depends on whether the process is used alone for complete treatment, or as a pre/ post treatment step (Frost et al., 1990; Loll, 1989).

Matsch and Drnevich (1977) recorded 30–40% volatile suspended solids (VSS) removal at 50°C and 4.2d retention time, while treating waste secondary sludge. Similar degradation was obtained by Jewell and Kabrick, (1980), while Burt et al. (1990a,b) achieved over 60% COD removal and 45% destruction of soluble solids during continuous operation of a pilot scale sewage sludge digestion at just under 50°C. Loll (1989), reported that TAD at over 50°C, and HRT of 0.5-3d when used in conjunction with a mesophilic anaerobic process led to a reduction in overall treatment time required in the anaerobic process by up to 10 days, in addition to pasteurisation of the sludge. TAD results in enhanced solids destruction, because of the high maintenance energy requirements of thermophiles (Brock, 1986). It has been reported that the settleability of sludge increases with increase in treatment temperature due to reduction in viscosity (Kambhu and Andrews, 1969). TAD also achieves odour reduction (resulting in sludge with acceptable odour) in sewage works at various retention times and loading rate (Murray et al., 1990).

Monitoring and Control of Aerobic Thermophilic Digestion

In spite of the advantages of TAD, comparatively little has been done to achieve good understanding of the process, particularly in the area of process monitoring and control (Maloney et al., 1997; Sonnleitner and Fiechter, 1983b). This is due, both to the relative newness of the process, and the diversity of possible applications, resulting in poorly defined process objectives and end points. Consequently, control of the process is still subject of trial and error, with no TAD specific control parameters in use. As a waste treatment process, control parameters have largely been based on those applied to activated sludge (from which it has evolved). Besides, much of the work carried out on the process have been limited to sewage sludge treatment (and hygeinisation), where classical control parameters have been conveniently, though inadequately applied (Kelly et al., 1993).

Based on the "degree-day" control process for activated sludge process (Koer and Marvinic (1977), Vismara (1985) recommended that treatment may be considered satisfactory if the product of operation temperature and time equal or exceed 250. As a complete waste treatment process applied to sewage sludge, EPA suggests that 40% of volatile suspended solids of sludge should be removed, while maintaining a minimum dissolved oxygen content of 2.0mg l^{-1} (Matsch and Drnevich, 1977). Additionally, sludge to be disposed of to agricultural land should have been held at 55°C for at least three days. The UK Department of the Environment (DoE) recommends that sludge to be disposed of to agricultural land should have been held at 55°C for a minimum of four hours at a retention time of 7 days (Frost et al., 1990).

Matsch and Drnevich (1977) suggested criteria based on the odour producing potential, provided that the parameters can be related to an easily measurable variable such as Oxygen Uptake Rate (OUR) or redox potential. Other control parameters that have been applied to TAD include, total/ viable microbial count and count of thermophiles at 65°C, ATP level and dehydrogenase activity, as well as lipid content, and changes in pH (Messenger et al., 1990; Burt et al., 1990a; Droste and Sanchez, 1983). These are in addition to classical waste quality parameters such as those related to profile of COD, BOD, TSS, and VSS. Although the latter parameters are acceptable where the target is the safe disposal of waste such as in sewage sludge, they are inadequate where TAD is part of an integrated process, and in particular during waste upgrading and recycling reactions.

The development of hydrolases has been studied in a number of waste treatment processes as monitoring parameters, including in landfill (Barlaz, 1997; Yamaguchi et al., 1991; Barlaz et al., 1989; Jones and Grainger, 1983a,b), anaerobic digestion (Sarada and Joseph, 1993a,b; Palmisano et al., 1993a,b; Godden et al., 1983) as well as in composting (Raut et al., 2008; Cayuela et al., 2008; Poulsen et al., 2008; Garcia et al., 1993). In these cases, different activities were considered reliable for following process performance. This is understandable, as the degradation of polymers is known to be a rate controlling step in waste digestion (Rivard et al., 1994; 1993; Mason et al., 1987, 1986; Eastman and Ferguson, 1981).

Though protease has been reported as the predominant activity during TAD of sewage sludge (Mason et al., 1992; Bomio et al., 1989), it has not been used as a monitoring tool nor has the profile of any other hydrolytic activities been used. Amylase, xylanase, cm-cellulase and proteases have also been shown to be related with the degradation and stabilization of waste and thermophilic population development during TAD of potato process waste

(Ugwuanyi et al., 2004ab). It was suggested that the profile of some of these enzymes may be used as process monitoring tools during TAD. This should be the case, since the degradation of the polymers determines the pace of waste stabilization in TAD processes. There is sparse information that describes hydrolytic activities or even polymer breakdown in TAD, although Bomio et al. (1989) suggested that up to 66% of the metabolic activities in TAD may be due to degradation of insoluble polymers.

Measurements of hydrolytic activities are considered economical, easy to execute and interpret, and may be process and waste type specific and adaptable. Whenever hydrolytic activities may be correlated with the degradation of polymers, it can become handy for monitoring the progress of waste stabilization. Ultimately, the monitoring and control of TAD will need to take into consideration the source and type of waste, and the aim of the treatment process, as a waste stream that is being reprocessed for use in animal nutrition will require different control parameters from wastes meant for land application and disposal. Similarly, a stand alone TAD process will require different control parameters from one in which it is part of an integrated process, as either pre- or post-treatment, pasteurisation or hydrolytic process. Ultimately, a process specific control fingerprint will need to be developed for any and each particular process.

Some Applications of Thermophilic Aerobic Digestion

TAD arose as a response to the need to find safe, economical and environmentally friendly means of disposing excess sewage sludge to land. This followed legislative restrictions on the disposal of sludge to sea, or of raw (unpasteurised) sludge to agricultural land (Kelly et al., 1993). It is understandable therefore, that virtually all early studies on TAD as well as more recent ones have focused on the use of this process in sewage sludge treatment and pasteurisation (Hawash et al., 1994; Chu et al., 1994; 1996; 1997). Notwithstanding, TAD has also been used for the treatment of a variety of other waste types, particularly those of agricultural and food industry origin intended for disposal. It has been deployed as the sole treatment method for different high strength wastes such as swine slurry (Beaudet et al., 1990), slaughter house effluent (Couillard et al., 1989), potato process waste (Ugwuanyi et al., 2005ab; Malladi and Ingham, 1993), and dairy, brewery/ distillery and other food industry wastes (Zvauya et al., 1994; Gumson and Morgan, 1982; Loll, 1976; Popel and Ohmnacht, 1972) spent liquor of pulp mills (Zentgraf et al., 1993), olive mill waste (Becker et al., 1999) and a wide variety of other waste, particularly those loaded with toxic organic and inorganic chemical (Adav et al., 2008). Sonnleitner and Fiechter (1983b,c) proposed that it may be used to treat virtually any kind of organic waste, provided that for particularly recalcitrant wastes enough time is allowed for the adaptation of the process, preferably in continuous processes. It has also been proposed that the metabolic versatility of TAD associated thermophiles may be exploited in the use of this process for the reprocessing and protein enrichment of a variety of wastes for use in animal nutrition, and in the production of high value bichemicals (Ugwuanyi, 1999).

Application of Thermophilic Aerobic Digestion in Waste Pasteurisation

Conventional waste treatment processes are accepted to be inefficient from a hygienic and epidemiological point of view (Pagilla et al., 1996; Plachy et al., 1995; 1993; Juris et al., 1993; 1992; EPA, 1992, 1990; De Bertoldi et al., 1988). Consequently, one of the major attractions of TAD has been the potential for exploiting its high temperatures to achieve the destruction of pathogenic microorganisms as well as protozoa, viruses and parasite eggs in treated wastes. These can all theoretically be destroyed at the temperatures typical of TAD (Wagner et al., 2008; Ugwuanyi et al., 1999; Mason et al., 1992; Kabrick and Jewell 1982).

Although various studies have been carried out on the inactivation of pathogens during waste treatment particularly in mesophilic processes and in composts (Gerba et al., 1995; Kearney et al., 1993ab; Carrington et al., 1991; Olsen and Larsen, 1987; Abdul and Lloyd, 1985), few have described the behaviour of pathogens and indicators in TAD (Borowski and Szopa 2007). As a result, various, often conflicting projections have been made on the capacity of this process to achieve pathogen destruction, based on the known sensitivity of indicator organisms to thermophilic temperatures (Ponti et al., 1995a; Messenger et al., 1993b). Jewell and Kabrick (1980) and Jewell (1991), reported that maintenance of the temperature of digesting sludge at 50°C for 24h destroys most pathogens including bacteria, viruses, protozoa and parasite (*Ascaris*) eggs, as does 60°C for 1h. Carlson (1982) reported that 24h was required to completely inactivate enteric bacteria in sewage sludge at 57°C, while Burt et al. (1990b) reported a reduction in populations of *E. coli* in sludge from 10^6-10^7 per gram to undetectable levels after digestion at 50°C, and a HRT of 24h. Efficient destruction of helminth eggs and ova as well as protozoan parasites, at different temperatures in TAD has also been reported (Whitmore and Robertson, 1995; Plachy et al., 1995,1993; Juris et al., 1992).

TAD has been reported to achieve better pasteurisation of waste than does composting (Paggila et al., 1996; Droffner and Brinton, 1995; Carrington et al., 1991), probably due to the presence of high levels of solids in composts, which may protect pathogens from thermal inactivation. Ponti et al. (1995a) applied TAD to achieve rapid inactivation of *E. coli* in sewage sludge at 55° and 57°C, and reported that increase in waste solid content considerably protected cells from thermal destruction. Morgan and Gumson (1989) reported that temperatures in excess of 50°C were required to eliminate *E. coli* and *Salmonella* in a continuous flow TAD with retention time of 8 days at Pontir Sewage Works, UK. Murray et al. (1990) reported efficient sludge pasteurisation in a small sewage treatment works even during periods of irregular sludge feed, while Spaull and McCormack (1988) employed TAD for the inactivation of plant pathogens and weed seeds with considerable success.

Though rapid destruction of mesophiles (*E. coli*) occurs between 54° and 62°C, recent studies have shown that mesophiles in compost may survive for very long periods, and may even multiply in the process, at temperatures comparable to what is likely to be employed in TAD (Droffner and Brinton, 1995). Besides, various pathogens can be expected to respond differently to the effect of heat at different pH, DO and waste solid contents (Plachy et al., 1995). Ugwuanyi et al., (1999) reported various D_{values} for different human and animal pathogens during TAD of agricultural waste. However, more studies are needed to be able to make projections on the survival of various pathogens in TAD, particularly given the variety of waste and digestion conditions likely to obtain in full scale processes. Since the survival of a single pathogen is often enough to cause re-infection of treated waste, Hamer, (1989) recommended that microbial death in waste treatment should be followed by plate count.

Although differences in design and assay procedures make comparisons between inactivation data in TAD difficult, the Environmental Protection Agency (EPA) recommends that sludge intended for spreading on agricultural and pasture land should be held at 55°C for 3 days to achieve pasteurisation ('class A' status (EPA, 1992)). While in the UK, the Department of the Environment (DoE) recommends that exposure of sludge to a temperature of 55°C for at least 4h, will reduce the number of viruses and pathogenic organisms to levels acceptable for disposal to agricultural land. Reduction of pathogens levels in wastes intended for land or even sea disposal is the subject of several legislations in Europe and North America, (Droffner and Brinton, 1995; EPA, 1992). This is a major driving force in the development, and acceptability of TAD. When this process is to be employed in waste reprocessing for animal feed use, new parameters related to control of infectious agents of both humans and animals will be expected to be given prominent consideration. This is particularly important with the increasing attention on the spread of zoonotic diseases.

Detoxification of Xenobiotics in TAD

The use of TAD for the destruction of xenobiotics and noxious chemicals that may contaminate otherwise recycleable waste biomass has received little attention (Laine and Jorgensen, 1996). The potential for application of TAD in this area derives from the ability of aerobic thermophiles to rapidly degrade a variety of chemicals due to their enzymatic versatility (Hernandez-Raquet et al., 2007; Moeller and Reeh 2003; Knudsen et al., 2000; Banat et al., 2000; Jones et al., 1998; Reinscheid et al., 1997) and removal efficiency may only be constrained by the bioavailability of the anthropogenic chemical (Patureau et al., 2008). Various enzymes such as oxygenases, ligninases and peroxidases, which are powerful and essential biocatalysts in xenobiotic degradation, remain the preserve of aerobic organisms (Adav et al., 2008; Verstreate et al., 1996). In TAD, these may play significant roles in the detoxification of xenobiotics, and a variety of plant toxins, such as linamarin in cassava, and related plant cyanogenic glycosides, whose presence in plants limit their use. The application of TAD in detoxification of xenobiotics and toxins will certainly increase its appeal as a waste management and recycling process.

Detoxification of cyanogenic glycoside (linamarin) associated with cassava and cassava process wastes using thermophiles isolated from TAD of agricultural waste has been demonstrated (Ugwuanyi et al., 2007). Although, little information currently exists on the use of either this process or any other for detoxification of waste in full scale operations, it is expected that the high temperatures of TAD and the selection of suitably adapted microbial population could lead to the breakdown of the toxins and noxious chemicals in various waste streams.

Protein Enrichment and Reprocessing of Waste by TAD

Various organic wastes, particularly the agro-food types, may be considered as resources insofar as their energy and mineral content are concerned. However, the recycling of these wastes is strictly tied to their microbiological and sanitary quality. Although in classical studies on the use of TAD waste upgrade and reuse were not always the intended endpoints, it has vigorously been proposed that TAD, as the vehicle for the protein enrichment of waste, may be applied to waste intended for upgrading and recycling as (components of) animal feed (Coulthard et al., 1981) given also that several agricultural wastes are currently being studied for upgrading and recycling, particularly in solid state fermentations including by silage (Villas-Boa et al., 2003; Laufenberg et al., 2003). The attraction for this application of TAD is buoyed by the fact that many agricultural and food industry wastes are already being employed in this capacity (Houmoumi et al., 1998; Krishna and Chandrasekaran, 1995; Coillard and Zhu, 1993; Barington and Cap, 1990). It is envisaged that carbohydrate-rich wastes, high-quality agricultural refuse and by-products, food industry wastes and other organic wastes may eventually be treated by this process, to achieve cost-effective pasterurisation, stabilisation and protein enrichment of zero-cost materials for animal feed use (Zentgraf et al., 1993; Coillard and Zhu, 1993; Barington and Cap, 1990). The emphasis here is on exploitation of the capacity of thermophilic populations to degrade carbohydrates and lipids (with loss of carbon as carbon dioxide) while accumulating nitrogen as high quality microbial protein (Ugwuanyi, 1999; Krahe et al., 1996; Bergquist et al., 1987), besides the selective conservation of waste protein under appropriate (elevated temperature) digestion conditions where nitrification is unlikely to occur.

The pattern of protein accumulation in TAD is influenced by the metabolism of intrinsic particulate and soluble waste proteins, the metabolism and disposition of microbial proteins, as well as the metabolism, particularly the conservation as microbial protein/biomass of intrinsic and extrinsic other nitrogenous compounds including ammonia in the waste slurry. Protein enrichment of wastes by TAD has considerable implication for global food security, especially in the tropics where, on account of scarcity, animals and humans often compete directly for the same sources of nutrition. Accumulation of protein in TAD-treated waste results from the selective degradation of carbohydrate, resulting in the loss of carbon (Ugwuanyi et al., 2006a). This leads to a relative accumulation of nitrogen as protein (possibly of waste origin but also, importantly, as microbial protein), and the relative increase in percentage of digested waste remaining as protein. If mineral nitrogen is provided, this is also converted to microbial biomass protein. The result of this nitrogen turnover is the conversion of low-quality waste biomass into higher-quality (microbial) protein-enriched digest. The protein-rich digest may then be economically applied to animal feeding. The extent of accumulation of protein is a function of waste digestion conditions, including temperature, aeration rate and pH (Ugwuanyi et al., 2006a).

Concluding Remarks and Safety Consideration

Global perception of waste in general and agricultural wastes in particular is changing rapidly in response to need for environmental conservation, sustainable agricultural productivity and global food security. Consequently, most of such wastes are currently seen more as resources in the wrong form and location that needs to be reprocessed and reused than as wastes to be disposed of. The need for appropriate technologis for the reprocessing of such resources has become more compelling as are the economic prospects arising from such efforts. Protein enrichment of waste for use in animal nutrition offers opportunities for the economic reuse of abundant agricultural wastes and refuse. Low tech, low cost biotechnological processes such as TAD offers great opportunities for the reprocessing and reuse of abundant agricultural wastes (particularly those generated as slurries or at elevated temperatures) in animal nutrition. The use of thermophilic organisms (perhaps also with GRAS status) to effect the protein enrichment, biomass production and detoxification reactions will help improve confidence in the final products derived from this process, and help drive its development and application in the valorization of agricultural wastes.

REFERENCES

Abdul, P. and Lloyd, D. (1985). Pathogen survival during anaerobic digestion: Fatty acids inhibit anaerobic growth of *Escherichia coli*. *Biotechnol Letters* 7 125-12

Abouelwafa, R., Amir S., Souabi, S., Winterton, P., Ndira, V., Revel, J-C., Hafidi, M. (2008). The fulvic acid fraction as it changes in the mature phase of vegetable oil- mill sluge and domestic waste composting. Bioresource Technology **99** 6112- 6118

Adams, J.D.W. and Frostick, L.E. (2008). Investigating microbial activities in compost using mushroom (Agaricus bisporus) cultivation as an experimental system. *Bioresource Technology* 99 1097–1102

Adav S.S, Lee D-J, Show K-Y. and Tay J-H (2008). Aerobic granular sludge: Recent advances. *Biotechnology Advances* 26 411–423

Agarwal S, Abu-Orf M. and Novak J.T. (2005). Sequential polymer dosing for effective dewatering of ATAD sludges. *Water Research* 39 (2005) 1301–1310

Ahn, J.-H. and Forster, C.F. (2002a). A comparison of mesophilic and thermophilic anaerobic upflow filters treating paper pulp liquors. *Process Biochemistry* 38 257-262

Ahn, J-H. and Forster, C.F. (2002b). The effect of temperature variations on the performance of mesophilic and thermophilic anaerobic filters treating a simulated paper mill wastewater. *Process Biochemistry* 37 589–594

Ahn, J-H., Forster C.F. (2000). A comparison of mesophilic and thermophilic anaerobic uplow filters *Bioresource Technology* 73 201-205

Ahring, B. K. (1994). Status on science and application of thermophilic anaerobic digestion. *Water Science and Technology* 30 241-249

Ahring, B. K. (1995). Methanogenesis in thermophilic biogas reactors. *Antonie van Leewenhoek* 67 91-102

Ahring, B.K., Ibrahim A.A. and Mladenovska, Z. (2001). Effect of temperature increase from 55° to 65°C on performance and microbial population dynamics of an anaerobic reactor treating cattle manure. *Water Research* 35, 2446–2452,

Aitken, M.D., Sobsey, M.D., Van Abel, N.A., Blauth, K.E., Singleton, D.R., Crunk, P.L., Nichols, C., Walters, G.W. and Schneider, M. (2007). Inactivation of Escherichia coli O157:H7 during thermophilic anaerobic digestion of manure from dairy cattle. *Water Research* 41 1659-1666

Aksu T, Byatok E. and Bolat D. (2004). Effects of bacterial silage inoculant on corn fermentation and nutrient digestibility. *Small Ruminant Research* 55: 249-252

Alfano, G., Belli, C., Lustrato, G. and Ranalli, G. (2008). Pile composting of two-phase centrifuged olive husk residues: Technical solutions and quality of cured compost. *Bioresource Technology* 99. 4694–4701

Alvarado-Lassman, A., Rustrian, E., Garcıa-Alvarado, M.A., Rodrıguez-Jimenez, G.C. and Houbron, E. (2008). Brewery wastewater treatment using anaerobic inverse fluidized bed reactors. *Bioresource Technology* 99 3009–3015

Andrews, S.A., Lee, H. and Trevors, J.T. (1994). Bacterial species in raw and cured compost from large scale urban composter. *Journal of Industrial Microbiology* 13 177-182

Angelidaki, I. and Ahring, B.K (1994). Anaerobic thermophilic digestion of manure at different ammonia loads: effect of temperatures. *Water Research* 28 727-731

Angelidaki, I. and Ahring, B.K.(1993). Effect of clay mineral bentonite on ammonia inhibition of anaerobic thermophilic reactors degrading animal waste. *Biodegradation* 3 409-415

Angelidakia, I. Chena, X., Cuia, J., Kaparajua, P. and Ellegaard, L. (2006). Thermophilic anaerobic digestion of source-sorted organic fraction of household municipal solid waste: Start-up procedure for continuously stirred tank reactor. *Water Research* 40 2621-2628

Appels, L., Baeyens, J., Degreve, J. and Dewil, R. (2008). Principles and potential of the anaerobic digestion of waste-activated sludge. Progress Energy Combust Sci, doi:10.1016/j.pecs.2008.06.002

Archer, D.B. and Kirsop, B.H. (1990). Microbiology and control of anaerobic digestion. In Wheatley, A. (ed.) *Anaerobic digestion; a waste treatment technology*. London, Elsevier Applied Science pp 43-92

Arvidsson K, Gustavsson A-M. and Martinsson K. (2008) Effect of conservation method on fatty acid composition of silage. Animal Feed Science and Technology doi:10.1016/j.anifeedsci.2008.04.003

Bampidis VA. and Robinson PH (2006) Citrus by-products as ruminant feeds: A review. *Animal Feed Science and Technology* 128: 175-217

Banat, F.A., Prechtl, S. and Bischof, F., (2000). Aerobic thermophilic treatment of sewage sludge contaminated with 4-nonylphenol. Chemosphere 41, 297–302.

Barlaz M.A. (1997) Microbial studies of landfills and anaerobic reefuse decomposition. In: Hurst CJ, Knudson GR, McInerney MJ, Stetzenbach LD, Walter MV (eds.) *Manual of Environmental Microbiology*. Washington, DC: ASM Press 541-557

Barlaz, M.A., Schaefer, D.M. and Ham, R.K. (1989). Bacterial population development and chemical characteristics of refuse decomposition in a simulated sanitary landfill. *Applied and Environmental Microbiology* 55 55-65

Baross, J.A. and Deming, J.M. (1983). Growth of 'black smoker' bacteria at temperature of at least 250C. *Nature* 303, 423-426

Baross, J.A., Lilley, M.D. and Gordon, L.I. (1982). Is the methane carbon-dioxide and hydrogen venting from submarine hydrothermal systems produced by thermophilic bacteria? *Nature* 298, 366-368

Barrington, S. F. and Cap, R. (1990) Thermophilic aerobic digestion of swine manures for feeding. Paper Abstract 90-115 p335. Canadian Society of Agricultural Engineering Conference, Penticton, BC, July 1990

Bath, C.A. (1991).Biomass. In Moss, V. And Cape, R.E. (eds.). *Biotechnology; The science and the business*. London, Harwood Academic Publisher pp 723-736

Beaudet, R, Gagnon C, Bisaillon JG and Ishaque M. (1990). Microbiological aspects of aerobic thermophilic treatment of swine waste. *Applied and Environmental Microbiology* 56: 971-976

Becker, P, Koster M, Popov MN, Markossian S, Antranikian G. and Markl H. (1999) The Biodegradation of olive oil and the treatment of lipid-rich wool scouring wastewater under aerobic thermophilic conditions. *Water Res* 33: 653–660.

Bergquist, P.L., Love, D.R., Croft, J.R., Streiff, M.B., Daniel, R.M., and Morgan, W.H. (1987). Genetics and potential application of thermophilic and extremely thermophilic microorganisms. *Biotechnology and Genetic Engineering Reviews* 5 199-224

Bewley, R.J.F., Sleat, R. and Rees, J.F. (1991) Waste treatment and pollution clean-up. In Moss, V. and Cape, R.E. (eds.). *Biotechnology; The science and the business*. London, Harwood Academic Publisher pp 507-519

Bhamidimarri, S.M.R. and Pandey, S.P. (1996). Aerobic thermophilic composting of piggery solid wastes. *Water Science and Technology* 33 89-94

Biddlestone, A.J. and Gray, K.R. (1985). Composting. In Robinson, C.W. and Howell, J.A. (eds.). *Comprehensive biotechnology; The principles, applications and regulations of biotechnology in industry agriculture and medicine vol. 4*. Oxford: Pergamon Press pp 1059-1070

Birou, B., Marison I.W. and Stocker von, U. (1987) Calorimetric investigation of aerobic fermentations. *Biotechnology and Bioengineering* 30 650-660

Bomio, M., Sonnleitner, B. and Fiechter, A. (1989). Growth and biocatalytic activities of aerobic thermophilic populations in sewage sludge. *Applied Microbiology and Biotechnology* 32, 356-362

Borowski, S. and Szopa J.S. (2007) Experiences with the dual digestion of municipal sewage sludge. *Bioresource Technology* 98 1199–1207

Bouallagui, H., Haouari, O., Touhami, Y., Cheikh, R.B., Marouani, L. and Hamdi, M. (2004). Effect of temperature on the performance of an anaerobic tubular reactor treating fruit and vegetable waste. *Process Biochemistry* 39 2143–2148

Brock, T.D. (1986). Introduction: An overview of the thermophiles. In Brock, T.D. (ed). *Thermophiles: general molecular and applied microbiology*. New York, John Wiley and Sons pp1-16

Bruce, A.M (1989). Other investigations on thermophilic aerobic digestion in the UK. In Bruce, A.M., Colin, F. and Newman, P.J. (eds.). *Treatment of sewage sludge: thermophilic aerobic digestion and processing requirements for landfilling*. London, Elsevier Applied Science pp39-43.

Buend, I.M., Fernandez, F.J., Villasenor, J. and Rodrıguez, L. (2008). Biodegradability of meat industry wastes under anaerobic and aerobic conditions. *Water Research* 42 3767- 3774

Burt, P., Littlewood, M.H., Morgan, S.F. Dancer, B.N. and Fry, J.C. (1990a). Venturi aeration and thermophilic aerobic sewage sludge digestion in small scale reactors. *Applied Microbiology and Biotechnology* 33 721-724.

Burt, P., Morgan, S.F., Dancer, B.N. and Fry, J.C. (1990b). Microbial population and sludge characteristics in thermophilic aerobic sewage sludge digestion. *Applied Microbiology and Biotechnology* 33 725-730

Cantrell K.B., Ducey, T., Ro, K.S. and Hunt, P.G. (2008). Livestock waste-to-bioenergy generation opportunities. *Bioresource Technology* 99 7941–7953

Carlson, C.H. (1982). Optimised aerobic and thermophilic treatment of municipal sewage sludges and night soils in a continuous operation. *Journal of Chemical Technology and Biotechnology* 32 1010-1015

Carrington, E. G., Pike, E. B., Auty, D. and Morris, R. (1991). Destruction of faecal bacteria, enteroviruses and ova of parasites in wastewatersludge by aerobic thermophilic and anaerobic mesophilic digestion. *Water Science and Technology* 24 377-380

Cayuela, M.L., Mondini, C., Sanchez-Monedero, M.A. and Roig, A. (2008) Chemical properties and hydrolytic enzyme activities for the characterisation of two-phase olive mill wastes composting. *Bioresource Technology* 99 4255–4262

Cecchi, F., Pava, P., Musacco, A., Mata Alvarez, J. and Vallini, G. (1993b). Digesting the organic fraction of municipal solid wastes: moving from mesophilic (37C) to thermophilic (55C) conditions. *Waste Management and Research* 11 403-414

Cecchi, F., Pava, P., Musacco, A., Mata Alvarez, J., Sana, C. and De Faveri, D. (1993a). Monitoring fast thermophilic re-start-up of a digester treating the organic fraction of municipal solid waste. *Environmental Technology* 14 517-530

Chang, J.I., Tsai, J.J. and Wu, K.U. (2006) Thermophilic composting of food waste *Bioresource Technology* 97 116–122

Chaudhry, S.M., Fontenot J.P. and Naseer Z. (1998) Effect of deep stacking and ensiling broiler litter on chemical composition and pathogenic organisms. *Animal Feed Science and Technology* 74: 155- 167

Chu, A, Mavinic, D.S., Kelly, H.G. and Guarnaschelli, C. (1997) The influence of aeration and solids retention time on volatile fatty acid accumulation in thermophilic aerobic digestion of sludge. *Environmental Technology* 18 731-738

Chu, A., Mavinic, D.S., Ramey, W.D. and Kelly, H.G. (1996). A biochemical model describing volatile fatty acid metabolism in thermophilic aerobic digestion of wastewater sludge. *Water Research* 30 1759-1770.

Chu, A., Mavinic, D.S.. Kelly, H.G. and Ramey, W.D. (1994). Volatile fatty acid production in thermophilic aerobic digestion of sludge. *Water Research* 28 1513-1522Chughtai and Ahmed, 1991

Cibis, E., Kent, C.A., Krzywonos, M., Garncarek, Z., Garncarek, B. and Miskiewicz, T. (2002). Biodegradation of potato slops from rural distillery by thermophilic aerobic bacteria. Bioresource Technology **85** 57- 61

Cibis, E., Krzywonos, M. and Miskiewicz, T. (2006). Aerobic Biodegradation of potato slops under moderate thermophilic conditions: effect of pollution load. Bioresource Technology **97** 679- 685

Colombatto, D., Mould F.L., Bhat M.K., Phipps R.H. and Owen E. (2004a) In vitro evaluation of fibrolytic enzymes as additives for maize (*Zea mays* L) silage: I. Effects of ensiling temperature, enzyme source and addition level. *Animal Feed Science and Technology* 111: 111-128

Colombatto, D., Mould F.L., Bhat M.K., Phipps R.H. and Owen E. (2004b) In vitro evaluation of fibrolytic enzymes as additives for maize (*Zea mays* L) silage: II. Effects of rate of acidification, fibre degradation during ensiling and rumen fermentation. *Animal Feed Science and Technology* 111: 129-143

Colombatto, D, Mould F.L., Bhat M.K., Phipps R.H. and Owen E. (2004c) In vitro evaluation of fibrolytic enzymes as additives for maize (*Zea mays* L) silage: III. Comparison of enzymes derived from psychrophilic, mesophilic or thermophilic sources. *Animal Feed Science and Technology* 111: 145-159

Coombs, J. (1990). The present and future of anaerobic digestion. In Wheatley, A. (ed.) *Anaerobic digestion; A waste treatment technology*. London, Elsevier Applied Science pp 43-92.

Coombs, J. (1994). Biomass goes to waste. *Renewable Energy* 5 733-740.

Cooney, C.L., Wang, D.I.C. and Mateles (1967). Measurement of heat evolution and correlation with oxygen consumption during microbial growth. *Biotechnology and Bioengineering* 11 269-281

Couillard, D. and Zhu, S. (1993). Thermophilic aerobic process for the treatment of slaughterhouse effluents with protein recovery. *Environmental Pollution* 79 121-126.

Couillard, D., Gariepy, S. and Tran, F.T. (1989). Slaughter house effluent treatment by thermophilic aerobic process. *Water Research* 23 573- 579

Coultate, T.P. and Sundaram, T.K. (1975). Energetics of *Bacillus stearothermophilus* growth: molar growth yield and temperature effect on growth efficiency. *Journal of Bacteriology* 121 55-64.

Coulthard, L.T., Townsley, P.M. and Saben, H.S. (1981). United States Patent No 4292328

Council of European Union (1999)

Cronan, J.E.Jr and Gelman, E.P. (1975). Physical properties of membrane lipids: Biochemical relevance and regulation. *Bacteriological Reviews* 39 232-256

Cuetos, M.J., Gomeza, X., Otero, M. and Moran, A. (2008). Anaerobic digestion of solid slaughterhouse waste (SHW) at laboratory scale: Influence of co-digestion with the organic fraction of municipal solid waste (OFMSW). *Biochemical Engineering Journal* 40 99–106

Cunha-Queda, A.C., Ribeiro, H.M., Ramos, A. and Cabral, F. (2007) Study of biochemical and microbiological parameters during composting of pine and eucalyptus bark *Bioresource Technology* 98 3213–3220

Davis, C.L., Hinch, S.A., Donkin, C.J. and Germishuizen, P.J. (1992). Changes in microbial population numbers during the composting of pine bark. *Bioresource Technology* 39 85-92

De Bertoldi, M., Zucconi, F. and Civilini, M. (1988) Temperature, pathogen control and product quality. *Biocycle* February 1988 43-50

Denek N. and Can A 2006 (Feeding value of wet tomato pomace ensiled with wheat straw and wheat grain for Awassi sheep. *Small Ruminant Research* 65: 260-265

Dolfing, J. (1992). The energetic consequences of hydrogen gradients in methanogenic ecosystems *FEMS Microbiology Ecology* 101 183-187

Droffner, M.L. and Brinton, W. F. (1995). Survival of E. coli and Salmonella populations in aerobic thermophilic composts as measured with DNA gene probes. *Zbl Hyg.* 197 387-397.

Droffner, M.L. Brinton, W. F. and Evans, E. (1995). Evidence for the prominence of well characterised mesophilic bacteria in thermophilic (50-70C) composting environments. *Biomass and Bioenergy* 8 191-1

Droste, R. L. and Sanchez, W.A. (1983). Microbial activity in aerobic sludge digestion. *Water Research* 17 975-983

Durak, Y. and Ozturk, C. (1993). Research of bacterial flora and microbial population in synthetic mushroom compost. *Turkey Journal of Biology* 17 201-209 during the composting of sludges from wool scour effluents. *Environ. Sci. Technol.* 32, 2623–2627.

Eastman, J.A. and Ferguson, J.F. (1981). Solubilisation of particulate organic carbon during the acid phase of anaerobic digestion. *Journal of the Water Pollution Control Federation* 53 352-366

Edginton, R. and Clay (1993). Evaluation and development of a thermophilic aerobic digester at Castle Donington sewage treatment works (IWEM 92 conference paper) *J. IWEM* 7, April 1993 149-155

Elango, D., Thinakaran, N., Panneerselvam, P. and Sivanesan, S. (2008). Thermophilic composting of municipal solid waste. *Appl Energ*, doi:10.1016/j.apenergy.2008.06.009

EPA (1990) Autothermal thermophilic aerobic digestion of municipal wastewater sludge. EPA/625/10-90/007 September, Washington, DC

EPA (1992) Control of pathogens and vector attraction in sewage sludge EPA/625/R-92/013 December, Washington, DC

Erses, S.A., Onay, T.T. and Yenigun, O. (2008). Comparison of aerobic and anaerobic degradation of municipal solid waste in bioreactor landfills. *Bioresource Technology* 99 5418–5426

Espinoza-Escalante, F.M., Pelayo-Ortız, C., Navarro-Corona, J., Gonzalez-Garcıa, Y., Bories, A. and Gutierrez-Pulido, H. (2008). Anaerobic digestion of the vinasses from the fermentation of Agave tequilana Weber to tequila: The effect of pH, temperature and hydraulic retention time on the production of hydrogen and methane

Farrand, S.G., Jones, C.W., Linton, J.D. and Stephenson, R.J. (1983). The effect of temperature and pH on the growth efficiency of the thermophilic bacterium *Bacillus acidocaldarius* in continuous culture. *Archives of Microbiology* 135 276-283

Ferrer, I., Ponsá S., Vázquezc, F. and Font, X. (2008) Increasing biogas production by thermal (70°C) sludge pre-treatment prior to thermophilic anaerobic digestion *Biochemical Engineering Journal* 42 186–192

Fezzani, B. and Cheikh, R.B. (2008). Optimisation of the mesophilic anaerobic co-digestion of olive mill wastewater with olive mill solid waste in a batch digester. *Desalination* 228 159–167

Fiechter, A. and Sonnleitner, B. (1989). Thermophilic aerobic stabilisation. In Dirkzwager, A.H. and L'Hermite, P. (eds.). *Sewage sludge treatment and use; new developments, technological aspects and environmental effects.* London, Elsevier Applied Science pp 291-301

Forster-Carneiro, T., Pérez, M. and Romero, L.I. (2008). Anaerobic digestion of municipal solid wastes: Dry thermophilic performance. *Bioresource Technology* 99 8180–8184

Forster-Carneiro, T., Pérez, M. and Romero, L.I. (2008) Thermophilic anaerobic digestion of source-sorted organic fraction of municipal solid waste. *Bioresource Technology* 99 6763–6770

Forster-Carneiro, T., Perez, M. and Romero, L.I. (2008). Influence of total solid and inoculum contents on performance of anaerobic reactors treating food waste. *Bioresource Technology* 99 6994–7002

Frost, R., Powsland, C., Hall, J.E., Nixon, S.C. and Young, C.P. (1990). Thermophilic aerobic digestion (TAD). In Review of sewage sludge treatment and disposal techniques. Report No PRD 2306-M/1 pp 81-91

Fujio, Y. and Kume, S. (1991). Isolation and characterisation of thermophilic bacteria from sewage sludge compost. *Journal of Fermentation Bioengineering* 72 334-337

Gandolla, M. and Aragno, M. (1992). The importance of microbiology in waste management. *Experientia* 48 362-366.

Garcia, C., Hernandez, T., Costa, C., Ceccanti, B., Masciandaro, G. and Ciardi, C. (1993). A study of biochemical parameters of composted and fresh municipal wastes. *Bioresource Technology* 44 17-23

Gardner N.J, Savard T., Obermeier P., Caldwell G. and Champagne C.P. (2001) Selection and characterization of mixed starter cultures for lactic acid fermentation of carrot, cabbage, beet and onion vegetable mixtures. *International J. Food Microbiology* 64: 261- 275

Gerba, C.P., Huber, M.S., Naranjo, J., Rose, J.B. and Bradford, S. (1995) Occurrence of enteric pathogens in composted domestic solid waste containing disposable diapers. *Waste Management and Research* 13 315-324

Godden, B., Pennickx, N., Pierard, A. and Lannover, R. (1983) Evolution of enzyme activities and microbial groups during composting of cattle manure. *European Journal of Applied Microbiology and Biotechnology* 17 306-310

Gomez J, de Gracia M, Ayesa E. and Garcia-Heras J.L (2007). Mathematical modelling of autothermal thermophilic aerobic digesters. *Water Research* 41 (2007) 959 – 968

Gould, M. and Drnervich, R.F. (1978). Autothermal aerobic digestion. *Journal of Environmental Engineering Division (ASCE)* 104 259-270

Grainger, J.M. (1987a). Microbiology of waste disposal. In Norris, J.R. and Pettipher, G.L. (eds.). *Essays in agricultural and food microbiology*. Chichester, John Wiley and Sons pp 105-134.

Grainger, J.M. (1987b). Methane from solid waste. In Sidwick J.M. and Holdom, R.S. (eds.). *Biotechnology of waste treatment and exploitation*. Chichester, Ellis Horwood Publishers pp 275-285.

Grueninger, H., Sonnleitner, B. and Fiechter, A. (1984). Bacterial diversity in thermophilic aerobic sewage sludge, iii. A source of organisms producing heat stable industrially useful enzymes, e.g. α-amylase. *Applied Microbiology and Biotechnology* 19 414-421.

Gumson, H.G. and Morgan, S.F. (1982). Aerobic thermophilic digestion of sewage sludge. *Master* 22, 319-320

Hamer, G. (1989). Fundamental aspects of aerobic thermophilic digestion. In Bruce, A.M., Colin, F. And Newman, P.J. (eds.). *Treatment of sewage sludge; thermophilic aerobic digestion and processing requirements for landfilling*. London, Elsevier Applied Science pp2-19

Hamer, G. and Bryers, J.D. (1985). Aerobic thermophilic sludge treatment, some biotechnological concepts. *Conservation and Recycling* 8 267-284

Hammoumi, A., Faid, M., El yachioui, M. and Amarouch, H. (1998) Characterisation of fermented fish waste used in feeding trials with broilers. *Process Biochemistry* 33 423-427

Hartmann, H. and Ahring B.K. (2005) Anaerobic digestion of the organic fraction of municipal solid waste: Influence of co-digestion with manure Water Research 39 1543–1552

Hassen A,. Belguith, A., Jedidi, N., Cherif A., Cherif M. and Boudabous A., (2001) Microbial characterisation during composting of municipal solid waste. *Bioresource Technology* 80: 217-225

Hawash, S., El Ibiari, N., Aly, F.H., El Diwani, G. and Hamad, M. A. (1994). Kinetic study of thermophilic aerobic stabilization of sludge. *Biomass and Bioenergy* 6 283-286

Heerden, van I., Cronje C, Swart S.H., Kotze J.M. (2002) Microbial, chemical and physical aspects of citrus waste composting. *Bioresource Technology* 81 71- 76

Heinonen-Tanski, H., Kiuru, T., Ruuskanen, J., Korhonen, K., Koivunen, J. and Ruokojarvi, A. (2005). Thermophilic aeration of cattle slurry with whey and or jam waste.Bioresource Technology **96** 247- 252

Hernandez-Raquet, G., Soef, A., Delgenes, N. and Balaguer, P. (2007). Removal of the endocrine disrupter nonylphenol and its estrogenic activity in sludge treatment processes. *Water Res.* 41, 2643–2651.

Hisset, R., Deans, E.A. and Evans, M.R. (1982). Oxygen consumption during batch aeration of piggery slurry at temperatures between 5 and 50C. *Agricultural Wastes* 4 477-487

Hobson, P.N. (1990). The treatment of agricultural wastes. In Wheatley, A. (ed.). *Anaerobic digestion, a waste treatment technology.* London, Elsevier Applied Science pp 93-138

Holms, W.H. and Bennett, P.M. (1971). Regulation of isocitrate dehydrogenase activity in *Escherichia coli* on adaptation to acetate. *Journal of General Microbiology* 65 57-68

Huang, G.F. Wu, Q.T. Wong, J.W.C. and Nagar B.B. (2006) Transformation of organic matter during co-composting of pig manure with sawdust. *Bioresource Technology* 97 1834–1842

Huang, R.T. (1980). *Compost engineering, principles and practice.* Lancaster: Ann Arbor Science Publisher (Technomic Publishing Co., Inc.) p 655

Inaba, N., Murayama, K., Miyake, M. and Fujio, Y. (1996). Composting of the processing waste of satsuma mandarin (*Citrus unshiu* Marc.) with a developed composting machine. *Journal of the Japanese Society for Food Science and Technology* 43 1205-1211

Jackson, M.L. (1964). Aeration in Bernoulli type devices. *Chemical Engineering Journal* 10 836-842.

Jacob, J., Rooss H.-J. and Siekman, K. (1989). Aerobic thermophilic method for disinfecting and stabilising sewage sludge. In Dirkzwager A.H. and L'Hermite , P. (eds.) *Sewage sludge treatment and use; new technological aspects and environmental effects.* London, Elsevier Applied Science pp 378-381

Jewell, W.J. (1991). Detoxification of sludges: autoheated aerobic digestion of raw and anaerobically digested sludges. In Freeman, H.M. and Sterra. R. (eds.). *Innovative hazardous waste treatment technology series vol. 3, Biological processes.* Lancaster, Technomic Publishing Company Inc. pp 79-90.

Jewell, W.J. and Kabick, R.M. (1980). Autoheated aerobic thermophilic digestion with aeration. *Journal of the Water Pollution Control Federation* 52 512-523.

Jewell, W.J., Kabrick, R.M. and Spada, J.A. (1982). Autoheated aerobic thermophilic digestion with aeration. In US Environmental Protection Agency report EPA-600/ S2-82-023 Project Summary

Jimenez, E.I. and Perez, G.V. (1991). Composting of domestic refuse and sewage sludge I. Evolution of temperature, pH, carbon-nitrogen ratio and cation exchange capacity. *Resource Conservation and Recycle* 6 45-60

Jones, K.L. and Grainger, J.M. (1983a). Characterisation of polysaccharidase activity optima in the anaerobic digestion of municipal solid waste. *European Journal of Microbiology and Biotechnology* 18 181-185

Jones, K.L. and Grainger, J.M. (1983b). The application of enzyme activity measurements to a study of factors affecting protease, starch and cellulose fermentation in a domestic refuse. Europ. *Journal of Applied Microbiology and Biotechnology* 18 242-245

Juris, P., Plachy, P., Dubinsky, P., Venglovsky, J. and Toth, F. (1993) Effect of aerobic stabilization of pig slurry on vitality of model pathogens. *Vet. Med.*, 9 553-558

Juris, P., Plachy, P., Toth, F. and Venglovsky, J. (1992) Effect of biofermentation of pig slurry on Ascaris suum eggs. *Helminthologia* 29 155-159.

Kabrick, R.M., and Jewell, W.J. (1982) Fate of pathogens in thermophilic aerobic sludge digestion. Water Research 16 1051-1060

Kambhu, K. and Andrew, J.F. (1969). Aerobic thermophilic process for the treatment of wastes- simulation studies. Journal of the Water Pollution Control Federation **41** R127-141

Kaparaju, P., Buendia, I., Ellegaard, L. and Angelidaki I., (2008). Effects of mixing on methane production during thermophilic anaerobic digestion of manure: Lab-scale and pilot-scale studies. *Bioresource Technology* 99 4919–4928

Kawada, N. and Nosoh, Y. (1981). Relation of Arrhenius discontinuities of NADH dehydrogenase to changes in membrane lipid fluidity of *Bacillus caldotenax*. *FEBS Letters* 124 15-18.

Kearney, T.E., Larkin, M.J. and Levett, P.N. (1993a) The effect of slurry storage and anaerobic digestion on the survival of pathogenic bacteria. *Journal of Applied Bacteriology* 74 86-93

Kearney, T.E., Larkin, M.J., Frost, J.P. and Levett, P.N. (1993b) Survival of pathogenic bacteria during mesophilic anaerobic digestion of animal waste. *Journal of Applied Bacteriology* 75 215-219

Kelly, H.G. (1991). Autothermal TAD of municipal sludges: conclusions of a one year full scale demonstration project. WPCF conference Toronto, October 1991

Kelly, H.G., Melcer, H. and Mavinic, D.S (1993). Autothermal thermophilic aerobic digestion of municipal sludges: A one year full scale demonstration project. *Water Research* 65 849-861

Kenkel, T. and Trela, J.M. (1979). Protein turnover in extreme thermophile *Thermus aquaticus*. *Journal of Bacteriology* 140 543-546.

Kim Y-K, Bae J-H, Oh B-K, Lee W. H. and Choi J-W (2002) Enhancement of proteolytic enzyme activity excreted from *Bacillus stearothermophilus* for thermophilic aerobic digestion process. *Bioresource Technoogy* 82 157-164

Knudsen, L., Kristensen, G.H., Jorgensen, P.E. and Jepsen, S.-E. (2000). Reduction of the content of organic micropollutants in digested sludge by a post-aeration process-a full-scale demonstration. *Water Sci. Technol.* 42 (9), 111–118.

Koers, D.A. and Mavinic, D.S. (1977). Aerobic digestion of waste activated sludge at low temperatures. Journal of the Water pollution Control Federation 49 460-468.

Koppar, A. and Pullammanappallil, P. (2008) Single-stage, batch, leach-bed, thermophilic anaerobic digestion of spent sugar beet pulp. *Bioresource Technology* 99 2831–2839

Kosseva, M.R., Kent, C.A. and Lloyd, D.R. (2001). Thermophilic bioremediation of whey: effect of physico-chemical parameters on the efficiency of the process. Biotechnology Letters **23** 1675- 1679

Krahe, M., Antranikan, G. and Mark, H. (1996). Fermentation of extremophilic organisms. *FEMS Microbiological Reviews* 18, 271-285.

Krishna, C. and Chandrasekaran, M. (1995) Economic utilization of cabbaeg wastes through solid state fermentation by native microflora. *Journal of Food Science and Technology* 32 199-201

Kuhn, H.J. Cometta, S. and Fiechter, A. (1980). Effects of growth temperature on maximal specific growth rate, yield, maintenance and death rate in glucose limited continuous culture of the thermophilic *Bacillus caldotenax*. *European Journal of Applied Microbiology and Biotechnology* 10 303-315.

Kuhn, H.J. Friedrich, U. and Fiechter, A. (1979). Defined minimal medium for a thermophilic *Bacillus* spp developed by a chemostat pulse and shift technique. *European Journal of Applied Microbiology and Biotechnology* 6, 341-349.

Kulikowska, D. and Klimiuk, E. (2008). The effect of landfill age on municipal leachate composition. *Bioresource Technology* 99 5981–5985

Laine, M.M. and Jorgensen, K.S. (1996). Straw compost and bioremediated soil as inocula for the bioremediation of chlorophenol-contaminated soil. *Applied and Environmental Microbiology* 62 1507-1513

Lansing, S., Botero, R. B. and Martin, J. F. (2008). Waste treatment and biogas quality in small-scale agricultural digesters. *Bioresource Technology* 99 5881–5890

Lee, M., Hidaka, T., Hagiwara, W. and Tsuno, H. (2008). Comparative performance and microbial diversity of hyperthermophilic and thermophilic co-digestion of kitchen garbage and excess sludge. *Bioresour. Technol*, doi:10.1016/j.biortech.2008.06.063

Lee, S. and Yu, J. (1997). Production of biodegradable thermoplastics from municipal sludge by a two stage bioprocess. *Resources Conservation and Recycling* 19 151-165

Lettinga, G. (1995). Anaerobic digestion and wastewater treatment systems. *Antonie van Leewenhoek* 67 3-28

Lettinga, G. and Hulshoff Pol, L.W. (1991). UASB Process design for various types of wastewaters. *Water Science and Technology* 24 87-107

Lettinga, G., Van Velson, A.F.M., Hobma, W., de Zeew, J. and Klapwijk, A. (1980). Use of the upflow sludge blanket (USB) reactor concept for biological waste water treatment, especially for anaerobic treatment. *Biotechnology and Bioengineering* 22 699-734.

Li, X., Zhang, R. and Pang, Y. (2008). Characteristics of dairy manure composting with rice straw. *Bioresource Technology* 99 359–367

Lieph R, Velos F.A. and Holmes D.S. (2006). Thermopiles like hot T. *Trends in Microbiology* 14: 423- 426

Lier van, J.B. (1996). Limitations of thermophilic anaerobic wastewater treatment and the consequences for process design. *Antonie van Leewenhoek* 69 1-14

Lier van, J.B., Boersma, F., Debets, M.M.W.H. and Lettinga, G. (1994). High rate thermophilic anaerobic wastewater treatment in compartmentalised up-flow reactors. *Water Science and Technology* 30 251-261

Lier van, J.B., Frijters, C., Grolle, C.T.M.J., Lettinga, G. and Stams, A.J.M. (1993b). Effects of acetate, propionate and butyrate on the thermophilic anaerobic degradation of propionate by methanogenic sludge and defined cultures. *Applied and Environmental Microbiology* 59 1003-1011

Lier van, J.B., Hulsbeek, J., Lettinga, G. and Stams, A.J.M. (1993a). Temperature susceptibility of thermophilic methanogenic sludge: implications for reactor start up and operation. *Bioresource Technology* 43 227-2235

Lier van, J.B., Sanz Martin, J.L. and Lettinga, G. (1995). Effect of temperatures on the anaerobic conversion of volatile fatty acids by dispersed and granular sludge. *Water Research* 30 199-207

Lindsay, J.A. (1995). Is thermophily a transferable property of bacteria? *CRC Critical Reviews in Microbiology* 21 165-174.

Linke, B (2006). Kinetic study of thermophilic anaerobic digestion of solid wastes from potato processing. *Biomass and Bioenergy* 30 892–896

Lipowsky, R. (1991). Conformation of membranes. *Nature* 349 475-481.

Loehr, R. C. (1977). *Pollution control for agriculture*. New York, Academic Press pp 383

Loll, U. (1976). Purification of concentrated organic waste waters from the foodstuffs industry by means of aerobic thermophilic degradation process. Progress in Water Technology 8 373-379.

Loll, U. (1989). Combined aerobic thermophilic and anaerobic digestion of sewage sludge. In Bruce, A.M., Colin, F. and Newman, P.J. (eds.). *Treatment of sewage sludge: thermophilic aerobic digestion and processing requirement for landfilling*. London, Elsevier Applied Science pp2-11

Lu, Y., Wu, X. and Guo, J (2008). Characteristics of municipal solid waste and sewage sludge co-composting. *Waste Management*, doi:10.1016/j.wasman.2008.06.030

Luostarinen S., Luste, S. and Sillanpää, M. (2009). Increased biogas production at wastewater treatment plants through co-digestion of sewage sludge with grease trap sludge from a meat processing plant. *Bioresource Technology* 100 79–85

Macias-Corral M., Samani Z., Hanson A., Smith G., Funk P., Yu H. and Longworth, J. (2008). Anaerobic digestion of municipal solid waste and agricultural waste and the effect of co-digestion with dairy cow manure. *Bioresource Technology* 99 8288–8293

Mackie, R.I. and Bryant, M.P. (1995). Anaerobic digestion of cattle waste at mesophilic and thermophilic temperatures. *Applied Microbiology and Biotechnology* 43 346-350

Maden, R.H. (1983). Isolation and characterisation of *Clostridium stecorarium* sp nov., cellulolytic thermophile. *International Journal of Systematic Bacteriology* 33 837-840

Malladi, B. and Iingham, S.C. (1993). Thrmophilic aerobic treatment of potato processing wastewater. *World Journal of Microbiology and Biotechnology* 9 45-49

Maloney, S.E., Marks, T.S. and Sharp, R.J. (1997). Detoxification of synthetic pyrethroid insecticides by thermophilic microorganisms. *Journal of Chemical Technology and Biotechnology* 68 357-360

Mari, I., Ehaliotis, C., Kotsou, M., Balis, C. and Georgakakis, D. (2003) Respiration profiles in monitoring the composting of by-products from the olive oil agro-industry. *Bioresource Technology* 87 331–336

Maso, M.A., and Blasi, A.B. (2008). Evaluation of composting as a strategy for managing organic wastes from a municipal market in Nicaragua. *Bioresource Technology* 99 5120–5124

Mason, C.A., Hamer G., Fleischman, T. and Lang, C. (1987) Aerobic thermophilic biodegradation of microbial cells. *Applied Microbiology and Biotechnology* 25 568-574

Mason, C.A., Hamer, G. and Bryers, J.D. (1986). The death and lysis of micro-organisms in environmental processes. *FEMS Microbiology Reviews* 39 373-401

Mason, C.A., Haner, A. and Hamer, G. (1992). Aerobic thermophilic waste sludge treatment. *Water Science and Technology* 25 113-118

Matsch, L.C. and Drnevich, R.F. (1977). Autothermal aerobic digestion. *Journal of the Water Pollution Control Federation* 49 296-310.

Matsche, N.F. and Andrew, J.F. (1973). A mathematical model for the continuous cultivation of thermophilic micro-organisms. *Biotechnology and Bioengineering Symposium* 4 77-90

McElhaney, R.N. and Souza, K.A. (1976). The relationship between environmental temperature, cell growth and the fluidity and physical state of the membrane lipids in *Bacillus stearothermophilus. Biochimica et Biophysica Acta* 444 359-359

McHugh, S., Carton, M., Mahony, T. and O'Flaherty, V. (2003). Methanogenic population structure in a variety of anaerobic bioreactors. *FEMS Microbiology Letters* 219 297-304

Mergaert, J. and Swings, J. (1996). Biodiversity of microorganisms that degrade bacterial and synthetic polymers. *Journal of Industrial Microbiology and Biotechnology* 17 463-469

Messenger, J.R. and Ekama, G.A. (1993a). Evaluation of dual digestion system: 3. Considerations in the process design of the aerobic reactor. *Water SA* 19 201-208

Messenger, J.R. and Ekama, G.A. (1993b). Evaluation of dual digestion system: Part 4: Simulation of the temperature profile in the batch fed aerobic reactor. *Water SA* 19 209-215

Messenger, J.R. de Villiers, H.A. and Ekama, G.A. (1990). Oxygen utilisation rate as a control parameter for the aerobic stage in dual digestion. *Water Science and Technology* 22 217-227

Messenger, J.R., de Villiers, H.A. and Ekama, G.A. (1993b) Evaluation of dual digestion system: Part 2: Operation and performance of pure oxygen aerobic reactor. *Water SA* 19 193-200.

Messenger, J.R., de Villiers, H.A., Laubscher, S. J. A., Kenmuir, K. and Ekama, G.A. (1993a). Evaluation of dual digestion system: Part 1: Overview of the Milnerton experience. *Water SA* 19 185-191.

Metcalf and Eddy Inc. (1991). *Wastewater engineering; treatment and disposal and reuse* (Tchobanoglous, G., ed.) 3rd edition. New York, McGraw-Hill pp765-926

Miller, F.C., Harper, E.R., Macauley, B.J. and Gulliver, A. (1990). Composting based on moderately thermophilic and aerobic conditions for the production of mushroom growing compost. *Australian Journal of Experimental Agriculture* 30 287-296

Moat, A.G. and Foster, J.W. (1995). *Microbial physiology, 3rd edition*, New York, John Wiley p580

Moeller, J. and Reeh, U. (2003). Degradation of nonylphenol ethoxylates (NPE) in sewage sludge and source separated municipal solid waste under bench-scale composting conditions. *Bull. Environ. Contam. Toxicol.* 70, 248–254.

Monte, M.C., Fuente, E., Blanco, A. and Negro, C. (2008). Waste management from pulp and paper production in the European Union. *Waste Management*, doi:10.1016/j.wasman.2008.02.002

Montero, B., Garcia-Morales, J.L., Sales, D. and Solera, R. (2008). Evolution of microorganisms in thermophilic-dry anaerobic digestion. *Bioresource Technology* 99 3233–3243

Morgan, A.M. and Gumson, H.G. (1981). Sludge digestion at elevated temperatures using air introduced by venturi aerators. *Water Industry* (1981) Brighton pp 482-487.

Morgan, A.M. and Gumson, H.G. (1989). The development of an aerobic thermophilic sludge digestion system in the UK. In Bruce, A.M., Colin, F. and Newman, P.J. (eds.). *Treatment of sewage sludge: thermophilic aerobic digestion and processing requirements for landfilling*. London, Elsevier Applied Science pp 20-28.

Morgan, S.F., Winstanley, R., Littlewood, M.H. and Gumson, H.G. (1986). The design of an aerobic thermophilic sludge digestion system. Institute of Chemical Engineering Symposium Series 96, 1-8.

Murray, K. C., Tong, A. and Bruce, A. M. (1990). Thermophilic aerobic digestion - a reliable and effective process for sludge treatment at small works. *Water Science and Technology* 22 225-232

Mutzel, A., Reinscheid, U.A., Antranikian, G. and Mueller, R (1996). Isolation and characterisation of a thermophilic bacillus strain that degrades phenol and cresol as sole carbon source at 70C. *Applied Microbiology and Biotechnology* 46 593-596

Nakasaki, K., Sasaki, K., Shoda, M. and Kubota, H. (1985C). Characteristics of mesophilic bacteria isolated during thermophilic composting of sewage sludge. *Applied and Environmental Microbiology* 49 42-45.

Nakasaki, K., Shoda, M. and Kubota, H. (1985a) Effect of temperature on composting of sewage sludge. *Applied and Environmental Microbiology* 50 1526-1530

Nguyen, P.H.L., Kuruparan, P. and Visvanathan, C. (2007). Anaerobic digestion of municipal solid waste as a treatment prior to landfill. *Bioresource Technology* 98 380–387

Noone, G.P. (1990). The treatment of domestic wastes. In Wheatley, A. (ed.) *Anaerobic digestion; a waste treatment technology.* London, Elsevier Applied Science pp 93-138

O'Sullivan, C., Burrell, P.C., Clarke, W.P. and Blackall, L.L. (2008) The effect of biomass density on cellulose solubilisation rates. *Bioresource Technology* 99 4723–4731

Oda Y, Saito K, Yamauchi H. and Mori M. (2002) Lactic acid fermentation of potato pulp by the fungus *Rhizopus oryzae*. *Curr. Microbiol.* 45: 1-4

Okine A, Aibibua H.Y. and Okamoto M. (2005) Ensiling of potato pulp with or without bacterial inoculants and its effect on fermentation quality, nutrient composition and nutritive value. *Animal Feed Science and Technology.* 121: 329-343

Olsen, J.E. and Larsen, H.E. (1987) Bacterial decimation times in anaerobic digestions of animal slurries. *Biological Wastes* 21 153-168

Ortega, L., Barrington, S. and Guiot, S.R. (2008) Thermophilic adaptation of a mesophilic anaerobic sludge for food waste treatment. *Journal of Environmental Management* 88 517–525

Pagilla, K.R., Craney, K.C. and Kido, W.H. (1996) Aerobic thermophilic pretreatment of mixed sludge for pathogen reduction and *Nocardia* control. *Water Environment Research* 68 1093-1098

Palmisano, A.C., Maruscik, D.A. and Schwab, B.S. (1993a). Enumeration of fermentative and hydrolytic microorganisms from 3 sanitary landfills. *Journal of General Microbiology* 139 387-391

Palmisano, A.C., Schwab, B.S. and Maruscik, D.A. (1993b). Hydrolytic enzyme activity in landfill refuse. *Applied Microbiology and Biotechnology* 38 828-832

Patureau, D., Delgenes, N. and Delgenes J-P. (2008) Impact of sewage sludge treatment processes on the removal of the endocrine disrupters nonylphenol ethoxylates. *Chemosphere* 72 586–591

Payne, W.J. (1970). Energy yield and growth of heterotrophs. Annual Review of Microbiology **24** 17- 52

Perez, M., Romero, L.I. and Sales D. (2001) Kinetics of thermophilic anaerobes in fixed-bed reactors. *Chemosphere* 44 1201-1211

Pirmohammadi, R., Rouzbehan, Y., Rezayazdi, K and Zahedifar, M. (2006) Chemical composition, digestibility and in situ degradability of dried and ensiled apple pomace and maize silage. *Small Ruminant Research Volume 66:* 150-155

Plachy, P., Juris, P. and Tomasovicova, O. (1993) Destruction of *Toxocara canis* eggs in wastewater sludge by aerobic stabilization under laboratory conditions. *Helminthologia* 30 139-142

Plachy, P., Placha, I. and Vargova, M. (1995) Effect of physico-chemical parameters of sludge aerobic exothermic stabilization on the viability of *Ascaris suum* eggs *Helminthologia* 32 233-237

Poh, P.E. and Chong, M.F. (2009) Development of anaerobic digestion methods for palm oil mill effluent (POME) treatment. *Bioresource Technology* 100 1–9

Ponti, C., Sonnleitner, B. and Feichter, A. (1995a) Aerobic thermophilic treatment of sewage sludge at pilot plant scale. 1. Operating conditions. *Journal of Biotechnology* 38 173-182.

Ponti, C., Sonnleitner, B. and Feichter, A. (1995b) Aerobic thermophilic treatment of sewage sludge at pilot plant scale. 2. Technical solutions and process design. *Journal of Biotechnology* 38 183-192

Popel, F. and Ohnmacht, C.H. (1972). Thermophilic bacterial oxidation of highly concentrated substrates. *Water Research* 6 807-815

Poulsen, P.H.B., Møller, J. and Magid, J. (2008) Determination of a relationship between chitinase activity and microbial diversity in chitin amended compost. *Bioresource Technology* 99 4355–4359

Raut, M.P., Prince William, S.P.M.., Bhattacharyya, J.K., Chakrabarti, T. and Devotta, S. (2008). Microbial dynamics and enzyme activities during rapid composting of municipal solid waste – A compost maturity analysis perspective. *Bioresource Technology* 99 6512–6519

Rees, J.F. (1985). Landfills for treatment of solid wastes. In Robinson, C.W. and Howell, J.A. (eds.) *Comprehensive biotechnology: The principles, applications and regulations of biotechnology in industry, agriculture and medicine vol.4*. Oxford: Pergamon Publisher pp 1071-1087

Reinscheid, U.M., Bauer, M.P. and Mueller, R. (1996-7). Biotransformation of halophenols by a thermophilic *Bacillus* sp. *Biodegradation* 7 455-461

Rimkus, R.R., Ryan, J.M. and Cook, E.J. (1982). Full scale thermophilic digestion at the west-south west sewage treatment plant. *Journal of the Water Pollution Control Federation* 47 950-961

Rivard, C.J., Nagle N.J. Adney, W.S. and Himmel, M.E. (1993). Anaerobic bioconversion of municipal solid wastes. Effect of total solids level on microbial numbers and hydrolytic enzyme activities. *Applied Biochemistry and Biotechnology* 39/40 107-117

Rivard, C.J., Nieves, R.A., Nagle N.J. and Himmel, M.E. (1994). Evaluation of discrete cellulase enzyme activities from anaerobic digester sludge fed a municipal solid waste feedstock. *Applied Biochemistry and Biotechnology* 45/46 453-462

Saludes, R.B., Iwabuchi, K., Miyatake, F., Abe, Y. and Honda, Y. (2008) Characterization of dairy cattle manure/wallboard paper compost mixture. *Bioresource Technology* 99 7285–7290

Sanchez-Arias, V., Fernandez, F.J., Villasenor, J. and Rodrıguez, L. (2008) Enhancing the co-composting of olive mill wastes and sewage sludge by the addition of an industrial waste. *Bioresource Technology* 99 6346–6353

Santana-Delgado, H., Avila, E. and Sotelo, A. (2008) Preparation of silage from spanish mackerel *comberomoru maculates* and its evaluation in broiler diets. *Animal Feed Science and Technology* 141: 129-140;

Sarada, R. and Joseph, R. (1993a). Profile of hydrolases acting on major macromolecules of tomato processing waste during anaerobic digestion. *Enzyme and Microbial Technology* 15 339-342.

Sarada, R. and Joseph, R. (1993b). Biochemical changes during anaerobic digestion of tomato processing waste. *Process Biochemistry* 28 461-466

Scerra, V., Caparraa, P., Fotia, F., Lanzab ,M. and Priolob, A. (2001). Citrus pulp and wheat straw silage as an ingredient in lamb diets: effects on growth and carcass and meat quality. *Small Ruminant Research* 40: 51-56

Schink, B. (1997). Energetics of syntrophic cooperation in methanogenic degradation. *Microbiology and Molecular Biology Reviews* 61 262-280

Schmidt, J.E. and Ahring, B.K. (1994). Extracellular polymers in granular sludge from different upflow anaerobic sludge blanket (UASB) reactors. *Applied Microbiology and Biotechnology* 42 457-462

Sellami, F., Jarboui, R., Hachida, S., Medhioub, K. and Ammar, E. (2008). Co-composting of oil exhausted olive-cake, poultry manure and industrial residues of agro-food activity for soil amendment. *Bioresource Technology* 99 1177–1188

Shaw, D.M., Narasimha, R.D. and Mahendrakar, N.S. (1998). Effect of different levels of molasses, salt and antimycotic agents on microbial profiles during fermentation of poultry intestine. *Bioresource Technol* 63: 237-241

Shuval, H., Jodice, R., Consiglio, M., Spagiarri, G. and Spigoni, C. (1991). Control of enteric microorganisms by aerobic thermophilic co-composting of waste water sludge and agro-industry wastes. *Water Science and Technology* 24 401-405

Siller, H. and Winter, J. (1998). Degradation of cyanide in agroindustrial or industrial wastewater in an acidification reactor or in a single-step methane reactor by bacteria enriched from soil and peels of cassava. *Applied Microbiology and Biotechnology* 50 384-389

Singh, S., Sandhu, M.S., Singh, M. and Harchand, R.K. (1991). Thermophilic actinomycetes associated with agro-environment of Punjab state India. *Journal of Basic Microbiology* 31 391-398

Slottner, D. and Bertilsson, J. (2006) Effect of ensiling technology on protein degradation during ensiling. *Animal Feed Science and Technology* 127: 101-111

Solera, R., Romero, L.I., and Sales, D. (2001). Determination of the Microbial Population in Thermophilic Anaerobic Reactor: Comparative Analysis by Different counting Methods. *Anaerobe* 07, 79–86

Sonnleitner B. and Feichter A (1983b). Bacterial diversity in thermophilic aerobic sewage sludge II. Types of organisms and their capacities. *European Journal of Applied Microbiology and Biotechnology* 18 174-180

Sonnleitner, B. (1983) Biotechnology of thermophilic bacteria- growth, products and application. *Advances in Biochemical Engineering and Biotechnology* 28 69-138

Sonnleitner, B. and Bomio, M. (1990) Physiology and performance of thermophilic microorganisms in sewage sludge processes. *Biodegradation* 1 133-146

Sonnleitner, B. and Feitcher, A. (1985).Microbial flora studies in thermophilic aerobic sludge treatment. *Conservation and Recycling* 8 303-313

Sonnleitner, B. and Fiechter, A. (1983a) Bacterial diversity in thermophilic aerobic sewage sludge: I. Active biomass and its fluctuations. *European Journal of Applied Microbiology and Biotechnology* 18 47-51

Sonnleitner, B. and Fiechter, A. (1983c). Advantages of using thermophiles in biotechnological processes; expectations and reality. *Trends in Biotechnology* 1 74-80.

Soto, M., Mendez, R. and Lema, J.M. (1992) Characterisation and comparison of biomass from mesophilic and thermophilic fixed bed anaerobic digesters. *Water Science and Technology* 25 203-212

Spaull, A.M. and McCormack, D.M. (1988). The incidence and survival of potato cyst nematodes (Globodera spp) in various sewage sludge treatment processes. *Nematologica* 34 452-461

Stams, A.J.M. (1994). Metabolic interaction between anaerobic bacteria in methanogenic environments. *Antonie van Leeuwenhoek* 66 271-294

Strom, P.F. (1985). Identification of thermophilic bacteria in solid- waste composting. *Applied and Environmental Microbiology* 50 906-913;

Subba Rao, N.S. (1982). Utilisation of farm wastes and residues. In Subba Rao, N.S. (ed.) *Advances in agricultural microbiology*. London, Butterworth Scientific pp 509-522

Sundaram, T.K. (1986). Physiology and growth of thermophilic bacteria. In Brock, T.D. (ed.) *Thermophiles: General molecular and applied microbiology*. New York, John Wiley and Sons pp 75-106

Surucu, G.A., Chian, E.S.K. and Engelbrecht, R.S. (1976). Aerobic thermophilic treatment of high strength wastewaters. *Journal of the Water pollution Control Federation* 48 669-679.

Surucu, G.A., Engelbrecht, R.S. and Chian E.S.K. (1975) Thermophilic microbiological treatment of high strength wastewaters with simultaneous recovery of single cell protein. *Biotechnology and Bioengineering* 17 1639-1662

Suutari, M. and Laakso, S. (1994). Microbial fatty acid and thermal adaptation. *CRC Critical Reviews in Microbiology* 20 285-328

Tatara, M., Makiuchi, T., Ueno, Y., Goto, M. and Sode, K. (2008) Methanogenesis from acetate and propionate by thermophilic down-flow anaerobic packed-bed reactor. *Bioresource Technology* 99 4786–4795

Thambirajah, J.J., Zulkali, M.D. and Hashim, M.A. (1995). Microbiological and biochemical changes during the composting of oil palm empty-fruit-bunches: effect of nitrogen supplementation on thesubstrate. *Bioresource Technology* 52 133-144

Tiquia, S.M., Tam, N.F.Y. and Hodgkiss, I.J. (1996). Microbial activities during the composting of spent pig manure sawdust litter at different moisture contents. *Bioresource Technology* 55 201-206

Tricase, C. and Lombardi, M. (2008). State of the art and prospects of Italian biogas production from animal sewage: Technical-economic considerations. *Renewable Energy*, doi:10.1016/j.renene.2008.06.013

Turner, C (2002). The thermal inactivation of E. coli in straw and pig manure. *Bioresource Technology* 84 57–61

Tyagi, R.D., Tran, F.T. and Agbebavi, T.J. (1990). Mesophilic and thermophilic aerobic digestion of sludge in air lift U-shaped bioreactor. *Biological Wastes* 31: 251-266

Uemura, S. and Harada, H. (1993). Microbial characteristics of methanogenic sludge consortia developed in thermophilic UASB reactors. *Applied Microbiology and Biotechnology* 39 654-660

Uemura, S. and Harada, H. (1995). Inorganic composition and microbial characteristics of methanogenic granular sludge grown in thermophilic upflow anaerobic sludge blanket reactor. *Applied Microbiology and Biotechnology* 43 358-364

Ugwuanyi, J. O., McNeil, B. and Harvey, L. M. (2009). Production of protein-enriched feed using agro-industrial residues. In Singh nee' Nigam, P. and Pandey, A. (eds.) Bioechnology for Agro-Industrial Residues Utilisation, DOI 10.1007/978-1-4020-9942-7-5, Netherlands, Springer Science+Business Media in press

Ugwuanyi, J.O. (1999). Aerobic Thermophilic Digestion of Model Agricultural Wastes. Ph.D. Thesis. University of Strathclyde, Glasgow

Ugwuanyi, J. O. (2008). Yield and protein quality of thermophilic *Bacillus* spp. Biomass related to thermophilic aerobic digestion of agricultural wastes for animal feed supplementation. *Bioresource Technology* 99: 3279- 3290

Ugwuanyi, J. O., Harvey, L. M. and McNeil, B. (1999). Effect of process temperature, pH and suspended solids content upon pasteurisation of a model agricultural waste during thermophilic aerobic digestion. *Journal of Applied Microbiology* 87 387-395

Ugwuanyi, J. O., Harvey, L. M. and McNeil, B. (2004). Development of thermophilic populations, amylase and cellulase activities during thermophilic aerobic digestion of model agricultural waste slurry. *Process Biochemistry* 39: 1661-1669

Ugwuanyi, J. O., Harvey, L. M. and McNeil, B. (2004). Protease and xylanase activities and thermophilic populations as potential process monitoring tools during thermophilic aerobic digestion. *Journal of Chemical Technology and Biotechnology* 79: 30 -38

Ugwuanyi, J. O., Harvey, L. M. and McNeil, B. (2005). Effect of aeration rate and waste load on evolution of volatile fatty acids and waste stabilization during thermophilic aerobic digestion of a model high strength agricultural waste. *Bioresource Technology* 96: 721 – 730

Ugwuanyi, J. O., Harvey, L. M. and McNeil, B. (2005). Effect of digestion temperature and pH on treatment efficiency and evolution of volatile fatty acids during thermophilic aerobic digestion of model high strength agricultural waste. *Bioresource Technology* 96: 707-719

Ugwuanyi, J. O., Harvey, L. M. and McNeil, B. (2006). Application of thermophilic aerobic digestion in protein enrichment of high strength agricultural waste slurry for animal feed supplementation. *Journal of Chemical Technology and Biotechnology* 81: 1641 – 1651

Ugwuanyi, J. O., Harvey, L. M. and McNeil, B. (2007). Linamarase activities in *Bacillus* spp responsible for thermophilic aerobic digestion of agricultural wastes for animal nutrition. *Waste Management* 27: 1501- 1508

Ugwuanyi, J. O., Harvey, L. M. and McNeil, B. (2008). Protein enrichment of corn cob heteroxylan waste slurry by thermophilic aerobic digestion using *Bacillus stearothermophilus*. *Bioresource Technology* 99: 6974- 6985

Ugwuanyi, J. O., Harvey, L. M. and McNeil, B. (2008). Diversity of thermophilic populations during thermophilic aerobic digestion of potato peel slurry. *Journal of Applied Microbiology* 104: 79- 90

Vázquez, J.A., Docasal, S.F., Prieto, M.A., Gonzalez, M.P. and Murado, M.A. (2008). Growth and metabolic features of lactic acid bacteria in media with hydrolysed fish viscera. An approach to bio-silage of fishing by-products. *Bioresource Technology* 99: 6246- 6257

Verstrate, W., de Beer, D., Pena, M., Lettinga, G. and Lens, P. (1996). Anaerobic processing of organic wastes. *World Journal of Microbiology and Biotechnology* 12 221-238

Vidotti R.M., Viegas E.M.M. and Carneiro D.J. (2003). Amino acid composition of processed fish silage using different raw materials. *Animal Feed Science and Technology* 105: 199-204

Vismara, R. (1985). A model for autothermic aerobic digestion; Effects of scale depending on aeration efficiency and sludge concentration. *Water Research* 19 441-447

Volanis M., Zoiopoulos P., Panagou E. and Tzerakis C. (2006). Utilization of an ensiled citrus pulp mixture in the feeding of lactating dairy ewes. *Small Ruminant Research* 64: 190-195

Wagner Otto A., Malin C. and Gstraunthaler, I.P. (2008). Survival of selected pathogens in diluted sludge of a thermophilic waste treatment plant and in NaCl-solution under aerobic and anaerobic conditions. *Waste Management*, doi:10.1016/j.wasman.2008.03.003

Waigant, W.M. and De Man, A.W.A. (1986). Granulation of biomass in thermophilic anaerobic sludge blanket reactors treating acidified wastewaters. *Biotechnology and Bioengineering* 28 718-727

Ward, R.L. and Ashley, C.S. (1978). Heat inactivation of enteric viruses in dewatered wastewater sludge. *Applied and Environmental Microbiology* 36 898-905

Wheatley, A.D. (1990). Anaerobic digestion; Industrial waste treatment. In Wheatley, A.D. (ed.). *Anaerobic digestion, a waste treatment technology.* London, Elsevier Applied Science pp171-224

White-Hunt, K. (1980). Domestic refuse—a brief history 1. *Solid Wastes* 70 609-615.

White-Hunt, K. (1981a). Domestic refuse—a brief history 2. *Solid Wastes* 71 159-166.

White-Hunt, K. (1981b). Domestic refuse—a brief history 3. *Solid Wastes* 71 284-292

Whitmore, T.N. and Robertson, L.J. (1995). The effect of sewage sludge treatment processes on oocyst of *Cryptosporidium parvum. Journal of Applied Bacteriology* 78 34-38

Williams, A.G., Shaw, M., Selviah, C.M. and Cumby, R.J. (1989). The oxygen requirements for deodorising and stabilizing pig slurry by aerobic treatment. *Journal of Agricultural Engineering Research* 43 291-311

Wolf, J. and Sharp, R.J. (1981). Taxonomic and related aspects of thermophiles within the genus *Bacillus* In Berkeley, R.C.W. and Goodfellow, M. (eds.). *The aerobic endospore-forming bacteria, classification and identification.* New York, Academic Press pp 251-296

Wolinski, W.K. and Bruce, A.M. (1984). Thermophilic oxidative sludge digestion; a critical assessment of performance and costs. E W P C A, Symposium Munich pp 385-408

Yamaguchi, M., Hake, J., Tanomoto, Y., Naritomi, T., Okamura, K. and Minami, K. (1991). Enzyme activity for monitoring the stability in a thermophilic anaerobic digestion of waste water containing methanol. Journal of Fermentation and Bioengineering 71 264-269

Yang, H.Y., Wang, X.F., Liu, J.B., Gao, L.J., Ishii, M., Igarashi, Y. and Cui, Z.J. (2006). Effect of water-soluble carbohydrate content on silage fermentation of wheat straw. *Journal of Bioscience and Bioengineering* 101: 232-237

Yilmaz, T., Yuceer, A. and Basibuyuk, M. (2008). A comparison of the performance of mesophilic and thermophilic anaerobic filters treating papermill wastewater. Bioresource Technology 99 156–163

Yu, S., Clark, O.G. and Leonard, J.J. (2008). Influence of free air space on microbial kinetics in passively aerated compost. *Bioresour. Technol.*, doi:10.1016/j.biortech.2008.06.051

Yun, Y-S., Park, J.I., Suh, M.S., Park, J.M. (2000). Treatment of food wastes using slurry-phase decomposition. *Bioresource Technology* 73 21 - 27

Zahiroddini, H., Baah, J., Absalom, W. and McAllister, T.A. (2004). Effect of an inoculant and hydrolytic enzymes on fermentation and nutritive value of whole crop barley silage. *Animal Feed Science and Technology* 117: 317-330

Zeikus, J.G., Hegge, P.W. and Anderson, M.A. (1979). *Thermoanaerobium brockii* gen. nov. and sp nov. A new chemoorganotrphic caldoactive anaerobic bacterium. *Archives of Microbiology* 122 41-48

Zentgraf, B., Gwenner, C. and Hedlich, R. (1993). Thermophilic bacteria from spent liquor of pulp mills. *Acta Biotechnol.* 13 83-87

Zhao, H-Z., Cheng, P., Zhao, B. and Ni, J-R. (2008). Yellow ginger processing wastewater treatment by a hybrid biological process. *Process Biochemistry* doi:10.1016/j.procbio.2008.07.019

Zinder, S.H. (1986). Thermophilic waste treatment systems. In Brock, T.D. (ed.). *Thermophiles; general, molecular and applied microbiology.* New York John Wiley and Sons pp257-277.

Zinder, S.H. (1990). Conversion of acetic acid to methane by thermophiles. *FEMS Microbiology Reviews* 75 125-138

Zupancic, G. D., Strazscar, M. and Ros, M. (2007). Treatment of brewery slurry in thermophilic anaerobic sequencing batch reactor. *Bioresource Technology* 98 2714–2722

Zvauya, R., Parawira, W. and Mawadza, C. (1994). Aspects of aerobic thermophilic treatment of Zimbabwean traditional opaque beer brewery wastewater. *Bioresource Technology* 48 273-274

In: Agricultural Wastes
Eds: Geoffrey S. Ashworth and Pablo Azevedo

ISBN 978-1-60741-305-9
© 2009 Nova Science Publishers, Inc.

Chapter 4

FLY ASH USE IN AGRICULTURE: A PERSPECTIVE

*Wasim Aktar**

Pesticide Residue Laboratory, Department of Agricultural Chemicals, Bidhan Chandra Krishi Viswavidyalaya, Mohanpur-741252, Nadia, West Bengal, India

INTRODUCTION

Fly ash has a potential in agriculture and related applications. Physically, fly ash occurs as very fine particles, having an average diameter of <10 mm, low- to medium-bulk density, high surface area and very light texture. Chemically, the composition of fly ash varies depending on the quality of coal used and the operating conditions of the thermal power stations. On average, approximately 95 to 99% of fly ash consists of oxides of Si, Al, Fe and Ca, and about 0.5 to 3.5% consists of Na, P, K and S. The remainder of the ash is composed of trace elements. In fact, fly ash consists of practically all of the elements present in soil except organic carbon and nitrogen (Table 1). Thus, it was discovered that this material could be used as an additive or amendment material in agricultural applications.

In view of the above, some agencies, individuals, and institutes at various locations conducted some preliminary studies on the effect and feasibility of fly ash as an input material in agricultural applications. Some amount of experience was gained in the country and abroad regarding the effect of fly ash utilisation in agriculture and related applications.

BENEFITS

I. Modification of Soil Texture

The addition of appropriate quantities of fly ash can alter the soil texture. Fly ash addition @ 70 t/ha has been reported to alter the texture of sandy and clayey soil to loamy (Fail and Wochock, 1977). Changes in soil texture could be expected in both agricultural soils as well as

---

[*] Correspondence to: wasim04101981@yahoo.co.in.

in strip-mined soils. It was generally observed that both sandy and clayey soils tend to become loamy in texture (Capp, 1978). In the U.S., the average silt content in fly ash is about 63.2% (Sharma et al., 1989), but in India this content ranges from about 16% (IIT, Kharagpur) to 45% (UAS, Raichur).

II. Modification of Bulk Density

The grain size distribution, especially the silt size range of fly ash, affects the bulk density of soil. Chang et al. (1977) observed that among five soil types, Reyes silty clay showed an increase in bulk density from 0.89 to 1.01 when the corresponding rates of fly ash amendment increased from 0 to 100%. But in soils with bulk densities varying between 1.25 and 1.60, a marked decrease in bulk density was observed by the addition of fly ash.

Page et al. (1979, 1980) reported that fly ash amendment to a variety of agricultural soils tends to decrease the bulk density. Optimum bulk density in turn improves the soil porosity, the workability of the soil, the root penetration and the moisture retention capacity of the soil.

III. Water-Holding Capacity of Soil

The application of fly ash has been found to increase the available water content of loamy sand soil by 120% and of a sandy soil by 67% (PAU, Ludhiana). RRL Bhopal reported that application of fly ash increase the porosity of black cotton soil and decreases the porosity of sandy soils and thereby saves irrigation water around by 26% and 30%, respectively.

Table 1. Physical and chemical characteristics of Indian fly ash and soil

Properties	Fly ash	Soil
BD (g cm-1)	<1.0	1.33
W.H.C. (%)	35–40	<20
Porosity (%)	50–60	<25
P (%)	0.004–0.8	0.005–0.2
K (%)	0.19–3.0	0.04–3.0
S (%)	0.1–1.5	0.01–0.2
Fe (%)	36–1333	10–300
Zn (ppm)	14–1000	2–100
Cu (ppm)	1–26	0.7-40
Mn (ppm)	100–3000	100–4000
B (ppm)	46–618	0.1–40

Source: CAS Raichur (1997).

Chang et al. (1977) reported that at an addition of 8% by weight of fly ash increased the water-holding capacity of soil. They also reported that soil hydraulic conductivity improved at lower rates of fly ash application but deteriorated when the rate of fly ash amendment exceeded 20% in calcareous soils and 10% in acidic soils. This improvement in water-holding capacity is beneficial to the plants, especially under rainfed agriculture.

IV. Soil pH

In India, most of the fly ash produced is alkaline in nature. Hence, an application of these to agricultural soil increases the soil pH. This property of fly ash can be exploited to neutralize acidic soils (Elseewi et al., 1978; Phung et al., 1978). Jastrow et al. (1979) reported that while addition of fly ash improves soil pH on one hand, it simultaneously adds essential plant nutrients to the soil on the other hand. Page et al. (1979) observed that experiments with calcareous and acidic soils revealed that fly ash addition increased the pH of the former from 8.0 to 10.8 and that of the latter from 5.4 to 9.9. It has also been reported that the use of excessive quantities of fly ash to alter pH can cause increase in soil salinity especially with unweathered fly ash (Sharma etal.1989). Some fly ashes are acidic which may be used for reclamation of alkaline soils. In one of the project sites of FAM at Phulpur, IFFCO has done some work on the reclamation of alkaline soils and observed that the pH of these soils could be brought to near neutral status using these acidic fly ashes.

V. Effect on Soil Crust

Fly ash application helps in reducing surface encrustation, which is a problem in red soils (CAS, Raichur). This effect in turn can enhance soil aeration and help in improvement of germination of plants grown on it.

VI. Effect on Growth and Yield of Crops

The positive impact of fly ash application on growth and yield of crops has been reported by various agencies. Some of which are given below:

Regional Research Laboratory (RRL) Bhopal reported that, on average, in comparison to control, there was around 50–60% more yield of Brinjal, around 45% more yield of potato and pea, around 40% more yield of tomato and around 29% more yield of cabbage were recorded in fly ash treated plot when fly ash was applied @25% of soil.

Punjab Agriculture University (PAU) observed that application of fly ash @10t/ha increased the yield of wheat from 21.5 q/ha to 24.1 q/ha and that of cotton from 1245 kg/ha to 1443 kg/ha. They have also been found that fly ash application @10% by weight increased the dry matter yield of moong from 3.80 gm to 7.36 gm and fly ash addition from 0 to 80 t/ha increased the yield of paddy from 61.82 q/ha to 63.58 q/ha. College of Agriculture, Raichur observed that the yield of groundnut was increased from 24.1 q/ha to 31.9 q/ha with the application of fly ash @20 t/ha.

View of groundnut crop grown on fly-ash–treated soil at IIT-Kharagpur.

Forestry and floriculture species on an ash pond (by TERI at BTPS Badarpur).

Promising indications were obtained from the preliminary research findings (highlights of which have been given above). A need was felt for collection of a set of empirical data through scientifically designed trials using standard procedures and protocols in a coordinated manner at the national level on the benefits and possible adverse effect of fly ash application in agricultural fields. Fly Ash Mission (FAM), along with its associate agencies, has taken up a large number of demonstrative trials (more than 50) at various sites at dispersed locations across the country under varied agro-climatic conditions on a spread of crops, forestry and horticulture species. These trials are being done with varied dose ranges based on the results of the part research experiences of respective centers at their sites. Even up to 100% ash bodies have been used to obtain the extreme effects. Tables 2 and 3 provide a picture of the locations, plant types and soil types and fly ash dose ranges that have been covered under Fly Ash Mission projects.

Table 2. Field crops and vegetables projects undertaken by FAM

S. no.	Soil	Fly ash doses range	Crops and no. of sites	Location	Executed by
1	Alluvial soil	0–200 t/ha	Rice, wheat (2)	Farakka	CFRI, Dhanbad
2	Alluvial soil	0–100 t/ha	Mustard, jute (1)	Farakka	CFRI, Dhanbad
3	Laterite soil	0–200 t/ha	Rice (5), wheat (4)	Bakreshwar	CFRI, Dhanbad
4	Laterite soil	0–100 t/ha	Mustard, potato, lentil (1)	Bakreshwar	CFRI, Dhanbad
5	Black soil	0–50 t/ha	Sugarcane	Chidambaram	Annamalai University
6	Laterite soil	0–150 t/ha	Groundnut	Neyveli	Annamalai University
7	Laterite soil	0–100 t/ha	Sugarcane	Neyveli	Annamalai University
8	Black soil	0–150 t/ha	Rice-green gram (1)	Sathamangalam	Annamalai University
9	Black soil	0–120 t/ha	Cotton-rice (1)	Vellampudugai	Annamalai University
10	Lateritic soil	0–10 t/ha	Rice-groundnut (3)	Kharagpur	IIT-Kharagpur
11	Lateritic soil	0–20 t/ha	Rice, groundnut-mustard (1)	Kharagpur	IIT-Kharagpur
12	Lateritic soil	0–30 t/ha	Mustard-rice (1)	Kharagpur	IIT-Kharagpur
13	Lateritic soil	0–10 t/ha	Rice (2)-mustard, groundnut, potato (1)	Balarampur, Gholghoria, Burari	IIT-Kharagpur
14	Lateritic soil (red)	0–80 t/ha	Sunflower-groundnut (2)	Raichur	CAS, Raichur
15	Black soil	0–80 t/ha	Sunflower-maize (2)	Raichur	CAS, Raichur
16	Alluvial soil	0–650 t/ha	Tomato (1), cabbage (1), potato (1), wheat (2), pea (1)- maize(6), wheat-maize (2)	Dhodhar, Nilgiri, Rihand Nagar	RRL, Bhopal
17	Alluvial soil	0–650 t/ha	Sunflower (1), tomato (1), potato (1), wheat (1), berseem (1), red gram (1), maize (4), rice (1)	Nilgiri, Rihand Nagar	RRL, Bhopal
18	Alluvial soil	0–40/0–80 t/ha	Rice-wheat (1), cotton-wheat (1), sunflower-maize (1), wheat-rice (1)	Ropar, Bhatinda	PAU Ludhaina
19	Alluvial soil	0–12 t/ha	Wheat	Ropar (Astalpur)	PAU Ludhaina
20	Alluvial soil	100% ash body with 7.5 cm soil cover	Arhar-wheat (1)	Bhatinda	PAU Ludhaina
21	Black soil	0–640 t/ha (residual effect)	Wheat-maize, soyabean-maize, lemon grass (1)	Sarni	RRL, Bhopal
22	Alluvial soil	0–640 t/ha	Maize-onion, rice-sunflower (1)	Angul	RRL, Bhopal

Source: Central Fuel Research Institute, Dhanbad (1999).

These trials have been scientifically designed to collect empirical data on effect of fly ash application on soil, plant and natural ground water near the application site. Indicative monitoring is being done of the soil, plant produce, macro and micro nutrient status including the trace and heavy metal status and also the changes in the natural radioactivity level (if any as a result of application of fly ash) on soil, plant and natural ground water near the trial site. The aspect of the nutritional quality of produce grown in fly ash treated soils if also being addressed in a special project by the National Institute of Nutrition. Institute of Physics, Bhubaneswar is testing the samples of all FAM project sites for some heavy metals and radionucleide levels. More than 1000 samples have been tested so far. At most places the levels appear to be in the normal range. Highlights of the some the important findings from these trials are as follows:

VII. Effect on Crop Growth and Yield

In rice-groundnut cropping system, application of Fly ash @ 10 t/ha to both the crops increased grain yield of rice on an average by 14% and pod yield of groundnut by 26% over control (IIT-Kharagpur). Application of fly ash @ 10 t/ha in combination with organic and inorganic sources either in one or both the seasons in rice-groundnut cropping system increased grain yield or rice and pod yield of groundnut significantly over application of only chemical fertilizer to both the crops (IIT-Kharagpur). The treatment combination fly ash @ 10 t/ha alongwith Paper Factory Sludge (@ 15 t/ha) and chemical fertilizer (CF) applied to rice and only CF to potato increased grain yield of Rice and tuber yield of Potato significantly over CF applied to both crops in Rice-Potato cropping system. (IIT-Kharagpur).

In the rice-mustard cropping system, application of pond ash @ 10 t/ha in combination with organic and inorganic sources to rice and subsequent mustard with CF alone increased yield of rice by 13–15% and seed yield by 15–18% as compared to the treatment of similar combination but without pond ash.(IIT-Kharagpur). Use of pond ash/fly ash either in splits or as one time application along with organic and inorganic sources were equally effective in increasing grain yield of rice and seed yield of mustard, as compared to the treatment without pond ash/fly ash. (IIT-Kharagpur). At the College of Agriculture–Raichur, yield of sunflower was increased by about 25% in red soil under rained as well as irrigated conditions when fly ash was applied @ 60T/ha along with 20t/ha FYM.

Table 3. Forestry, land reclamation projects undertaken by FAM

S. no.	Soil/land type	Fly ash doses range	Tree species and no. of sites	Location	Executed by
1	Laterite Soil	0–240 t/ha	Eucalyptus (1)	Chaudwar, Cuttak	TCRDC, Patiala
2	Laterite Soil	0–24% of pit volume	Eucalyptus, Acacia auriculiformis, Casurina equisetifolia, Acacia mangium (1)	Durga Prasad, Cuttack	TCRDC, Patiala
3	Alkali-saline eroded land (in arid zone)	0–20% v/w	Eucalyptus, Zizyphus, Jojoba (1)	Jaipur	TERI, New Delhi
4	Ash pond	–	Melia azadirach, Delbergia Sisso, Eucalyptus sp.,	Badarpur	TERI, New Delhi

S. no.	Soil/land type	Fly ash doses range	Tree species and no. of sites	Location	Executed by
			Populus deltoides (1)		
5	Low Fertile Soil	1/3 pit volume	Ceiba pentandra, Melia azadirach, Cassia siamea, Erythrina indica, Cassia glauca, Bauhinia purpurea, Putranjiva, Pongamia glabra, Thevetia elifera (1)	New Delhi	TERI, New Delhi
6	Usar	0–5%	Rice, wheat (1)	Dailapur	IFFCO, Phulpur
7	Usar	0–5%	Rice, mustard (1)	Tardih	IFFCO, Phulpur
8	Usar	0–5%	Rice, wheat (1)	Yakubpur	IFFCO, Phulpur
9	Usar	0–6%	Rice, wheat (1)	Purisudi	IFFCO, Phulpur
10	Usar	0–6%	Rice, wheat (1)	Parasinpur	IFFCO, Phulpur
11	Usar	0–6%	Rice, wheat (1)	Mobarukpur	IFFCO, Phulpur
12	Ash pond	–	Rajnigandha, Tagetus, Carnation, Palmarosa, Sunflower (1)	Badarpur, New Delhi	TERI, New Delhi

Source: RRL Bhopal (1999).

More than 70% increase in yield of groundnut was observed when fly ash was applied @ 30 t/ha along with FYM @ 20 t/ha at CAS Raichur. The yield of maize also increased by about 35% of present when fly ash was applied @ 30 t/ha along with FYM @ 20 t/ha (CAS Raichur).

The performance efficiency of both dry fly ash and pond ash in respect of crop growth parameters (yield and improvement in physical and chemical properties of soil was found to be similar by CFRI, Dhanbad in their studies at Farakka and Bakeraswar). Best grain and straw yield of both paddy and wheat crops were observed 200 t/ha pond ash dose (20–40% increases were observed). Paddy and Wheat crops grown with fly ash showed early maturing tendencies at Farakka. Annamalai University, in their trials found that the application of 100 t/ha of Lignite Fly Ash (LFA) did not adversely affect the germination of seeds of rice, green-gram, groundnut, sugarcane or cotton. Satisfactory levels of germination were observed and these were found to be at par with those in control plots (where no fly ash/pond ash was applied).

Annamalai University also reported yield increases due to LFA application in eight out of twelve of the field trials (with LFA application between 4-120 t/ha) showing a range of 8% to 36% increase in yield of produce (over control). Significant yields increases were seen in 2nd and 3rd crop of Groundnut at 10t/ha LFA. Increasing the dose of LFA to 100 t/ha significantly increased pod yield in the first crop itself. (Annamalai University).

Fly ash for reclaiming saline alkaline soil—rice crop at IFFCI, Phulpur.

Flower at ash pond (by TERI at Badarpur).

VIII. Effect on Soil Health

In rice based cropping system, application of fly ash/pond ash @ 10 t/ha alongwith organic and inorganic sources wither in one season or in consecutive seasons improved physical properties of soil through decreasing its bulk density and increasing its water holding capacity and porosity (IIT-Kharagpur). Being alkaline in nature, application of fly ash/pond ash @ 10 t/ha in combination with organic and inorganic sources increased pH of acid-latertic soil to a considerable extent (IIT-Kharagpur). In rice based cropping system, repeat application of fly ash/pond ash @ 10 t/ha in combination with organic and inorganic sources raised the fertility status of soil, as compared to the treatment of similar combination but without fly ash/pond ash (IIT-Kharagpur). In general, the available heavy metal status of soil decreased under fly ash/pond ash based treatments (IIT-Kharagpur). Application of fly ash/pond ash increased soil dehydrogease activity, which was more discernible in aerobic condition than flooded/reduced soil condition (IIT-Kharagpur). RRL, Bhopal (in the project site Dhodhar, Rihand Nagar) found that the nutrient availability was enhanced in soil where 25% coal ash was applied at one time and in those plots wherein addition to the 25% ash was made initially and 5% addition was made every year. At CAS, Raichur pH and Electrical Conductivity of soil did not differ significantly due to application of different fly ash levels. CAS, Raichur also found that the combined application of fly ash and FYM had a beneficial effect on the fertility status of soil the content of total lead, arsenic and selenium did not change significantly due to application of recommended dose(upto 60 t/ha) of fly ash. However, at higher rate of applications the contents of these toxic elements increased marginally. CAS, Raichur found that the content of toxic elements was lesser in red soils as compared to black soils. RRL, Bhopal in its trials at Nilgiri, Rihand Nagar (landfill site) found that the primary and secondary nutrients were increased in ash filled plots. The heavy metals like Co, Ni, Cr, Pb, Cd where found to be below detectable limits.

CFRI, Dhanbad found that the application of alkaline fly ash (pH 8.3) helped in neutralizing the acidic red soil (pH 5.01) making it more productive and suitable for cultivation. It also helped in improving the utilisation efficiency of NPK fertilizer. Annamalai University found that application of lignite fly ash in various soil types showed the following types of soil modifications: 1) neutralising soil pH, 2) increasing EC, and 3) increasing available levels of potassium, sulphur and boron.

IX Effect on Quality of Yield and Uptake of Nutrients and Toxic Elements

In rice based cropping system, application of fly ash/pond ash @ 10 t/ha in combination with organic and inorganic sources increased the concentration of macronutrients (N, P, CA and Mg) in rice grain and edible part of the subsequent crops (groundnut, potato and mustard) as compared to the treatment of similar combination, but without fly ash/pond ash (IIT-Kharagpur). Decrease in heavy metal concentration in grain or edible part of the different crops under fly ash/pond ash based treatment is due to dilution effect of these elements through increased grain/edible yield (IIT-Kharagpur). Radionucleide levels in grain/edible part of rice, groundnut, potato and mustard varied under fly ash/pond ash based treatments as compared to the treatments without it and indicated no adverse effect of fly ash/pond ash (IIT-Kharagpur). The nutritional value of agricultural produce grown on ash-filled land-fill sites in terms of protein and carbohydrates were found to be comparable with the National Institute of Nutrition-Standards (RRL, Bhopal- trials at Nilgiri, Rihand Nagar).

X. Effect on Ground Water

Application of fly ash/pond ash in combination with organic and inorganic sources released lower quantity of Fe and Mn to ground water as compared to the treatment without fly ash/pond ash (IIT-Kharagpur). In ground water samples the level of ^{286}Ra was decreased, ^{228}Ac remained unchanged under fly ash/pond ash based treatment as compared to the treatments without it, indicating poor leaching of radionucleide to ground water. Thereby the ground water quality due to application of fly ash/pond ash remained unaffected with respect to radionucleide contamination (IIT-Kharagpur).

XI. Other Effects

The crops grown under fly ash/pond ash based treatment were observed to be resistant to disease, insect, and pest infestation as compared to the crops grown without fly ash/pond ash IIT-Kharagpur). At Bakreshwar, in farmers' field trials the farmers have observed that the crops grown in fly Ash treated plots were relatively more resistant to pest attack in compared to those in control plots. Farmers were enthusiastic about taking groundnut crop in rabi using fly ash compared to the traditional boro rice crop at Kharagpur for a better cost-benefit ratio.

USE OF FLY ASH AS A MINE SOIL AMENDMENT

The physical effects of fly ash additions on soils were discussed earlier, but relatively high loading rates (> 100 tons per acre) are generally required to significantly influence soil physical properties such as water holding capacity and aggregation. In most instances, fly ash is added to soils primarily to affect chemical properties such as pH and fertility, and loading rates are limited by chemical effects in the treated soils. Plant growth on fly ash-amended soils is most often limited by nutrient deficiencies, excess soluble salts and phytotoxic B levels (Page et al., 1979; Adriano et al., 1980). Fly ash usually contains virtually no N and has little plant-available P. However, newer power plants may be adding ammonia as a flue gas conditioner to limit NOX emissions which may lead to some plant-available N. Application of fly ash to soil may cause P deficiency, even when the ash contains adequate amounts of P, because soil P forms insoluble complexes with the Fe and Al in more acidic ashes (Adriano et al., 1980) and similarly insoluble Ca-P complexes with Class C ashes. Amendment of K-deficient soil with fly ash increases plant K uptake, but the K in fly ash is apparently not as available as fertilizer K, possibly because the Ca and Mg in the fly ash inhibit K absorption by plants (Martens et al., 1970).

In some cases, soils have been amended with fly ash in order to correct micronutrient deficiencies. Acidic-to-neutral fly ash has been found to correct soil Zn deficiencies, although alkaline fly ash amendment can induce Zn deficiency because Zn becomes less available with increasing pH (Schnappinger et al., 1975). Fly ash application has also been shown to correct B deficiencies in alfalfa (Plank and Martens, 1974). In some cases, plant yields after fly ash application have been reduced because of B toxicity (Martens et al., 1970; Adriano et al., 1978). Soil amendment with fly ash to alleviate B deficiencies should be carefully monitored in order to avoid B toxicity. Fly ash often contains high concentrations of potentially toxic trace elements. Plants growing on soils amended with fly ash have been shown to be enriched in elements such as As, Ba, B, Mo, Se, Sr, and V (Furr et al., 1977; Adriano et al., 1980). Although trace amounts of some of these elements are required for plant and animal nutrition, higher levels can be toxic. Highly phytotoxic elements often kill plants before the plants are able to accumulate large quantities of the element; which limits their transfer to grazing animals. Elements such as Se and Mo, however, are not particularly toxic to plants and may be concentrated in plant tissue at levels that cause toxicities in grazing animals. Soils amended with high rates of fly ash may accumulate enough Mo to potentially cause molybdenosis in cattle (Doran and Martens, 1972; Elseewi and Page, 1984).

Finally, amendment of soil with fresh fly ash may increase soil salinity (reported as soluble salts or electrical conductance-EC) and associated levels of soluble Ca, Mg, Na, and B. Incorporation of 80 T/A unweathered fly ash from a Nevada power plant increased soil salinity 500 to 600% and also caused a significant increase in soluble B, Ca, and Mg (Page et al., 1979). Fly ash that has been allowed to weather and be leached by rainfall for several years generally has much lower soluble salt and soluble B concentrations and is more suitable for use as a soil amendment (Adriano et al., 1982). In general, ashes which have been wet-handled in the plant and stored in ponds will be much lower in soluble salts and B than dry-collected ashes.

USE OF FLY ASH IN ACIDIC SPOIL AND COAL REFUSE REVEGETATION

Alkaline fly ash can aid in the reclamation of acidic spoils and refuse piles, although one-time ash applications do not appear to be effective in maintaining increased pH if pyrite oxidation is not completely stopped and neutralized. The pH of an extremely acidic surface mine soil and a coal refuse bank in West Virginia was initially raised to near neutral by application of high rates of alkaline (pH 11.9) fly ash. Soil pH dropped 1 to 2 units over the next two growing seasons, however, presumably because of continued pyrite oxidation in the spoils and leaching of Ca and Mg oxides from the fly ash (Adams et al., 1972). Jastrow et al. (1981) used fly ash as an alternative to lime in a greenhouse experiment involving acidic coal refuse. The initial pH of the refuse was 3.5.

Amendment with fly ash raised the pH to 4.8, but it dropped to 4.2 by the end of one growing season. In another greenhouse experiment, the application of fly ash to extremely acidic coal refuse resulted in a higher pH and significantly increased barley yields (Taylor and Schumann, 1988). Boron toxicity has been observed in plants grown on fly ash-amended mine spoils, although in some cases toxicity symptoms were apparent but yields were not reduced (Adams et al., 1972; Keefer et al., 1979; Taylor and Schumann, 1988). Jastrow et al. (1981) implicated Mn, Zn, and V toxicity as possible factors in reduction of tall fescue yields on fly ash-amended coal refuse. Coal refuse often contains high levels of trace elements and fly ash application can raise the concentrations of these elements to toxic levels, especially if pH is not controlled.

STUDIES ON POSSIBLE NEGATIVE EFFECTS OF FLY ASH APPLICATION

I. Ground Water

Fly ashes contain a small amount of trace and heavy metals which may percolate down and pollute ground water. The solubility of these elements is <10% (Rohriman, 1971). Natusch (1975) observed in a laboratory experiments on leaching potential that 5 to 30% of toxic elements especially Cd, Cu and Pb are leachable. Gralloway et al. (1976) observed that atleast 10% of total Cd would be solubilized in the acidic pH range of 3 to 5. It is unlikely that these will have any major effect on the quality of ground water. However, monitoring of this aspect may be advisable. At Central Fuel Research Institute (CFRI), Dhanbad it was observed that the quality of ground water did not change with the application of fly ash and all the parameters including the trace and toxic metal contents were within the permissible limits. Some other research organisations also observed that fly ash has no significant polluting effect on ground water.

II. Uptake of Heavy Metals and Toxic Elements by Plants

Fly ash has a ppm level concentration of heavy metals; when applied to soil these elements may get absorbed by plants grown in it which may ultimately enter the food chain. However, the absolute quantities of these elements in fly ashes are low which may not result into negative effect. The data on trace element uptake and accumulation by plant are limiting. Despite fairly intensive research over the last 25 years, the data on trace element accumulation are rather sketchy and inconsistent. Boron in fly ash is readily available to plants and investigators consider B to be limiting factor in unweathered fly ash utilisation (Townsend and Gillham, 1975; Elseewi et al., 1978; Ciravolo and Adriano, 1979). RRL, Bhopal conducted a study regarding the uptake of heavy and trace metals by some vegetable crops and it was observed that the uptake is quite low and remains within the normal range. Central Fuel Research Institute, Dhanbad observed that there is no significant differences in uptake of trace and heavy metal between control and fly-ash–treated plots. Although fly ash contains a moderate amount of trace and heavy metals, the uptake and accumulation of these by plants is very negligible.

III. Radionuclides

There have been several reports in the literature on the presence of radionuclides in fly ash, but studies on their impact have been few (Coles et al., 1978; Gowiak and Pacynas, 1980). The radiochemical pollution of Uranium and Thorium series is always present in fly ash (Eisenbud and Petrow, 1964). The concentration of natural Uranium varies from 14 to 100 ppm although in exceptional cases it may be as high as 1500 ppm whereas that of Thorium is less than 10 ppm. The fly ash concentrates besides other gaseous and trace metal oxides, several radioactive contaminants like $222Ru$ and $220Ru$ (Sharma et al., 1989). Bhaba Atomic Research Centre, Bombay is of the opinion that most of the Indian coals has very low levels of radioactivity which is well below the hazardous limit. Hence radioactivity of fly ash may not be a limiting factor for its application for agriculture purposes. Central Fuel Research Institute, Dhanbad observed that there is no significant uptake of radioactive elements by plants and also that there was negligible cumulative build up of these contaminants in soil when fly ash applied for agriculture purposes.

CONCLUSION

The potential of fly ash as a resource material in agriculture and related areas is now a well-established fact, and more and more researchers and users are becoming convinced regarding its utility potential in this field. The major attribute that makes fly ash suitable for agriculture is its texture and the fact that it contains almost all of the essential plant nutrients except organic carbon and nitrogen. Although fly ash cannot replace the need for chemical fertilizers or organic manure, it can be used in combination with these (or in some cases may substitute in part their requirement) to get additional benefits in terms of improvement in soil physical characteristics, increased yields, etc. As in the case of fertilizers and any other agriculture input, the amount and method of fly ash application would vary with the type of soil, the crop to be grown, the prevailing agroclimatic condition and the type of fly ash available.

Although fly ash has many benefits as an input material for agriculture applications, in view of the fear in the minds of many (regarding the levels of natural radioactivity in fly ash and the characteristic presence of some amounts of heavy and toxic elements in it) there may be some precautions that have to be taken for the time being while using fly ash in agriculture. From the information available until now, there appears to be little ground for concern on these accounts (heavy metals, radioactivity, etc.); however, further confirmatory studies at the ICAR centers would be helpful in establishing recommendations in this field. Meanwhile, there appears to be sufficient ground now for the cautious and judicious application of this useful material, which is otherwise being wasted and underutilized.

REFERENCES

Arthur, M.F., Zwick, T.C., Tolle, D.A., and Van Varis, P. (1984) Effects of fly ash on microbial Co2 evolution from our agricultural soil. Water Air Soil Pollut., 22, 209.

CAS Raichur(1997) Interim report of Fly Ash Mission sponsored project "Utilization Of Fly Ash in Agriculture " submitted to Fly Ash Mission

Capp, J.P. (1978) Power Plant flyash utilisation for land reclamation in the eastern United States, in Reclamation of Drastically Disturbed Lands. Schaller, F.W. and Sutton, P., Eds., Sol. Sci. Soc. of Am., Madison, WI, 339.

Central Fuel Research Institute, Dhanbad(1999) Draft report Of Fly Ash Mission sponsored project "Utilization of Fly Ash in Agriculture" submitted to Fly Ash Mission

Ciravolo, T.G. and Adriano, D.C. (1979) Utilisation of Coal ash by crops under green house conditions, in Ecology and Coal Resources Development, Wali, M.,Ed., Pergamon Press, New York, 958.

Chang, A.C., Lund, L.J., Page, A.L. and Warneke, J.E. (1977) Physical properties of flyash amended soils. J. Environ Qual. 6(3), 267.

Eisenbud, M.and Petrow, H.C. (1964) Radioactivity in the atmospheric effluents of power palnts that use fossil fuel. Science 144, 288.

Elseewi, A.A., Binghman, F.T. and Page, A.L.(1978) Growth and mineral composition of lettuce and swiss chard grown on flyash amended soils, in Environmental Chemistry and

Cycling processes, Conf. 760429, Adriano, D.C. and Brisbin, I.L.,Eds., U.S. Department of Commerce, Springfield, VA, 568.

Faculty of Agriculture, Annamalai University (1999). Interim Report of Fly Ash Mission sponsored project "Selected Technology Project for Fly Ash Disposal and Utilization in Agriculture" (10-03).

Fail, J.L. amd Wochok, Z.S. (1977) Soyabean growth on flyash amended strip mine soils. Palnt Soil, 48, 473.

Gowiak, B.J. and Pacyna, J.M. (1980) Radiation dose due to atmospheric releases from coal-fired power stations. Int.J. Environ. Stud. 16,23.

Gralloway, J.N., Likens, G.E. and Edgeston, E.S.(1976) Acid rain precipitation in the north eastern United States; pH and acidity, Science 194, 722,

IIT Kharagpur (1999) Draft report Of Fly Ash Mission sponsored project " Utilisation Of Fly Ash And Organic Wastes In Restoration Of Crop Land Ecosystem " submitted to Fly Ash Mission

Jastrow, J.D., Zimmerman, C.A., Dvorak, A.J. and Hinchman, R.R.(1979) Comparison of Lime and Flyash as Amendments to Acidic Coal Mine Refuse: Growth Responses and Trace Element Uptake of Two Grasses. Argonne National Laboratory, Argonne, IL, 43.

Kumar, V. (1996) Fly Ash Utilisation: A Mission Mode Approach in Ash Ponds and Ash disposal Systems. Raju, V.S., Dutta, M., Seshadri, V., Agarwal, V.K. and Kumar, V., Eds. Narosa Publishing House, New Delhi, 365.

Kumar V, Goswami G and Zacharia K A (1999). Fly Ash: Its Influence on Soil Properties. Indian Society Soil Sciences Workshop, 18-21st October 1999, Calcutta

Kumar V, Goswami G and Zacharia K A (1998). Fly Ash Use in Agriculture: Issues and Concern. International Conference on Fly Ash Disposal and Utilisation, 20-22nd January, New Delhi.

Natusch, D.F.S. (1975) Characteristics of pollutants from coal combustion and conversion process, in Toxic Effects on the Aquatic Biota from Coal and Oil Shale Development, Quarterly Progress Rep. Oct.- Dec., Natural Resources Ecology Laboratory, Colorado State University, Fort Collins, 73, 1975

Padmakaran, P. etal.(1994) Fly ash and its utilisation in industry and agricultural land development. Research and Industry, 40, 244-250.

Page, A.L., Elseewi, A.A. and Straughan, I.R. (1979) Physical and Chemical Properties of flyash from coal-fired plants with reference to environmental impacts. Residue Rev., 7, 83.

Page, A.L., Elseewi, A.A., Lund, L.J., Bradford, G.R., Mattigod, S., Chang, A.C. and Bingham, F.T. (1980) Consequences of Trace Element Enrichment of Soils and Vegetation from the Combustion of Fuels Used in Power Generation. University of Claifornia, Riverside, 158.

Phung, H.T., Lund, I.J. and Page, A.L. (1978) Potential use of flyash as a liming material in Environmental Chemistry and Cycling Processes, Conf. 760429, Adriano, D.C. and Brisbin, I.L., Eds. U.S. Department of Energy, 504.

Punjab Agriculture University (1993) Utilisation of flyash in agriculture and revegetation of dumping sites. Annual progress report.

Rohriman, F.A.(1971) Analysing the effect of flyash on water pollution. Power, 115, 76.

RRL Bhopal (1999) Interim report Of Fly Ash Mission sponsored project "Long Term Effect Of Fly Ash On Soil Fertility And Crop Yield" submitted to Fly Ash Mission

Sharma, S. etal. (1989) Flyash dynamics in soil-water systems. Critical Reviews in Environmental Control 19(3), 251-275.

Townsend, W.N. and Gillham, E.W.F. (1975) Pulverised fuel ash as a medium for plant growth, in The Ecology and Resource Degradation and Renewal, Chadwick, M.L. and Goodman, G.T., Eds., Blackwell Scientific, Oxford, 287.

Vijayan, V. and Ramamurthy, V.S. (1995) Measurement of indoor radon levels in Bhubaneshwar. Bulletin of Radiation Protection, vol (18) No. 1 and 2.

Zacharia, K.A.; Kumar, V. and Velayutham, M. (1996) Fly Ash Utilisation in agriculture towards a holistic approach. National Seminar on Fly Ash Utilisation, Neyveli Lignite Corporation Limited, Neyveli.

In: Agricultural Wastes
Eds: Geoffrey S. Ashworth and Pablo Azevedo

ISBN 978-1-60741-305-9
© 2009 Nova Science Publishers, Inc.

Chapter 5

PESTICIDES: USE, IMPACT AND REGULATIONS FOR MANAGEMENT

Vandita Sinha[*], Vartika Rai and P.K. Tandon

Department of Botany, University of Lucknow
Lucknow-226007, Uttar Pradesh, India

ABSTRACT

Modern farming employs many chemicals to produce and preserve large quantities of high-quality food. Fertilizers, pesticides, cleaners and crop preservatives are the major categories that are now abundantly used in agriculture for increasing production. But each of these chemicals poses a hazard— unforeseen side effects such as toxicity to non target organisms, development of resistance in pests to materials used and environmental contamination with the potential to affect the entire food chain, are some major drawbacks thereby causing serious ecological imbalance. However, in many countries, a range of pesticides has been banned or withdrawn for health or environmental reasons, and their residues are still detected in various substances such as food grains, fodder, milk, etc. The majority of chemical insecticides consist of an active ingredient (the actual poison) and a variety of additives that improve efficacy of their application and action. All of these formulations degrade over time. The chemical by-products that form as the pesticide deteriorates can be even more toxic than the original product.

Often stockpiles of pesticides are poorly stored and toxic chemicals leak into the environment, turning potentially fertile soil into hazardous waste. Once a pesticide enters soil, it spreads at a rate that depends on the type of soil and pesticide, moisture and organic matter content of the soil and other factors. A relatively small amount of spilled pesticide can, therefore, create a much larger volume of contaminated soil. The International Code of Conduct on the Distribution and Use of Pesticides states that packaging or repackaging of pesticides should be done only on licensed premises where staff is adequately protected against toxic hazards. Now, many agencies have come forward to prevent the contamination and accumulation of pesticides in the environment—for example, the issuing of the International Code of Conduct on the Distribution and Use of Pesticides by the United Nations Food and Agriculture Organization (FAO). In addition, the

---

[*] E-mail: sinha_vandita@yahoo.com , vandita999@gmail.com

organization works to improve pesticide regulation and management in developing countries. In order to prevent accumulation of pesticides, the WHO works to raise awareness among regulatory authorities and helps to ensure that good regulatory and management systems for the health sector are in place. The United Nations Industrial Development Organization (UNIDO) is supporting cleaner and safer pesticide production with moves toward less hazardous products based on botanical or biological agents. Wider use of these products will result in reductions in the imported chemicals that contribute to obsolete pesticide stockpiles. The World Bank has established a binding safeguard policy on pest management that stipulates that its financed projects involving pest management follow an Integrated Pest Management (IPM) approach.

INTRODUCTION

Pesticides are indispensable to modern agriculture. Establishing a balance between the demand and supply of food for the unstoppable growing population, modern farming has inclined to the use of varied chemicals to produce and preserve large quantities of high-quality food. Today, the food produced in one part of the world is being used in another region, which involves storage and transportation. During all of these stages, fertilizers, pesticides, cleaners and crop preservatives are the major categories that are now used abundantly in agriculture. A pesticide is a substance or mixture of substances used to kill a pest [1]. A pesticide may be a chemical substance, biological agent (such as a virus or bacteria), antimicrobial, disinfectant or device used against any pest. Pesticides are defined under the Federal Environmental Pesticide Control Act as "any substance or mixture of substances intended for preventing, destroying, repelling, or mitigating any pest" including insects, rodents, nematodes, fungus, weeds, other forms of terrestrial or aquatic plant or animal life or microorganisms on or in living humans or animals [2]. There are many kinds of pesticides, such as insecticides, fungicides, herbicides, larvicides, acaricides, rodenticides, molluscides, nematocides and aphicides, etc.

The most common mode of delivery of pesticide in agriculture is spraying although some highly toxic compounds are formulated as slow release granules for direct application to soil. Overspraying of surface water, run off from agricultural land and movement through soil into water courses can lead to appreciable concentrations in river and estuaries. There potential to cause adverse effects to human and wildlife populations has been the subject of intense study and has led to the development of increasingly stringent and encompassing regulations for the risk assessment of novel formulations and to control the use of existing compounds [3].

Pesticides that are being actively used to boost agriculture to meet the demand of the growing population have both positive and negative effects. Pesticide use, is not without problems, unforeseen side effects such as toxicity to non target organisms, development of resistance in pests to materials used and environmental contamination with the potential to affect the entire food chain, are some major drawbacks. In developing countries additional problems are a lack of understanding of their proper use, non availability of suitable application equipment, inadequate storage conditions and high prices [4].

It is evident from the biological monitoring studies that farmers are at higher risk for acute and chronic health effects associated with pesticides due to occupational exposure. Furthermore, the intensive use of pesticides (higher sprays more than the recommended dose) in cotton areas involves a special risk for the field workers, pickers, and of an unacceptable residue concentration in cottonseed oil and cakes [5].

Besides being toxic to the target as per design, pesticides have other effects on these as well as on non target organisms. These effects are genetic effects recognized for their long term value. All these effects may threaten the genetic health of current and future generation. In developing countries where users are often ill trained and lack appropriate protective devices the risk of pesticide's poisoning are magnified [6].

Pesticides are known polluters of the environment, and their improper usage under exploitative agriculture, particularly in vegetable and cotton cultivation, has created havoc. Fruits and vegetables generally contain pesticide residues even after being washed or peeled. By eating food that has been grown with pesticides, we are exposed to pesticides. Many fungicides, herbicides and insecticides have toxic effects on germination and seedling growth. Seed germination is inhibited, growth of plants is greatly reduced, and a decrease in protein content has been observed due to the phytotoxicity of pesticides [7, 8]. There is growing demand for organic fruits and vegetables having no blemishes. Also there is an increasing pressure by consumers for "clean" and uncontaminated foods. These concerns are one reason for the organic food movement. This in turn is putting increased demand on the insecticide industry to produce chemicals with low mammalian toxicity that can be used at low doses with little environmental impact. There is an increasing cost in the production of insecticides based on finite supplies of oil and increasing requirements for data to prove their environmental safety. Most of the assessments of the benefits of insecticide use are based on direct crop yields and economic returns, termed risk/benefit analysis. However, these may not take into account of the indirect costs associated with insecticide use in terms of environmental impact.

Impact of Pesticides on Humans

Widespread use of pesticides over the years has resulted in problems caused by their interaction with biological systems in the environment. The enormous and continued use of pesticides has added to the environment pollution to such an extent that human health is adversely affected and ecosystems are endangered [9]. Improper use of pesticides endangers the health of users and consumers of agricultural products, Farmers and farm workers in developing countries have been reportedly threatened by the disastrous effect of the pesticides. For example, breast milk samples from women in cotton-producing regions of developing countries have some of the highest levels of DDT ever recorded in humans, and the illness and mortality rates from pesticide poisoning in these areas approach to those of major diseases. Practices such as spraying of broad-spectrum pesticides on a frequent calendar schedule threatens the health of farm workers, impairs the quality of drinking water supplies, and threatens the ecological integrity and economic productivity of coastal and other aquatic ecosystems [10].

Additionally, many studies have indicated that pesticide exposure is associated with long-term health problems such as respiratory problems, memory disorders, dermatologic conditions [11,12], neurological deficits [13,14] miscarriages, and birth defects. Summaries of peer-reviewed research have examined the link between pesticide exposure and neurologic outcomes and cancer, perhaps the two most significant things resulting in organophosphate-exposed workers [15, 16].

Use of the pesticides is still the dominant pest control mean across the globe while the pattern of pesticide use is alarming. Among pesticide, insecticides share is over 90% [17].

Exposure of multiple pesticides for prolonged period has affected the normal functioning of different organ systems and possibly produced characteristics clinical effects such as hepatitis, dyspnea and burning sensation in urine [18]. A significant association between use of organochlorine and organophosphate pesticides and incidence of diabetes has been reported [19]. Rapidly increasing resistance against every type of insecticide along with an acute off target neurotoxicity make further insecticide application an impending threat to ecology and human health [20].

Exposure routes other than consuming food that contains residues, in particular pesticide drift, are potentially significant to the general public. The Bhopal disaster occurred when a pesticide plant released 40 tons of Methyl IsoCyanate (MIC) gas, a chemical intermediate in the synthesis of some carbamate pesticides. The disaster immediately killed nearly 3,000 people and ultimately caused at least 15,000 deaths [21].

Pesticides only work temporarily and may be re-applied often for ridding school buildings from rodents, insects, pests, etc. There is increasing pressure from national and international pesticide registration authorities, on insecticide manufacturers to provide comprehensive data about the environmental behavior of insecticides and on acute toxicity of their chemicals to humans, rats, fish, aquatic crustaceans and also in plants [22].

Impact of Pesticides on Plants

Pesticide residue refers to the pesticides that may remain on or in food after they are applied to food crops. Terminal residues can be defined as the chemicals, which accumulate in biological material, in the environment as a result of the introduction of pesticides. To a large extent, pesticide residues on plants are the result of direct application or especially in case of root vegetables, the result of absorption from soil. In many countries range of pesticides has been either banned or withdrawn for health and environment reasons; even though their residues are detected till date in various kinds of materials like food grains, fodder, milk etc. However some pesticides are systemic and can penetrate into plant tissue and thus may be metabolized to other compounds, or remains as such in the plant. Accumulation of pesticide residues occurs when a molecule is selectively directed or attracted towards specific surfaces or into specific cells or cell products from a reservoir and is accumulated faster than it is released. Such accumulation occurs with natural substances such as the essential mineral elements, amino acids, fat and other building blocks used for life processes [23]. The United Nations Codex Alimentarius Commission has recommended international standards for Maximum Residue Limits (MRLs), for individual pesticides in food [24].

Many Pesticides are toxic to plants and cause severe injuries to plants. Toxic effects of many fungicides, herbicides and insecticides on germination and seedling growth have been reported by many workers. It is possible that plants of different families may differ in this respect [25], sometimes even plants of the same genus [26] may differ due to the difference in pesticide residue translocation [27]. Visual phytotoxicity symptoms such as yellowing, epinasty, hyponasty, scorching, necrosis and death of plants were observed due to the toxicity of pesticides [28]; inhibition in seed germination by organophosphate pesticides [7, 8, 29]; and inhibition in reducing sugar content and amylase activity due to pesticide treatments [29] was also observed. A decrease in plant growth and protein content in a pulse crop by an organophosphate pesticide was observed by some workers [7, 8], along with visual crop injury by other pesticides [28, 30, 31].

CHALLENGES AND MANAGEMENT OF PESTICIDES AS AGRICULTURAL WASTE

In the early years of rapid expansion of the use of insecticides, the effectiveness of these chemicals on a wide range of insect pests was so spectacular that they were applied widely and often indiscriminately in most countries. However, later on there was anxiety concerning possible human, ecological or environmental hazards. These concerns include the acute and chronic toxicity of many insecticides to humans, domestic animals and wildlife; their phytotoxicity to plants; the development of new pest species after extensive pesticide use; the development of resistance to these chemicals by pests; the persistence of many insecticides in soil and water; and their capacity for global transport and environmental contamination [22]. Most of the developed countries have banned majority of the organochlorine insecticides due to long term contamination of the environment and hazard to the environment. Many of the organophosphates are highly toxic to mammals [15, 16] and sometimes cause severe local environmental problems, particularly in the contamination of water and local kills of wildlife, although they can contaminate human food if suitable regulatory precautions are not observed [22]. The carbamates are used principally to control insect pest of agricultural and horticultural crops [32]. Some of them have potential for considerable environmental impact, particularly in soils as they tend to be rather more persistent than the organophosphates in soil [22]. However, synthetic pyrethroids are readily biodegradable and have short persistence.

The majority of chemical insecticides consist of an active ingredient (the actual poison) and a variety of additives, which improve efficacy of its application and action. The formulation of an insecticide will influence the method by which it is applied, its persistence in the field and also its toxicity [32]. Some of the older types which are persistent, accumulate in farm land and crops, and contaminate environment are now banned, however newer types tend to be less persistent. Indiscriminate use of pesticides threatens the health and contaminate agricultural land also. Over-dosage and accumulation of pesticide in crops and farm land are not sustainable practices.

Pesticides residue on agricultural lands, soil, plants and water is a matter of great concern. Besides that stocks of pesticides which are of no use or banned or become obsolete have created a major problem today. Food and Agricultural Organization (FAO)'s Programme on the Prevention and Disposal of Obsolete Pesticides is working to inform the world about the dangers of obsolete pesticide stocks [33]. It collaborates with developing countries to prevent more obsolete pesticides from accumulating and assists them dispose of their existing stockpiles. Discussed below is the FAO programme on obsolete pesticides, their prevention and disposal [33].

Pesticides that can no longer be used for any purpose are dangerous toxic waste. Pesticides become obsolete when their expiry date has passed or when bans are placed on them for environmental and public health reasons. Some stockpiles contain products that are still usable but are not wanted, because of excess of stock or a greatly reduced pest problem or any other reason. Unwanted pesticides may not be obsolete but run a high risk of becoming obsolete as a result of prolonged storage. Pesticide are not only toxic chemicals but also perishable and are effective only for a certain period of time. Their shelf life depends on their active ingredients and the type of formulation. If not stated otherwise on the label, products normally have a shelf life of two years from the date of manufacture, provided that they are stored according to instructions stated on the label. Often it is obvious that a pesticide whose expiry date has passed has become obsolete. Clear liquid formulations may have formed flakes or crystals, emulsions and powders may have solidified. It may be more difficult to identify the products whose chemical properties have changed, while the visible physical properties remain the same [33].

Without proper testing in the field it is impossible to know whether a pesticide formulation that has been proven effective against pests in the donor or manufacturing country will work under conditions in developing countries. If no field-testing is done beforehand, the product may be found to be unsuitable for use and will remain in storage and become obsolete [33].

In many countries, when a range of products has been banned or withdrawn for health or environmental reasons, the fate of existing stocks is often given scarce condition. Stocks remain where they are stored and eventually deteriorate. Good practice in such cases requires pesticide regulatory authorities to allow a phase out period when products are banned or restricted so that existing stocks can be used up before the restriction is fully applied. There are many cases where highly hazardous pesticides, which are not permitted for use in industrialized countries, are exported to developing countries. For a pesticide to be banned, it has to be registered first. Some pesticide companies have not registered or re-registered products which they knew would have not been authorized in their own country but continue to produce and export the same products to developing countries There are also cases of pesticide manufacturers increasing exports of products that have been banned or restricted in their own countries, possibly in order to use up existing stocks or to compensate for depleted local markets. The argument is put forward that developing countries are demanding these hazardous pesticides because less toxic products are often too expensive [33].

Pesticides stocks will go "bad" quickly if they are not properly transported, stored and managed. Proper care during transport is vital not just for maintaining the quality of pesticide products, but for ensuring public health as well. Pesticides should never be transported along with other products especially foods. During transport, containers and other packaging materials are often handled very roughly leading to accelerated corrosion and leakage. Also, long periods of exposure to direct sun light during transit can cause both the container and its contents to deteriorate. Improper storage affects not just the quality of pesticide formulations, it seriously damages the pesticide containers and the contents may leak. One leaking container can contaminate many other products or cause more metal containers to corrode, creating a situation where the entire stock becomes unusable. Often stockpiles of old pesticides are poorly stored and toxic chemicals leak into the environment, turning potentially fertile soil into hazardous waste. Liquid pesticides can leak out of corroded drums into the soil and groundwater and end up polluting local lakes and rivers. The wind can spread pesticide powders over a wide area. Once pesticides enter soil they spread at rates that depend on the type of soil and pesticides, moisture and organic matter content of the soil or agricultural land and other factors. A relatively small amount of spilled pesticides can therefore create a much larger volume of contaminated soil. For example, approximately 30 tonnes of pesticides buried on a site in Yemen in the 1980s contaminated over 1500 tonnes of soil. Cleaning up contaminated water and soil is a desirable part of any obsolete pesticide disposal operation.

But dealing with contaminated soil is a costly, technically complex and difficult task. Limited funds for cleanup usually focus first on removal of the source of contamination (the pesticides themselves) and decontamination of soil and water is generally addressed on the basis of risk analysis when additional funds are available [33].

Prevention of Accumulation of Obsolete Pesticide Stocks

Food and Agriculture Organization of the United Nations (FAO) prepared guidelines, published as provisional, to enhance the formulation of policies and procedures aimed at prevention of such accumulation. The guidelines analyse the causes of this accumulation and recommend how it can be prevented. The guidelines are considered generally applicable and of interest to many countries, aid agencies and the pesticide industry [34]. Some of the important suggestions are mentioned below:

It was recommended that governments and other large-scale users examine critically their policies on pesticide management, plant protection and vector control; their procedures for assessing pesticide requirements; and their procedures for procurement of pesticides. It may moreover be necessary to revise pesticide management regulations and/or provide training for government and non-government staff responsible for stock keeping at pesticide stores.

Stocking more than one season's pesticide requirements should be avoided because storage under tropical conditions may further reduce the already short shelf-life of many products. Large quantities of pesticides should not be stocked if there are plans to review, reduce or abolish subsidies or preferential tariffs for pesticides. Responsibility for pesticide stocks lies primarily with the owner of the pesticides. The owner should manage pesticide stocks in a proper, safe and environmentally sound manner and take the necessary precautions to prevent stocks from becoming obsolete.

It was suggested to reduce the amount of pesticides by careful selection of products. In many cases, conventional pesticides can be replaced by more modern products (such as biological pesticides and growth inhibitors) that are more selective, less dangerous to human beings and animals and of which much smaller volumes are required.

Pesticide procurement should be based on what is actually and immediately required Products that have not been tested in trials conducted under conditions comparable to those of the intended use should not be procured and accepted.

Detailed specifications should be provided when procuring or requesting pesticides. The following factors should be taken into consideration: Formulation should be suitable for application equipment available. Package type should be durable enough to stand foreseen transport and storage conditions and storage period. Package size should be suitable/practical/affordable for the end-user. Label should contain a batch number and date of manufacture/release.

Pesticide consignments that have not been requested, or that deviate from the requested specifications or quantity, should not be accepted.

Material safety data sheets should be provided with each consignment. Provision of additional information on product stability under tropical circumstances and storage and stacking recommendations would be useful.

For surplus products that can still be used, agrochemical companies should assist governments in identifying potential users in other countries.

Agrochemical companies should establish effective delivery systems to provide products at short notice so that large stocks do not need to be held in the country.

Return services should be developed to take back unused quantities of pesticides, in particular unwanted products that can be reformulated, and surplus stocks that can be used elsewhere. As an incentive for such services, the fact that companies are prepared to accept the return of unused products might be taken into consideration by governments and aid agencies when they select a supplier.

REGULATIONS FOR PESTICIDE USE

In most countries, before a pesticide product can be marketed and used to manage a pest problem, the product must be registered with a government agency responsible for regulating the sale, distribution and use of pesticide products. Initially, registration of pesticide products was required to protect the consumer from fraudulent claims and limited attention was given to the impact of the product on consumer safety or the environment. As awareness of the potential impact of pesticides on the user, on the consumer, and on the environment developed, the registration of pesticide products became the predominant method for regulating the use of pesticide products. As requirements for registration of pesticide products expanded, the product label became the bottom line of the registration process. Every specific statement on the label had to be supported by evidence that no adverse effect would be caused to man or the environment if the product was used according to instructions specified on the product label [35].

Although a pesticide product may be approved to control a specific pest problem on a given host crop or agricultural land, problems may still occur if the applicator does not follow the instructions specified on the label or fails to use sound judgment when exceptional situations occur. As a result, it has been recognized over time that distribution and use of some pesticide products need to be restricted to applicators or users having the training or expertise to use the pesticide product in a manner that no adverse harm will occur to man or the environment [35].

Regulating applicators of pesticide products includes two steps: (1) pesticide products designated for restricted use must be labeled accordingly during registration process, and (2) a system of pesticide applicator training and certification must be implemented to ensure that only trained applicators are granted a license to purchase and use pesticide products labeled for restricted use. Some pesticides are considered too hazardous for sale to the general public and are designated restricted use pesticides. Only certified applicators who have passed an examination may purchase or supervise the application of restricted use pesticides [35].

Promotion of Integrated Pest management (IPM) by eminent institutes and international funding agencies like FAO, ADB, and UNDP etc has resulted in increase of the use of biopesticides and biocontrol agents. It is high time that the IPM modules should be propagated to the masses. The promotion of IPM will not only contribute to the protection of the environment but also raise the yields and quality of produce, thereby raising the socioeconomic levels of the farmers [36]. Biopesticides and biocontrol agents are going to play a key role in IPM modules. There is a need to promote IPM and fine-tune it at local levels. Many biopesticides and biocontrol agents, like neem oil, Neem Kernel Extracts (NSKE),

Trichogramma cards, *Trichoderma viride, Bacillus thuringenesis*, and nuclear polyhedrosis virus are now commercially available [7]. Use of biopesticides and non chemical alternatives are promoted by many international agencies also. The adoption of the IPM approach in pest management programs has been accepted as a policy which will reduce problems associated with pesticides. In recent years, the United States Department of Agriculture (USDA) and Environmental Protection Agency (EPA) have promoted an initiative to achieve a national goal of IPM on 75% of crop acres by the year 2000. The Federal Insecticide, Fungicide, and Rodenticide Act (FIFRA), a United States federal law that set up the basic US system of pesticide regulation to protect applicators, consumers and the environment, forbids the use of a pesticide in a manner inconsistent with its label and denies registration of pesticides that may have unreasonable adverse effects to man or to the environment [35].

Food and Agricultural Organisation (FAO) Programme on the Prevention and Disposal of Obsolete Pesticides is working with developing countries, donor agencies and industry to deal with the environmental and public health implications of this aspect of agriculture. The FAO Programme has developed an extensive training programme, published guidelines and other resources for countries to effectively and safely conduct the necessary activities to dispose of stockpiles and implement specific measures to avoid their re-accumulation. In this regard, the programme also provides technical and legal advice to make sure that obsolete pesticide projects are carried out effectively [33].

The cost of disposal of obsolete pesticides varies with the location, condition and type of waste and the methods used for its destruction. The problem is compounded by inadequate infrastructure, the wide dispersal of obsolete stocks and their deteriorated condition. The options available for the management or destruction of obsolete pesticides in a safe and environmentally acceptable manner are extremely limited. They are described in some detail in the FAO provisional guidelines on the disposal of bulk quantities of obsolete pesticides [37]. The technical, economic and political complexities are discussed in several papers by Greenpeace and the United Kingdom Pesticide Action Network (PAN) [38, 39].

EFFORTS AT THE INTERNATIONAL LEVEL

Though pesticide regulations differ from country to country, pesticides and products on which they were used are traded across international borders. To deal with inconsistencies in regulations among countries, delegates to a conference of the FAO adopted an International Code of Conduct on the Distribution and Use of Pesticides in 1985 to create voluntary standards of pesticide regulation for different countries [35]. The Code was updated in 1998 and 2002. [40] The FAO claims that the code has raised awareness about pesticide hazards and decreased the number of countries without restrictions on pesticide use [41].

Efforts by some of the international organizations and other agencies as discussed in FAO Pesticide Disposal Series 11—Country Guidelines [42] are mentioned below:

Food and Agricultural Organisation (FAO) FAO-led activities on obsolete pesticides include: organizing and running workshops and consultation meetings to raise awareness and generate action on obsolete pesticides in affected countries and regions. Initiating and coordinating completion of national inventories of obsolete pesticide stockpiles. Initiating and formulating disposal projects for FAO member countries. Supervision, monitoring and follow-up of disposal and prevention operations in the field. Liaison with donors and industry to generate support for disposal and prevention operations. Public outreach to raise awareness of the problems of obsolete pesticides globally. FAO has also issued the International Code of Conduct on the Distribution and Use of Pesticides.

Empty pesticide containers and small quantities of unused or unwanted pesticides constitute hazardous waste and mechanisms need to be put in place to facilitate sound management of this waste. These mechanisms should be compliant with FAO guidelines [43] based on the principle of removing hazardous waste from end-users—who lack the resources and expertise to manage it properly and returning it to the supplier.

United Nations Institute for Training and Research (UNITAR) UNITAR has prepared guidelines for the preparation of National Profiles and can assist in this process. The important benefit of these profiles is that they encompass all issues concerning the management of chemicals and highlight gaps and priorities for action. In some cases obsolete pesticides are identified as a priority, but in many countries other chemical management issues take precedence. A National Profile is a starting-point for all action associated with chemical management; from it countries are then able to develop an action plan. This plan should be transparent and have well-identified targets, and it can be used to mobilize resources. UNITAR can assist with and is developing guidelines for this process.

United Nation Environment Programme (UNEP) Chemicals The main focus of UNEP Chemicals activities in relation to obsolete pesticides relates to the implementation of the requirements of the Stockholm Convention on Persistent Organic Pollutants (POPs) and raising awareness with regard to obsolete pesticides in the framework of the POPs negotiation process. UNEP is exploring possible funding sources that may also support obsolete pesticide management. UNEP is assisting the lead agencies in this work: FAO for agricultural pesticides and WHO for health pesticides. UNEP Chemicals take part in the Inter-Organization Programme for the Sound Management of Chemicals (IOMC) coordinating group on obsolete pesticides.

World Health Organisation (WHO) The health sector is a major user of pesticides but is often overlooked in processes related to the better management of pesticides. WHO is particularly concerned about this and proposes that *national health authorities always be included in training and awareness-raising exercises dealing with obsolete pesticides.* DDT, which is one of the POP pesticides and is also commonly found in obsolete pesticides stockpiles, is still used in many developing countries for the control of malaria vector mosquitoes. While WHO is leading efforts in the search for alternatives to DDT through its DDT working group, and the effective control of malaria through the Rollback Malaria Programme, it is also aware that DDT continues to be used. Destruction technology for obsolete pesticides, POPs and other hazardous materials is also its concern.

United Nations Industrial Development Organization (UNIDO) UNIDO helps developing countries to develop the infrastructure for hazardous waste management. The thrust of UNIDO's approach is to minimize waste production by applying clean production methods and better controls. Where waste exists or continues to be generated, reuse and recycling are promoted. The organization advocates waste treatment close to the source whenever possible. Its programmes are based on waste minimization. UNIDO helps to promote cleaner and safer pesticide production and is also supporting the production of botanical pesticides such as neem, which has the potential to generate income and replace imported chemical pesticides in marginal areas. *Bacillus thuringiensis* (Bt) and other biopesticide productions are also promoted, as is the development and use of safer pesticide application technology.

The Secretariat of the Basel Convention has established regional centres that aim to support Parties to the Convention in its implementation. They are able to offer guidance and advice on hazardous waste management issues. They will also direct Parties to other sources of advice and information such as FAO on pesticides or UNIDO on industrial processes.

The International Maritime Dangerous Goods (IMDG) Code was developed as a uniform international code for the transport of dangerous goods by sea. It covers such matters as packing, container traffic and stowage, with particular reference to the segregation of incompatible substances. The Code lays down basic principles; detailed recommendations for individual substances, materials and articles; and a number of recommendations for good operational practice, including advice on terminology, packing, labelling, stowage, segregation and handling, and emergency response action.

Inter-Organization Programme for the Sound Management of Chemicals (IOMC) Inter-Organization Programme for the Sound Management of Chemicals (IOMC), was established in 1995 to strengthen cooperation and increase coordination in the field of chemical safety. It plays an important role in coordination of scientific and technical work carried out by participating organizations. The seven Participating Organizations (POs) of the IOMC are: the Food and Agriculture Organization of the United Nations (FAO), the International Labour Organization (ILO), the Organisation for Economic Co-operation and Development (OECD), the United Nations Environment Programme (UNEP), the United Nations Industrial Development Organization (UNIDO), the United Nations Institute for Training and Research (UNITAR) and the World Health Organization (WHO) [44].

The International Code of Conduct on the Distribution and Use of Pesticides was one of the first voluntary Codes of Conduct in support of increased food security, while at the same time protecting human health and the environment. The Code established voluntary standards of conduct for all public and private entities engaged in, or associated with, the distribution and use of pesticides, and since its adoption it has served as the globally accepted standard for pesticide management. The Code through supplementary technical guidelines, has been instrumental in assisting countries to put in place or strengthen pesticide management systems. The basic function of the Code remains to serve as a framework and point of reference for the judicious use of pesticides for all those involved in pesticide matters, particularly until such time as countries have established adequate and effective regulatory infrastructures for the sound management of pesticides [41].

The Intergovernmental Forum on Chemical Safety (IFCS) discussed the issue of obsolete pesticides and chemicals and states "The identification, neutralization, and safe disposal of obsolete stocks of pesticides and other chemicals (especially polychlorinated biphenyl [PCB]) must be urgently facilitated particularly in developing countries and countries with economies in transition." As well, future stockpiling of other obsolete pesticides and chemicals must be prevented [45].

The Global Crop Protection Federation (GCPF) GCPF members are committed to a process of product stewardship that effectively means implementation of the International Code of Conduct on the Distribution and Use of Pesticides [46].

CONCLUSION

Proper waste control requires a plan to reduce the amount of waste generated or the toxicity of the waste produced. However, the most environmentally sound and economically efficient way to manage any waste is to not generate it in the first place (source reduction).

Agricultural wastes should be restricted from land disposal to protect human health and the environment, and proper treatment should be given to diminish the toxicity of these wastes prior to disposal. To further improve the environmental performance and waste management standards on farms so that the agricultural land and human health are better protected, we should concentrate on the following measures:

- guiding farmers on hazardous waste controls and managing their wastes
- stopping all waste activities that could give rise to pollution or harm, in particular the burning of waste and the mis-description and poor management of hazardous waste
- better regulation of pesticides in developing countries, including better control over imports, adherence to packaging and labeling requirements, quality controls and product registration
- better management of pesticides, including storage and handling
- awareness of the problems of obsolete pesticides and accumulated hazardous chemical waste
- fewer inappropriate practices, such as burial, dumping or uncontrolled incineration of obsolete pesticides

The rate of accumulation of obsolete pesticide stocks is now very slow; however, in some countries where obsolete stocks have already been disposed of, new obsolete stocks are accumulating. Although the causes of accumulation are known, and in many cases have been or are being addressed, there are cases in which these practices continue. These need to be identified and stopped. The international community should put pressure on donors, whose donated pesticides have become obsolete, to maintain their responsibility for these products and fund their disposal. Assurances must be provided to affected countries that their economies and food security will not suffer in the event of a locust outbreak. Once the situation is under control, measures must be put in place to prevent future accumulation of obsolete pesticide stockpiles.

Sometimes obsolete and even banned pesticides are delivered for use against pests for which no alternative control is known locally, although alternatives may be identified if efforts are directed towards seeking alternatives. Sometimes alternatives to chemical pesticides already exist, yet support is needed for research and implementation to increase confidence in their effectiveness and their availability.

The adverse effects of a chemical depend on its toxicity, how people are exposed to the chemical, and each person's individual susceptibility. Some of them have potential for considerable environmental impact, particularly in soils. Before a product can be sold, it must be tested, usually under both laboratory and field conditions, including degradation studies, to determine what may happen to the active ingredient in soil, water, sediment and plants. Exposure to chemical agents can lead to a wide range of health effects that may be expressed immediately or take years to develop. All parties, including IGOs, governments, development agencies and donors, NGOs and researchers, should collaborate to develop and promote pest management methods that reduce reliance on chemical pesticides. Reducing the use of pesticides may reduce the risks that they place on society and the environment.

REFERENCES

[1] US Environmental Protection Agency, About Pesticides http://www.epa.gov/pesticides/about/index.htm
[2] Hayes, J.W. (1975). In: *Toxicology of Pesticides*. The William and Wilkins Company, Baltimore, U.S.A., Wavely Press Inc., U.S.A.
[3] Galloway, T. and Handy, R. (2003). Immunotoxicity of Organophosphorus Pesticides. *Ecotoxicology* 12: 345-363
[4] Dimetry, N.Z. (1996). Neem as a safe trend for pest control in the future. In: Neem and Environment vol. I World Neem Conference. (eds: Singh, R.P., Chari, M.S., Raheja, A.K. and Krause W.) Oxford and IBH Publishing Co. Pvt. Ltd., New Delhi p. 51-60
[5] Tariq, M.I., Shahzad Afzal, S., Hussain, I. and Sultana, N. (2007). Pesticides exposure in Pakistan: A review. *Environment International* 33(8): 1107-1122
[6] Srivastava, A.K., Singh, P. and Singh, A.K. (2008). Sensitivity of the mitotic cells of barley (*Hordeum vulgare* L.) to insecticides on various stage of cell cycle. *Pesticide Biochemistry and Physiology* 91: 186-190
[7] Sinha Vandita (2005). Effects of Chemical and Biopesticide (Neem) and their interactive Effects with Fe and Zn on Urd, Gram and Moong Plants, Mphil Thesis, Department of Botany, University of Lucknow, Lucknow, India.
[8] Thirumaran, D. and Xavier, A. (1987). Effect of methyl parathion (Metacid 50) on growth, protein, free amino acid and total phenol content of black gram (*Vigna mungo* L.) seedlings. *Indian J. Plant Physiol.* 30(3): 289-292
[9] Kanekar Pradnya, Bhadbhade Bharatij, Deshpande Neelima M. and Sarnaik Seema S. (2004). Biodegradation of Organophosphorus Pesticides. *Proc.Indian Natn. Sci. Acad. B70 (.1):* 57-70
[10] Jayaraj, S. (2000). Perspective on horticultural development in India: Need for efficient IPM Strategies. In: *Innovative pest and disease management in horticultural and plantation crops.* (eds: Narasimhan, S., Suresh, G. and Daniel, S.), Allied Publishers Ltd. p. 6-15
[11] Arcury, T.A., Quandt, S.A., Mellen, B.G. (2003). An exploratory analysis of occupational skin disease among Latino migrant and seasonal farmworkers in North Carolina. *Journal of Agricultural Safety and Health*, Volume 9, Issue 3, Pages 221–232.
[12] O'Malley, M.A. 1997. Skin reactions to pesticides. *Occupational Medicine* Volume 12, Issue 2, Pages 327–345.
[13] Kamel, F., et al. (2003). "Neurobehavioral performance and work experience in Florida farmworkers". *Environmental Health Perspectives* 111: 1765–1772.
[14] Firestone, J.A., Smith-Weller, T., Franklin, G., Swanson, P., Longsteth, W.T., Checkoway, H. 2005. Pesticides and risk of Parkinson disease: a population-based case-control study. *Archives of Neurology* 62(1):91–95.

[15] Alavanja, M.C., Hoppin, J.A., Kamel, F. (2004). Health effects of chronic pesticide exposure: cancer and neurotoxicity. *Annu Rev Public Health* 25:155–197.

[16] Kamel, F., Hoppin, J.A. (2004). Association of pesticide exposure with neurologic dysfunction and disease. *Environ Health Perspect* 112:950–958.

[17] Ramarethinam, S., Marimuthu, S. and Murugesan, N.V. (2008). Potential Effect of Nimbicidine 0.03% EC Formulation in Improving the Performance of Insecticide on *Spodoptera litura. Pestology* 32(7): 13-18

[18] Azmia, M.A., Naqvia, S.N.H., Azmib, M.A. and Aslamb, M. (2006). Effect of pesticide residues on health and different enzyme levels in the blood of farm workers from Gadap (rural area) Karachi—Pakistan. *Chemosphere* 64(10): 1739-1744

[19] Jamshidia, H.R., Ghahremania, M.H., Ostada, S.N., Sharifzadeha, M., Dehpourb, A.R. and Abdollahia, M. (2009). Effects of diazinon on the activity and gene expression of mitochondrial glutamate dehydrogenase from rat pancreatic Langerhans islets. *Pesticide Biochemistry and Physiology* 93(1): 23-27

[20] Rembold, H. (1996). Neem and its general development for pest control In: Neem and Environment (eds: Singh, R.P., Chari, M.S., Raheja, A.K. and Krause, W.) Oxford and IBH publishing Co. Pvt. Ltd. New Delhi. p.1-10

[21] "1984: Hundreds die in Bhopal chemical accident".*On This Day:3December*.BBC News. http://news.bbc.co.uk/onthisday/hi/dates/stories/december/3/newsid_2698000/2698709.stm

[22] Edwards, C.A. (2000). Ecological based use of insecticides. In: *Insect pest management techniques for environmental protection.* (eds: Rechigl and Rechigl) Lewis Publishers, London. p.103-109

[23] Kenaga, E.E. (1972). Chlorinated hydrocarbon insecticides in the environment. In: *Environmental toxicology of pesticides.* (eds: Matsumura, F., Bousch, G.M. and Misato, J.), Academic Press, New York. p. 193-224

[24] Codex Alimentarius Commission, Code of Ethics for International Trade in Food. CAC/RCP 20-1979 (Rev. 1-1985).

[25] Wyrill, J.B. III and Burnside, O.C. (1976). Absorption, translocation and metabolism of 2,4-D and glyphosphate in common milkweed and hemp dogbane. *Weed Sci.* 24: 557-566

[26] Anderegg, B.N. and Lichtenstein, E.P. (1981). A comparative study of water transpiration and the uptake and metabolism of ^{14}C Phorate by C_3 and C_4 plants. *J.Agric. food chem.* 29 : 733-738

[27] Stoller, E.W. (1970). Mechanism of isopropyl 3-chlorocarbanilate by soyabean plants. *J. Agric. Food Chem.*, 17: 1017-1020

[28] Anandhakrishnan, B. and Jayakumar, R. (2003). Bioefficacy and phytotoxicity of Triclopyr Butotyl Ester (TBE) in transplanted rice *Oryza sativa* L. *Pestology* 29(9) : 10-14

[29] Prasad, B.N. and Mathur, S.N. (1983). Effect of metasystox and cuman-L on seed germination, reducing sugar content and amylase activity in *Vigna mungo* (L) Hepper *Ind. J. Plant Phy.* 26(2): 209-213

[30] Miller, S.D., Dalrymple, A.W. and Krall, J.M. (1990). Weed control in pinto beans with preplant incorporated or complementary preplant incarporated/post emergence treatments. *West. Soc. Weed Sci. Prog. Rep.* p. 257-258

[31] Soltani, N., Shropshire, C., Cowan, T. and Sikkema, P. (2003). Tolerance of Cranberry beans (*Phaseolus vulgaris*) to soil applications of S-metolachor and imazethapyr. *Can. J. Plant Sci.* 83: 645-648

[32] Dent, D. (2000). Insecticides. In: *Insect Pest Management 2nd Edition*, (ed: Dent, D.), ABI Publishing, New York. p. 81-121

[33] Prevention and Disposal of Obsolete Pesticides http://www.fao.org/waicent/faoinfo/agricult/agp/agpp/pesticid/disposal/en/index.html

[34] FAO 1995. Prevention of accumulation of obsolete pesticide stocks. Provisional guidelines. FAO Pesticide Disposal Series No. 2. Rome. 31 pp. http://www.fao.org/waicent/faoinfo/agricult/agp/agpp/pesticid/disposal/common/ecg/103807_en_v7460e.pdf

[35] Willson, H.R. 23 February, 1996. Pesticide Regulations. In: E. B. Radcliffe, W. D. Hutchison & R. E. Cancelado [eds.], Radcliffe's IPM World Textbook, URL: http://ipmworld.umn.edu, University of Minnesota, St. Paul, MN (retrieved on 17 October, 2008)

[36] Kaushik, N. (2000). Integrated Pest Management: Status and Concerns. In: *Innovative Pest and Disease Management in Horticultural and Plantation Crops* (eds: Narsimhan, S., Suresh G. and Wesley, Daniel S) Allied Publishers Ltd. p. 124-129.

[37] FAO (1996). Disposal of bulk quantities of obsolete pesticide in developing countries. Provisional technical guidelines. FAO Pesticide Disposal Series No. No. 4. Rome. 44 pp.

[38] Greenpeace International. 1998. *Technical criteria for the destruction of stockpiled persistent organic pollutants,* by P. Costner, M. Simpson and D. Luscombe. Amsterdam, the Netherlands. 39 pp.

[39] UNEP (1999). *Technology options for the management and destruction of obsolete or unwanted pesticides in developing countries,* by M. Davis. Subregional Expert Meeting on Technologies for Treatment/Destruction of PCBs and Obsolete Pesticides. Golitsino, Moscow, Russian Federation, 6-9 July 1999. UNEP Chemicals/CIP State Committee of the Russian Federation on Environmental Protection.

[40] FAO, Programmes: International Code of Conduct on the Distribution and Use of Pesticides. www.fao.org/ag/agp/agpp/pesticid/code/pm_code.htm

[41] FAO (2002). International Code of Conduct on the Distribution and Use of Pesticides. www.fao.org/waicent/faoinfo/agricult/agp/agpp/pesticid/code/download/code.pdf

[42] FAO Pesticide Disposal Series 11 Country guidelines ftp://ftp.fao.org/docrep/fao/005/y2566E/y2566E00.pdf

[43] FAO (1999). Guidelines for the management of small quantities of unwanted and obsolete pesticides. FAO Pesticide Disposal Series No. 7. 25 pp.

[44] Inter-Organization Programme for the Sound Management of Chemicals (IOMC) http://www.who.int/iomc/en/

[45] IFCS (2000). IFCS priorities for action beyond 2000. Prepared by Sweden and Hungary in collaboration with the Forum Standing Committee, IFCS/FORUM III/09w. March.

[46] GCPF (2000). Obsolete stocks of crop protection products. Global Crop Protection Federation (GCPF) position update, July. Brussels. (www.gcpf.org)

In: Agricultural Wastes
Eds: Geoffrey S. Ashworth and Pablo Azevedo

ISBN 978-1-60741-305-9
© 2009 Nova Science Publishers, Inc.

Chapter 6

CARBONIZATION OF RICE HUSK TO REMOVE OFFENSIVE ODOR FROM LIVESTOCK WASTE AND COMPOST

Seiji Kumagai[1,], Koichi Sasaki[2] and and Yukio Enda[3]*

[1]Department of Machine Intelligence and Systems Engineering
Akita Prefectural Univeristy
84-4 Tsuchiya-aza-ebinokuchi, Yurihonjo, 015-0055, Akita, Japan
[2]Livestock Experiment Station
Akita Prefectural Agriculture, Forestry and Fisheries Research Center
13-3 Kaisonumayachi, Jinguji, Daisen, 019-1701, Akita, Japan
[3]Akita Research and Development Center
4-11 Arayamachi-aza-sanukiyori, Akita, 010-1623, Akita, Japan

ABSTRACT

An attempt was made to convert the agricultural waste of rice husk (RH) into an adsorbent to remove the offensive odor released from livestock waste and compost. The ammonia gas adsorption of the RH carbonized at 400°C was much faster than those of several commercial deodorants as well as those of carbonized wood wastes. Acidic functional groups remaining at 400°C were useful to promote adsorption of basic ammonia gas. The actual compost was covered with or mixed with the RH carbonized at 400°C. The covering method reduced the concentration of ammonia gas emitted from the compost much faster than the mixing method, which was connected to volatilization of ammonia gas lighter than ambient air. Wetting the carbonized RH was also effective in reducing the ammonia gas concentration. An assorted feed to which was added the RH carbonized at 400°C at the level of 2 mass% was given to growing pigs. The addition of the carbonized RH reduced about 80% of the concentrations of hydrogen sulfide and mercaptans emitted from the pig dung. The removal of acidic gases of hydrogen sulfide and mercaptans was suggested to result from basic inorganic matter of K, Ca and P, which

---

[*] Corresponding author: Tel/Fax: +81-184-27-2128, Email: kumagai@akita-pu.ac.jp.

were intrinsically composed in RH. The testing results showed that the RH carbonized at 400°C was a promising material for removing the offensive odor produced by the livestock industry.

1. INTRODUCTION

Rice is a staple food of the Japanese people. About 2 million tons of rice husk (RH) are produced every year in the process of rice threshing in Japan, and 63 mass% of RH is recycled in agricultural, livestock and other industries [1]. The most common method for disposal of residual RH had been incineration on farms. However, this practice is now prohibited in Japan because it produces ash, fumes, and toxic organic gases, leading to serious air pollution. Thus, recycling of the residual RH is now a socially important subject, especially in the rice-growing regions of Japan.

There is much literature describing methodologies to convert RH into functional materials, e.g., liquid adsorbents [2-12], gas adsorbents [13-17], and high-strength mechanical materials [18-21]. However, the cost to produce these functional materials is higher than that of conventional products. On the other hand, RH can be used as fuel in place of oil and petroleum gas, resulting in the focus of research on RH combustion [22-24]. However, RH contains high-content inorganic matter (about 20 mass% in dried state), resulting in lower heating value and in more ash production compared with other agricultural wastes, such as woods. Nowadays, wood-based wastes are instead used as fuel in Japan.

A total of 43 mass% of RH produced in Japan is recycled in the livestock industry, in which 21 mass% is used as paving materials in livestock housing and 22 mass% is used for producing compost [1]. Hence, an increase in the use of RH in the livestock industry is a realistic solution to increase the recycling rate of RH in Japan. Livestock housing and compost factories are often adjacent to rice farms. Offensive odor pollution produced by the livestock industry is now a public concern even in suburban areas, which is related to the spread of residential districts. It is now very important for the sustainable livestock industry to address the problems related to offensive odor.

A conversion of RH into an adsorbent to remove the offensive odor seems to be a promising recycling methodology for both rice and livestock farmers. Carbonization is an effective method in enabling RH to remove the offensive odor, from the viewpoints of mass production and production cost. Modern RH carbonization plants in Japan can produce a large quantity of carbonized RH using a tiny quantity of oil for ignition and power for furnace control [25]. Carbonization of RH now has few economical, energy and environmental disadvantages.

Ammonia gas is a principal component of the offensive odor produced by livestock waste and compost. In this chapter, the ammonia gas adsorption of RH carbonized at different temperatures was evaluated and was compared to several commercial deodorants as well as carbonized wood wastes in enclosed bags, correlating their ammonia adsorption with their pore structures and chemical natures. Field tests were also conducted. The carbonized RH was exposed to ammonia gas emitted from actual fermenting compost. Effective methods for enhancing the ammonia adsorption of the carbonized RH were proposed. An assorted feed to which carbonized RH was added was given to growing pigs. The effect of the feed with the carbonized RH on the reduction of the offensive odor emitted from pig's dung was evaluated.

2. MATERIALS AND METHODS

2.1. Material and Carbonization

Two types of RH samples were obtained by threshing Akita Komachi rice grown in Akita Prefecture, Japan. One was harvested in Ogata-mura in the autumn of 2003, and the other was harvested in Nishiki-mura in the autumn of 2005. The wood wastes used were Japanese cedar stem, Japanese cedar bark and black pine stem. All of these were collected in Honjo city in the autumn of 2004. The RH samples whose dimensions were 6–8 mm (L) × 2–4 mm (Dia.) were subjected to tests without any mechanical treatments. Japanese cedar and black pine stems were chopped in a chipping machine, but were not separated into sap and heart wood. The chips whose dimensions were 40–60 mm (L) × 8-12 mm (W) × 6–10 mm (T) were selected. Japanese cedar was chopped with a debarking machine. The bark was cut into pieces of dimensions 40–60 mm (L) × 8–12 mm (W) × 3–6 mm (T). The carbonization was carried out using electrical furnaces. The ambient atmosphere of the sample during carbonization is vacuum or nitrogen. The vacuum pressure was maintained to be 500 Pa using a vacuum pump system. In the case of nitrogen atmosphere, nitrogen gas flowed in a stainless steel cylinder at the rate of 0.7–1.0 L/min. The furnace temperature was increased linearly from room temperature to the desired temperature in 1 h; it was then maintained at the desired temperature for 3 h for vacuum atmosphere or for 1 h for nitrogen atmosphere. The furnace was then cooled naturally to room temperature. Several commercial deodorants were also prepared for a comparison purpose. Table 1 shows the used commercial deodorant products.

2.2. Material Characterizations

The sample was dried at 105°C in air for 3 h prior to analyses and tests. It was verified beforehand that the above periods were sufficient to saturate the samples' weights in drying. After the drying process, the sample was cooled at 22–24°C and 30-40 % RH, and then weighed 5 min later. In evaluating the moisture content of the sample, the weight in the dry state is determined to be 100 mass%. The inorganic content of the raw RH (0.1 g) was determined from the residual-ash ratio after incineration at ca. 800°C for 1 h in air. The ash composition was determined using a fluorescent X-ray analyzer (XRF-1700, Shimadzu Corp., Japan). A CHN/S analyzer (2400 I, PerkinElmer Inc., USA) was used to determine the hydrogen, carbon and nitrogen contents, as well as those of other elements. The composition of the raw RHs and their ach composition are shown in Tables 2 and 3, respectively. The ash content of the used RHs was found to be ca. 20 mass%. It is noteworthy that major component of the ash is SiO_2 (>90 mass%) and K, Ca and P, which are related to soil and fertilizers, are contained in the RH samples.

A gas-adsorption analyzer (Autosorb, Quantachrome Instruments, USA) was used to study the sample porosity, correlating it with the ammonia adsorption property. N_2 adsorption isotherms were obtained, providing the specific surface area based on the BET theory [26], and the total pore volume at a relative pressure of 0.995. The samples (ca. 0.02 g) were degassed at 200°C for > 4 h prior to isotherm measurement. The BET specific surface area (S_{BET}) was calculated using the N_2 adsorbed volume at relative pressures (P/P_0) of 0.1–0.3.

Table 1. Commercial deodorants used for comparison with the carbonized RHs

Sample ID	Product	Primary ingredient	Manufacturer
Activated carbon	Refrigerator deodorizer	Coconut shell activated carbon	Kobayashi Pharmaceutical Co., Ltd.
Silica gel	Deodorant sand	Silica gel (SiO_2)	Unicharm Corp., Japan
Wood chip pellet	Antibacterial (deodorant) chips	Raw woods in Germany	Kao Corp., Japan

Table 2. Dry base composition of raw husks of Akita Komachi rice (unit: mass%)

Production site	Ash	H	C	N	O* (by difference)
Ohgata-mura	21.0	5.1	38.6	0.6	34.7
Nishiki-mura	17.1	5.2	40.5	0.4	36.8

* Except for that in ash.

Table 3. Composition of the ash from raw RH (unit: mass%)

Production site	SiO_2	K_2O	P_2O_5	CaO	Detectable others
Ohgata-mura	94.5	3.3	1.1	1.1	2.1
Nishiki-mura	91.1	5.3	0.6	1.6	1.4

The liquid N_2 volume, referring to N_2 adsorbed volume at P/P_0 0.995, was determined to be the total pore volume (V_t). The volume of micropores (V_m) was obtained following the t-method [27], giving the volume of mesopores and macropores ($V_e = V_t - V_m$).

2.3. Ammonia Gas Adsorption in Enclosed Bag

The raw RH, the carbonized RH, the commercial deodorant products, and the carbonized wood wastes were subjected to ammonia gas in enclosed bags. The dried sample was enclosed quickly in a gas sampling bag made of polyvinyl fluoride film (Tedlar®; DuPont, USA) and residual air in the bag was degassed using a small pump for 5 min. Ammonia gas adjusted at the concentration of 100 vol. ppm (dry air based, <1% RH at 22–24°C) provided by Taiyo Nippon Sanso Corp., Japan, was injected 3 L to the degassed bag. The contact time started when the ammonia injection started. Kitagawa-type detecting tubes targeting ammonia gas (Komyo Rikakagaku Kogyo K.K., Japan) were used to measure the ammonia concentration in the bag. The gas of 100 mL in the bag was evacuated for every concentration measurement. The ammonia concentration was evaluated at 22–24°C as a function of the contact time.

2.4. Removal of Ammonia Gas from Fermenting Compost

The sample which had shown the best performance of ammonia adsorption in the above bag test was subjected to offensive odor released from actual fermenting compost produced in the Livestock Experiment Station, Akita Prefectural Agriculture, Forestry and Fisheries Research Center, Japan in autumn of 2007. The major component of the offensive odor was confirmed to be ammonia gas. In order to find the best adding method to reduce the ammonia concentration, the fermenting compost, including moisture and weighing 200 g, was covered with or was mixed with the sample at the level of 0-5 % to the compost weight. The compost with the sample was placed in an enclosed vessels of 800 mL. Figure 1 shows the experimental setup of this ammonia adsorption test. The contact time started when the vessel was enclosed. Kitagawa-type detecting tubes targeting ammonia gas (Komyo Rikakagaku Kogyo K.K., Japan) was inserted into the vessel. The tube tip was placed in top space in the vessel and 100 mL of gas was evacuated to measure the ammonia concentration at the time of 1, 3 and 24 h.

The hole opened for this measurement was sealed using a scotch tape. The humid and water-sprayed samples were also tested to study the influence of adsorbed moisture on the ammonia adsorption. The humid sample was prepared in a humid chamber of which temperature and humidity were controlled to be respectively 25°C and 95% RH for 24 h. The moisture content of the spayed sample was set to be 100 mass%.

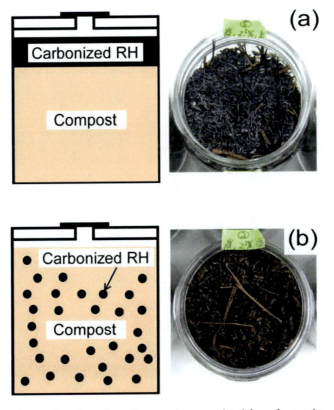

Figure 1. Experimental setup for adsorption of ammonia gas emitted from fermenting compost. (a) covering method, (b) mixing method.

2.5. Giving Carbonized RH to Growing Pigs to Reduce Offensive Odor from Dung

A commercial assorted feed (Smooth, Kanematsu Corp., Japan) added with the carbonized RH, which had shown the best performance in the bag test, was provided to castrated three-way crossbred pigs (Landrace × Large White × Duroc). Before this feeding test, the pigs solely and freely feed for seven days. During the feeding test, an assorted feed (4 % of the pig's weight) with and without the carbonized RH was given to 4 and 4 pigs, respectively, at every morning and evening. The addition level of the carbonized RH was 2 mass% to the assorted feed. Figure 2 shows photos of the assorted feed added with the carbonized RH and the used pigs. The pigs freely drunk water and they solely fed during the test. At the start of the feeding test, the pigs were 109 days old. Average weight ± standard deviation of the pigs to which the feed with 0 and 2 mass% carbonized RH was given were 49.3 ± 6.7 kg and 50.4 ± 5.1 g, respectively. In the 5th day and 6th testing days, all the dung was collected. Then, the collected dung was preserved in the pig house at 15–20°C and at 50% RH for 24 h. Then, 200 g of the dung was encased in the vessel of 800 mL and was preserved for 90 min. The gas concentrations of ammonia, hydrogen sulfide, mercaptans and lower fatty acids in the vessel were measured using Kitagawa-type detecting tubes (Komyo Rikakagaku Kogyo K.K., Japan). In this article, mercaptans are defined as C_nH_m-SH and are mostly composed of methylmercaptan. Lower fatty acids are C_nH_m-COOH (C <= 12) and their major components are propionic acid, n-butyric acid, n-valeric acid and iso-valeric acid.

Figure 2. Assorted feed with the carbonized RH at the level of 2 mass% and the pig in the testing chamber.

3. RESULTS AND DISCUSSION

3.1. Material Properties of Carbonized RH

Changes in the composition and yield of the Ohgata-mura RH which was carbonized in vacuum at 300–800°C for 3 h are shown in Table 4. It is found that the yield of the carbonization process decreased up to 600°C and remained at a similar level of 38–39 mass% at higher temperatures. The carbon content of the carbonized RH changed only slightly. The hydrogen and oxygen contents of the RHs carbonized at 300 and 400°C were higher than those of the RHs carbonized at 600 and 800°C. It is also noteworthy that about half of the weight of the RH carbonized at >300°C is attributable to ash of which major component was SiO_2.

The pore structure of the carbonized RH was also analyzed. All the nitrogen adsorption isotherms except for that of the raw RH were found to belong to Type II, as categorized by the BDDT classification [27]. The BET specific surface area (S_{BET}) and the pore volume information (V_t, V_m and V_e) were obtained. Table 5 shows the pore characteristics of the raw and carbonized RHs. Between 300–600°C, S_{BET} and V_t increased with the carbonizing temperature. At higher temperatures, S_{BET} and V_t decreased with the carbonizing temperature. The most porous structure was observed on the RH carbonized at 600°C. Micropores of which pore diameter is < 2 nm were produced in the RH carbonized at 600°C.

Table 4. Compositional change and yield of the carbonized Ohgata-mura RH (unit of the composition and the yield is mass%)

Carbonizing temperature (°C)	Ash	H	C	N	O[*]	Yield
300	37.2	3.5	41.2	0.4	17.7	56.5
400	46.9	2.3	43.0	0.4	9.2	44.8
600	54.8	1.1	42.2	0.3	1.6	38.3
800	54.4	0.5	42.0	0.5	2.6	38.6

[*] Oxygen content was calculated: O = 100 - (C + H + N + ash).

Table 5. Pore characteristics of the raw and carbonized Ohgata-mura RHs

Carbonizing temperature (°C)	S_{BET} (m²/g)	V_t (mL/g)	V_m (mL/g)	V_e (mL/g)
Raw	1.7	0.012	0	0.012
300	8.7	0.029	0	0.029
400	52.2	0.083	0	0.083
600	170.0	0.165	0.038	0.127
800	59.9	0.101	0	0.101

3.2. Ammonia Gas Adsorption Performance of Carbonized RH in Enclosed Bag and Comparison with Commercial Deodorants and Carbonized Wood Wastes

The carbonized RHs were subjected to 100 vol. ppm ammonia gas in enclosed bags. Figure 3 shows the ammonia adsorption properties of the raw RH and the RH carbonized at 300–800°C, of which weights were all adjusted to 1.00 g. Data are average of three separate measurements. The control indicates the ammonia concentration in the gas in which no sample was enclosed. The results showed that a slightly higher time-rate of the ammonia adsorption was observed on the RH carbonized at 300 °C than on the raw RH. With increasing carbonizing temperature, the time-rate of the ammonia adsorption increased at 300–400°C, whereas it decreased at 400–800°C. The RH carbonized at 600–800°C required a much longer time to reduce the ammonia concentration (<10 vol. ppm) than the raw RH did. The fastest uptake appeared on the RH carbonized at 400°C, which reduced the ammonia concentration <5 vol. ppm in 0.5 h. High carbonizing temperature >600°C was found to be unnecessary to enhance the ammonia uptake. The most porous structure appeared on the RH carbonized at 600°C. Thus, the porosity probably did not play a predominant role in determining their ammonia adsorption performance. That fact implies that hydrogen and oxygen much remaining at the RHs carbonized at 300 and 400°C might be related to enhancement of the ammonia adsorption.

Different types of the commercial deodorants which are useful for adsorption of ammonia gas were evaluated similarly and were compared with the RH carbonized at 400°C, which had displayed the best performance in the bag test. Figure 4 shows the ammonia gas adsorption properties of the commercial deodorants and the RH carbonized at 400°C. Data are average of three separate measurements. It is obvious that the carbonized RH adsorbed ammonia gas in a much shorter time than the commercial deodorants. The carbonized RH required 0.5 h to reduce the ammonia concentration <5 ppm, while the silica gel, showing the fastest ammonia adsorption in all the commercial deodorants, required 4 h. Those results demonstrate that the ammonia adsorption performance of the carbonized RH was superior to several types of commercial products.

The RH and wood wastes which were carbonized at 400°C in a similar manner were given to the ammonia adsorption test in the enclosed bags. Figure 5 shows the ammonia gas adsorption properties of the carbonized wood wastes and the RH carbonized at 400°C. Data are average of three separate measurements. All the samples reduced the ammonia concentration quickly compared to the above commercial deodorants. It is observed that the carbonized RH and Japanese cedar bark displayed the greater and similar ammonia adsorption performance. The ash content of the raw wood wastes was confirmed to be less than 3 mass%. Considering that wood wastes can be used as fuel sources, RH should be utilized as an adsorbent of ammonia gas.

The ammonia adsorption test using an enclosed bags suggested that the RH carbonized at 400 °C was a promising material for the removal of ammonia gas rather than commercial deodorant products and carbonized wood wastes. It was also shown that the carbonizing temperature strongly influenced the ammonia adsorption of RH. The pore structure in the carbonized RH had a tenuous relation to their ammonia adsorption.

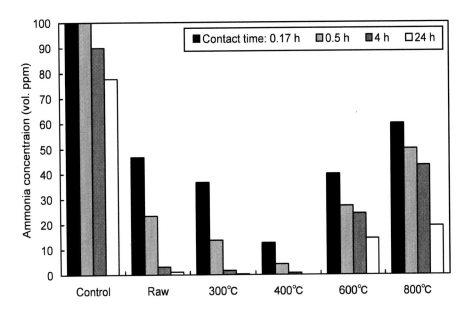

Figure 3. Ammonia adsorption of the raw RH and the RH carbonized at 300–800°C as a function of contact time.

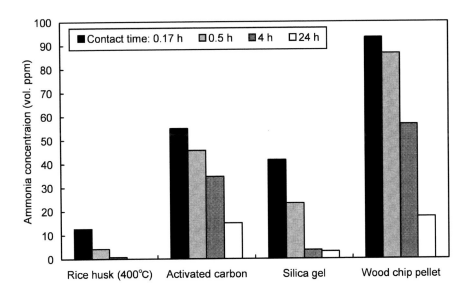

Figure 4. Ammonia adsorption of the commercial deodorants and the RH carbonized at 400°C.

3. 3. Mechanisms of Ammonia Adsorption on Carbonized RH

The above results showed that the carbonizing temperature around 400°C could enhance ammonia adsorption performance of RH. The RHs carbonized at 300 and 400°C, which showed the faster adsorption of ammonia gas, allowed higher contents of hydrogen and oxygen than other carbonized RHs.

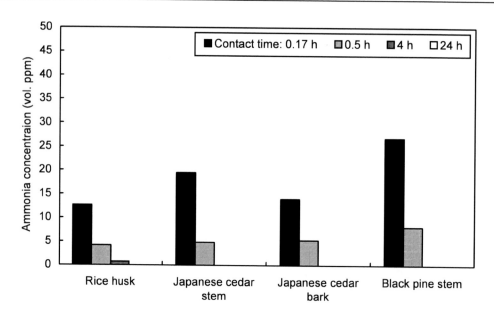

Figure 5. Ammonia adsorption of wood wastes carbonized at 400°C and RH carbonized at 400°C.

Functional groups such as hydroxyl (-OH), carboxyl (-COOH), and carbonyl (>C=O) groups were likely to be rich in these carbonized RH. It was pointed out in two other studies that acidic functional groups produced by carbonization at 400°C can promote ammonia gas adsorption in other species of biomass (e.g., Japanese cedar and Japanese cypress) [28, 29]. Acidic carboxyl groups can promote adsorption of ammonia molecules through a neutralization process as described in Equation 1.

$$\text{-COOH} + NH_3 \rightarrow \text{-COONH}_4 \tag{1}$$

On the other hand, it should be recognized that the above polar functional groups can adsorb ambient moisture. Equation 2 shows that ammonia molecules can be dissolved into moisture.

$$H_2O + NH_3 \rightarrow OH^- + NH_4^+ \tag{2}$$

The ammonia adsorption test was implemented in dry air (<1 % RH). However, residual moisture might be involved in this process. An evaluation of hygroscopic properties of the carbonized RH samples provides helpful information connected to the mechanism of their ammonia adsorption. The raw and carbonized RH samples (1.00 g) were dispersed on glass petri dishes (60 mm Dia.). They were then placed in a chamber (50 × 50 × 50 cm^3) in which the temperature and the humidity were controlled to be 25–27°C and 78–82% RH, respectively. The weight changes of the samples resulting from exposure to humid air are shown in Figure 6. The sample weight prior to the test (dried state) was 1.00 g. Data are results of single measurement. Exceptionally, the raw RH adsorbed much moisture. The largest amount of moisture uptake in the carbonized RH appeared for the RH carbonized at 400°C. Less moisture uptake was apparent for the RHs carbonized at 600 and 800°C than for the RHs carbonized at 300 and 400°C.

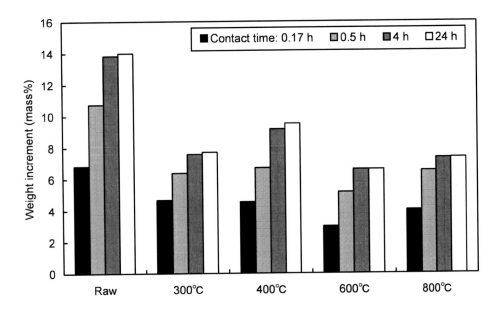

Figure 6. Moisture uptake of raw RH and RH carbonized at 300–800°C in a humid environment (25–27°C and 78–82% RH) as a function of contact time.

The weakest moisture uptake appeared at 600°C, which was commonly observed in other carbonized biomass species [30]. A stronger moisture uptake of the RHs carbonized at 300 and 400°C seems to be consistent with their faster ammonia adsorption. However, the raw RH, showing the greatest moisture uptake, was shown to adsorb less ammonia than the RH carbonized at 400°C. The raw RH and the RH carbonized at 800°C should have adsorbed more ammonia gas if moisture uptake were a dominant mechanism of ammonia adsorption. The lesser moisture uptake on the RH carbonized at 300°C than that on the raw RH seems to be attributable to the loss of wood constitution. Formation of polar functional groups is likely to regain the moisture uptake ability, which is relevant to the change in hygroscopic nature of the RH carbonized at 400°C.

It is described in one study that acidic carboxyl groups, rather than basic carbonyl groups, are readily produced at lower temperatures during biomass pyrolysis (carbonization) [31]. It was shown that the hydrogen content of the RHs carbonized at 600 and 800°C decreased to lower levels (ca. 1 mass%), implying that basic carbonyl groups were richer than acidic carboxyl groups in this temperature range. Consequently, the fastest ammonia adsorption of the RH carbonized at 400°C was deemed to result from formation of acidic carboxyl groups, in which Equation 1 should have proceeded predominantly. The higher carbonizing temperature (>400°C) reduced carboxyl groups and might produce carbonyl or other functional groups instead. Hence, the ammonia adsorption of the carbonized RH diminished with increasing carbonizing temperature.

The ammonia adsorption of the raw RH is inferred to be the result of a strong hygroscopic nature derived from intrinsic celluloses. In this adsorption, Equation 2 probably describes the primary process.

The role of SiO_2 in the ammonia adsorption by carbonized RH should also be discussed. The compositional analyses indicate that the SiO_2 content increased with increasing the carbonizing temperature. It was further demonstrated that the ammonia adsorption was not enhanced simply with the carbonizing temperature. Therefore, the contribution of SiO_2 in RH to its ammonia adsorption was minor, which was consistent with the result of the ammonia adsorption test using the carbonized wood wastes..

3.4. Removal of Ammonia Gas from Fermenting Compost by Carbonized RH

The RH carbonized at 400°C, which had shown the best ammonia adsorption performance in the bag test, was employed for the removal of ammonia gas emitted from actual fermenting compost. The RH produced in Nishiki-mura was carbonized at 400°C for 1 h in nitrogen gas flow. The compost of 200 g was covered or mixed with the carbonized RH at the levels of 0, 1.0, 2.5 and 5.0 mass% in dry state to the compost. Figure 7 shows the concentration of ammonia gas in the vessel in which the compost added with the carbonized RH was enclosed, as a function of contact time. Data are results of one single measurement. The control indicates the ammonia concentration in the gas in which no carbonized RH was added. The ammonia concentration of the control decreased with the contact time, implying that gas in the vessel was evacuated and fermentation of the compost weakened. It is clearly found that increasing the addition level of the carbonized RH decreased the ammonia concentration in the vessel for both the covering and mixing methods. Covering method effectively reduced the ammonia concentration than the mixing method. Ammonia gas is lighter than ambient air, indicating that the carbonized RH covering the compost trapped efficiently the ammonia gas released from the whole region of the compost. It is also found that covering the compost with the carbonized RH at 5 mass% reduced the ammonia gas concentration from 200 to <1 vol. ppm. For the case of high-level addition (2.5 and 5.0 mass%) in the covering method, the ammonia concentration rather increased with the contact time. This is probably due to the adsorption saturation of the carbonized RH.

In order to more enhance the ammonia adsorption of the carbonized RH, humid and water-sprayed carbonized RHs were tested in a similar manner, with a comparison to humid and water-sprayed raw RH. Figure 8 shows the concentration of ammonia gas emitted from the compost covered with the dried and wetted samples, as a function of contact time. The quantity of the covered samples was 1.0 mass% in dry state to the compost weight. It is found that wetting the samples resulted in a more decrease in the ammonia concentration for both the carbonized and the raw RHs. Water-sprayed samples, into which moisture of 100 mass% was added, displayed the lowest ammonia concentration. The impact of wetting was found to be larger for the raw RH. However, wetting sample had no influence on the ammonia concentration at the contact time of 24 h. The enhancement of the ammonia adsorption owing to wetting was temporal.

Because it was confirmed that humidity in the top space of the vessel in which the compost and the sample was encased was very high (>90% RH), uptake and release of moisture on the RH carbonized at 400°C and the raw RH in a humid environment were evaluated. Figure 9 shows the moisture content of the dried and the water-sprayed samples in a humid testing chamber (20°C and 90% RH). The moisture content of the dried samples increased and that of the water-sprayed samples decreased with the contact time. The moisture content of the dried samples was saturated in 1 h, while that of the water-sprayed samples was saturated in 4 h. The raw RH showed the higher moisture content at the longer contact time, which is consistent with results of the moisture uptake test shown in Figure 6.

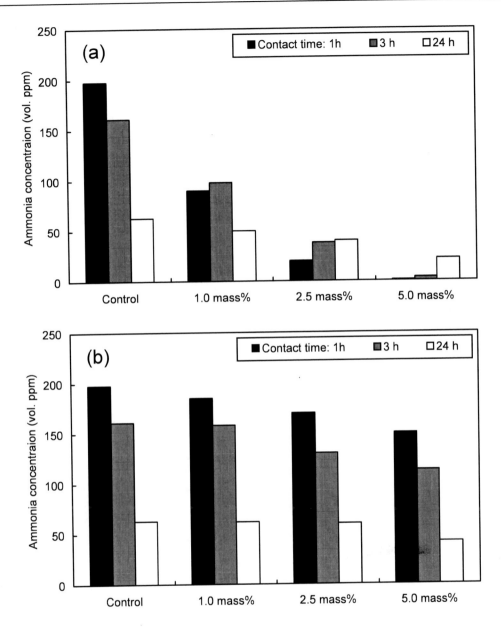

Figure 7. Concentration of ammonia gas emitted from the compost covered or mixed with different amounts of RH carbonized at 400°C as a function of contact time. (a) covered, (b) mixed.

The moisture content of the water-sprayed carbonized RH at the contact time of 24 h was almost similar to that of the dried carbonized RH, which was similarly observed on the raw RH. The enhancement mechanism of the ammonia adsorption on the carbonized and raw RHs is explainable from the solution of ammonia into moisture, as shown in Equation 2. Thus, the enhanced ammonia uptake at the contact time of 1 and 3 h should have been attributed to the residual moisture on the RH samples. The enhanced uptake imparted by water-spraying was found to diminish with a release of moisture from the RHs.

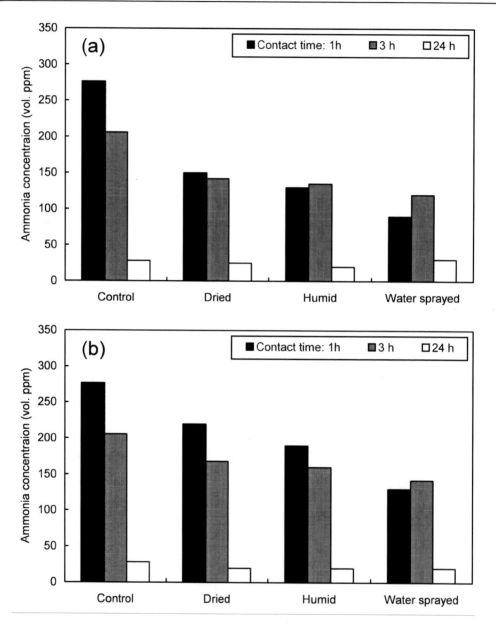

Figure 8. Concentration of ammonia gas emitted from the compost covered with dried and wetted samples as a function of contact time. (a) RH carbonized at 400°C, (b) raw RH.

The above results suggested that the best way to prevent ammonia gas to be released from compost to ambient environment was to cover compost with the RH carbonized at 400°C and then spraying water to the carbonized RH. In addition to this, the carbonized RH needs not to be removed from the compost because it includes useful inorganic fertilizer components such as K, Ca and P as well as produces void in the compost for promoting the ferment.

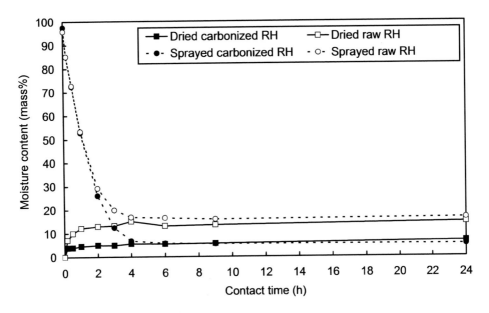

Figure 9. Moisture content of dried and water-sprayed RH samples in a humid environment (20°C and 90% RH) as a function of contact time.

3.5. Reducing Offensive Odor from Pig Dung by Carbonized RH

A commercial assorted feed added with the RH carbonized at 400°C at the level of 2 mass% to the feed was provided to growing pigs. Figure 10 shows the concentrations of several types of offensive gases emitted from the pigs' dung. Ammonia gas was attempted to be measured, but its concentration was under the detectable level (<0.1 vol. ppm). Data are average of 4 separate measurements with standard deviations indicated as an error bar. It is found that an addition of the carbonized RH to the feed reduced the concentrations the offensive gases emitted from the dung, in which the significances between the two testing sections (0 and 2 mass%) were verified by the t-test (P<0.05). In particular, sulfur-related compounds of hydrogen sulfide and mercaptons were notably (ca. 80%) reduced. The effects of the carbonized RH on the health and physical conditions of the growing pigs were evaluated (see Table 6). Noticeable changes in the health and physical conditions resulting from the carbonized RH were not observed, suggesting that the addition of the carbonized RH was useful to reduce offensive odor emitted from the dung of pigs, even if only the short-term (ca. 1 week) effect was verified.

Two mechanisms on the reduction of offensive odor emitted from the dung are available. One is direct gases adsorption of the carbonized RH in the bowel of pig. It was stated above that acidic functional groups formed on the RH carbonized at 400°C were useful to promote adsorption of basic gas of ammonia. However, hydrogen sulfide, mercaptans and lower fatty acids are acidic gases [32]. Therefore, the adsorption of these acidic gases was likely to be attributed to basic inorganic matters of K, Ca and P, which were intrinsically composed in RH. The other mechanism is to change the activity of enteric bacteria [33]. However, further discussion on the mechanism of the offensive odor reduction is now impossible referring to the present results.

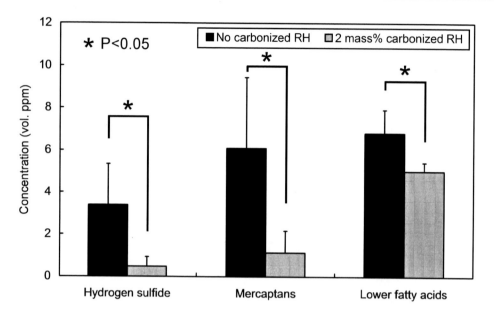

Figure 10. Concentrations of hydrogen sulfide, mercaptans and lower fatty acids emitted from the dung of pigs given assorted feed with and without RH carbonized at 400°C.

Table 6. The condition of pigs given an assorted feed with and without added RH carbonized at 400°C at 2 mass%

	Health condition	Weight of pig (g)	Dung amount (g)	Moisture content of dung (mass%)
No addition	Good	51.4 ± 6.7	1600 ± 437	72.9 ± 1.2
Carbonized RH added	Good	52.2 ± 5.8	1648 ± 372	73.6 ± 3.7

Average ± standard deviation.

More research is being undergone by the authors to determine the optimum adding level of the carbonized RH, to verify the long-term health and physical conditions of pigs, to evaluate the quality of the produced flesh, and to elucidate the detailed mechanism.

4. CONCLUSIONS

RH was carbonized at different temperatures to test it as a deodorizing adsorbent of offensive odor produced by the livestock industry. The carbonized RH was exposed to ammonia gas at a concentration of 100 vol. ppm in an enclosed bag. The RH carbonized at 400°C reduced the ammonia gas concentration in the bag the fastest. The deodorization performance was much better than those of several commercial deodorization adsorbents as well as those made of wood wastes (Japanese cedar and black pine) carbonized at 400°C. Pore and chemical characteristics of the carbonized RH were examined to correlate them with the ammonia adsorption property, suggesting that acidic functional groups remaining at 400°C were useful to promote adsorption of basic ammonia gas.

The performance of the RH carbonized at 400°C was evaluated with respect to the adsorption of ammonia gas produced from actual fermenting compost in an enclosed vessel. The actual compost was covered with or was mixed with the carbonized RH at the level of 0–5 mass% to the compost. The covering method reduced the ammonia concentration in the vessel much faster than the mixing method did, which was attributed to volatilization of ammonia gas which is lighter than ambient air. Covering the compost with the carbonized RH at 5 mass% reduced the ammonia gas concentration from 200 to <1 vol. ppm. Wetting the carbonized RH (mass similar to water was provided to the carbonized RH) was found to be effective in reducing the concentration of ammonia gas emitted from the compost, even if this effect was temporal up to the completion of moisture release.

An assorted feed to which was added RH carbonized at 400°C at 2 mass% was given to growing pigs. The addition of the carbonized RH reduced about 80% of the concentration of hydrogen sulfide and mercaptans emitted from the pigs' dung. Notable changes in the health and physical conditions of pigs were not observed. The uptake of hydrogen sulfide and mercaptans on the carbonized RH was likely to be related to the basic inorganic matter of K, Ca and P, which were intrinsically composed in RH.

Finally, the above laboratory and field tests showed that the RH carbonized at 400°C can be used as a promising material to remove the offensive odor produced by the livestock industry. This also contributes to an efficient circulation of waste in agriculture and livestock.

REFERENCES

[1] New energy and industrial technology development organization (NEDO). Estimated amount of existent and available biomass in Japan: GIS database. 2008. Available from: http://app1.infoc.nedo.go.jp/kinds/no2.pdf (in Japanese).

[2] Proctor, A., Palaniappans, S. Soy oil lutein adsorption by rice hull ash. *Journal of the American oil chemists' society* 1989, 66, 1618-1621.

[3] Proctor, A., Palaniappans, S. Adsorption of soy oil free fatty acids by rice hull ash. *Journal of the American oil chemists' society*, 1990, 67, 15-17.

[4] Proctor, A. X-ray diffraction and scanning electron microscope studies of processed rice hull silica. *Journal of the American oil chemists' society*, 1990, 67, 576-584.

[5] Kim, K.S., Choi, HC. Characteristics of adsorption of rice-hull activated carbon. *Water science and technology*, 1998, 38, 95-101.

[6] Guo, Y. Qi, J. Yang, S. Yu, K. Wang, Z. Xu, H. Adsorption of Cr(VI) on micro-and mesoporous rice husk-based active carbon. *Materials chemistry and physics*, 2002, 78, 132-137.

[7] Ajmal, N., Rao, R.A.K., Anwar, S., Ahmad, J., Ahmad, R. Adsorption study on rice husk: Removal of Cd(II) from waste water. *Bioresource technology*, 2003, 86, 147-149.

[8] Bishnoi, N.R., Bajaj, M., Sharma, N., Gupta, A. Adsorption of Cr(VI) on activated rice husk carbon and activated alumina. *Bioresource technology*, 2004, 91, 305-307.

[9] Kennedy, L.J., Mohan das, K., Sakeran, G. Integrated biological and catalytic oxidation of organics/inorganics in tannery wastewater by rice husk based mesoporous activated carbon-Bacillus sp. *Carbon*, 2004, 42, 2399-2407.

[10] Chandrasekhar, S., Pramada, P.N. Rice husk ash as an adsorbent for methylene blue: Effect of ashing temperature. *Adsorbent*, 2006, 12, 27-43.

[11] Kumagai, S., Noguchi, Y., Kurimoto, Y., Takeda, K. Oil adsorbent produced by the carbonization of rice husks. *Waste management*, 2007, 27, 554-561.

[12] Kalderis, D., Koutoulakis, D., Paraskeva, P., Diamadopoulos, E., Otal, E., Olivares del Valle, J., Fernández-Pereira, C. Adsorption of polluting substances on activated carbons prepared from rice husk and sugarcane bagasse. *Chemical engineering journal*, 2008, 144, 42-50.

[13] Pendyal, B., Johns, M.M., Marshall, W.E., Ahmedna, M., Rao, R.M. The effect of binders and agricultural by-products on physical and chemical properties of granular activated carbons, *Bioresource technology*, 1999, 68, 247-254.

[14] Yalcin, N., Sevinc, V. Studies of the surface area and porosity of activated carbons prepared from rice husks, *Carbon*, 2000, 38, 1493-1945.

[15] Guo, Y., Yu, K., Wang, Z., Xu, H. Effects of activation conditions on preparation of porous carbon from rice husk. *Carbon*, 2000, 41, 1645-1687.

[16] Amaya, A., Medero, N., Tancredi, N., Siva, H., Deiana, C. Activated carbon briquettes from biomass materials. *Biresource technology*, 2007, 97, 1635-1641.

[17] Kumagai, S., Sasaki, K., Shimizu, Y., Takeda, K. Formaldehyde and acetaldehyde adsorption properties of heat-treated rice husks. *Separation and purification technology*, 2008, 61, 398-403.

[18] Oya, A., Kishimoto, N., Mashio, S., Kumakura, K., Suzuki, T., Serrano-Talavera, B., Linares-Solano, A. Structure and properties of a molded carbon derived from rice hull. *Journal of materials science*, 1995, 30, 6249-6252.

[19] Unuma, H., Niino, K., Sasaki, K., Shibata, Y., Iizuka, H., Shikano, S., Nakamura, T. Preparation and characterization of glass-like carbon/silica composite from rice hull and phenolic resin. *Journal of materials science*, 2006, 41, 5593-5597.

[20] Watari, T., Nakata, A., Kiba, Y., Torikai, T., Yada, M. Fabrication of porous SiO_2/C composite from rice husks, *Journal of the European ceramic society*, 2006, 26, 797-801.

[21] Yoshida, K., Iizuka, H., Shibata, Y., Unuma, H., Nakamura, T., Shikano, S. Water resistance of porous carbon materials made from rice hull. *International journal of the society of materials engineering for resource,* 2006, 13, 49-53.

[22] Liou, T.H. Evolution of chemistry and morphology during the carbonization and combustion of rice husk. *Carbon*, 2004, 42, 785-794.

[23] Kwong, P.C.W., Chao, C.Y.H., Wang, J.H., Cheung, C.W., Kendall, G. Co-combustion performance of coal with rice husks and bamboo. *Atmospheric environment*, 2007, 41, 7462-7472.

[24] Tsai, W.T., Lee, M.K., Chang, Y.M. Fast pyrolysis of rice husk: Product yields and compositions. *Bioresource technology*, 2007, 98, 22-283

[25] KANSAI Corporation. Characteristics of ARHC. 2008. Available from http://www15.ocn.ne.jp/~kansai/e-index2.html

[26] Bansal, R.C. Goyal, M. *Activated carbon adsorption.* 2005, Boca Raton, FL, CRC Press, Inc.

[27] Do, D.D. *Adsorption Analysis: Equilibria and Kinetics*, 1998, London, Imperial College Press.
[28] Asada, T., Yamada, A., Ishihara, S., Komatsu, T., Nishimatsu, R., Taira, T., Oikawa, K. Countermeasure against indoor air pollution using charcoal board. *Tanso*, 2004, 211, 10-15.
[29] Hitomi, M., Kera, Y., Tatsumoto, H., Abe, I., Kawafune, I., Ikuta, N. Preparation of charcoals from Cryptomeria and Chamaecyparis and their properties, *Tanso*, 1993, 160, 247-254 (in Japanese).
[30] Mori, M., Saito, Y., Shida, S., Arima, T. Adsorption properties of charcoals from wood-based materials. *Mokuzai Gakkaishi*, 2000, 46, 355-362 (in Japanese).
[31] Yanai, H., Ishizaki, N. *Kasseitan Tokuhon*, 1996, Tokyo, Nikkankogyo Shimbunsya (in Japanese).
[32] Yasugi, R., Koseki, H., Furuya, M., Hidaka, T. (Editors) *Iwanami Seibutsugaku Jiten 4th Edition*, 1996, Tokyo, Iwanami Syoten (in Japanese).
[33] Oshida, T., Kakiichi, T., Haga, K. *Chikusan Kankyo Hozenron*, 1998, Tokyo, Yokendo (in Japanese).

In: Agricultural Wastes
Eds: Geoffrey S. Ashworth and Pablo Azevedo

ISBN 978-1-60741-305-9
© 2009 Nova Science Publishers, Inc.

Chapter 7

BIO (SINGLE CELL) PROTEIN: ISSUES OF PRODUCTION, TOXINS AND COMMERCIALISATION STATUS

Ravinder Rudravaram[1], Anuj Kumar Chandel[2], Linga Venkateswar Rao[1], Yim Zhi Hui[3] and Pogaku Ravindra[3,]*

[1] Department of Microbiology, Osmania University, Hyderabad-500 007, India
[2] Department of Biotechnology, Jawaharlal Nehru Technological University, Hyderabad-500 007, India
[3] School of Engineering and Information Technology, Universiti Malaysia Sabah, Kota Kinabalu, 88999, Malaysia

ABSTRACT

The alarming rate of population growth and a regular depletion in food production and food resources are important factors in the present dire need to find new viable options for food and feed sources. Based on scientific developments, particularly in industrial microbiology, one feasible solution could be the consumption of microorganisms as human food and animal feed supplements. Humans have used microbial-based products—like alcoholic beverages, curd, cheese, yogurt, and soya—even before the beginning of civilization. Due to research developments in the scientific arena in the last two decades, (Bio) single cell protein (SCP) has drawn new attention towards its use as supplement in human food, animal feed or staple diet. There are several benefits to using SCP as food or feed, viz. its rapid growth rate and high protein content. The microorganisms involved as SCP have the ability to utilize cheap and plentiful available feedstock for their growth and energy, making them an attractive option. However, in spite of laboratory-based success stories, only a limited number of commercial SCP production plants have been seen worldwide. This review analyzes the possibility of SCP production, various raw materials for its production, available microorganisms with cultivation methods, toxicity assessment

---

[*] Corresponding author: Dr. Pogaku Ravindra, School of Engineering and Information Technology, Universiti Malaysia Sabah, Kota Kinabalu, 88999, Malaysia. Email: dr_ravindra@hotmail.com. Telephone: 006 088 320000 ext 3048. Mobile: 006 013 87666634. Fax: 006 088 320348.

and their removal. Also, new developments and risk assessment using SCP along with worldwide industrial SCP production are discussed.

Key words: single cell protein (SCP), solid state fermentation (SSF), toxins, nucleic acid

1. INTRODUCTION

Single cell proteins are microbiologically-produced proteins, identified and named at Massachusetts Institute of Technology (MIT) in 1966 by Carrol Wilson. It can be derived from a variety of microorganisms, both unicellular and multicellular—namely, bacteria, yeast, fungi, or microscopic algae. These potentially important food substances are not pure proteins but are, rather, dehydrated cells consisting of mixtures of proteins, lipids, carbohydrates, nucleic acids and a variety of other non-protein nitrogenous compounds, vitamins, and inorganic compounds (Anupma and Ravindra, 2000).

Earth has always undergone fast and drastic changes in the name of progress. It has fulfilled all our needs and necessities, but the changes are so swift that the resources provided by it are falling short. One reason, first and foremost, is the geometric progression of the population explosion (Kapitza, 1996). Population growth has ill effects on the economy of the country, and in social and political aspects as well. Further, it has a direct impact on the availability of food (Levy-Costa, Sichieri, dos Santos, Pontes and Monteiro, 2005).

At the start of the 20th century, the population of the world was less than two billion people, and the population had grown to six billion when the century ended. The incremental growth of the human population in the last half-century is unprecedented. In the present scenario, world population is projected to continue to rise exponentially for next 50 years (Smil, 2005). India is the second most populous country in the world after China. Over the past four decades, the size of India's population has more than doubled and is projected to reach 1.22 billion by 2015 and 1.40 billion by 2030, despite the considerable slowdown in fertility and population growth rates.

Regardless of which projection one uses, it is clear that at least 1.3 billion people will be added to the world population during the next 25 years (Figure 1). A population spurt is seen both in developing and developed countries, although with a difference in rate of explosion (Chamie, 2004). The developing countries owe about 98% of the total population increase, while developed countries add the remaining 2% difference. The population growth diverts our thinking towards the availability of basic needs for this enormous number.

Apart from increased population, rapid urbanization and fast reduction of area under cultivation have further aggravated the problem of availability of food.

Proteins are useful for body building, repair and maintenance of tissues, osmotic pressure, and chemical coordination (in the form of hormones). In extreme conditions protein may also be used for the production of energy in the human body (ZoBell, Olson, Wiedmeier and Stonecipher, 2004). The nutritive value of proteins depends on its amino acid composition. Amino acids are the bricks with which tissue protein is built and replaced. There are some 20 amino acids commonly found in dietary proteins, out of which 10 amino acids cannot be synthesized by human beings and they have to be supplied through diet. These are called essential amino acids (Koehnle, Russell and Gietzen, 2003).

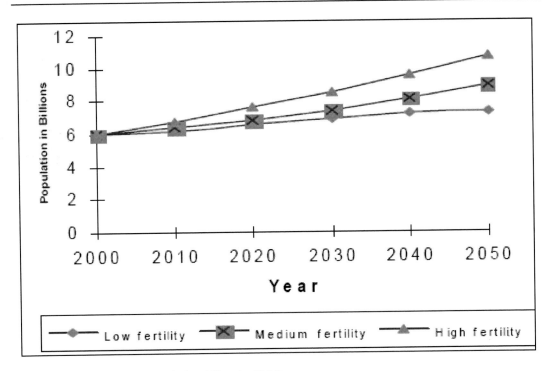

Figure 1. Projected world population (Chamie, 2004).

Table 1. Daily requirement of essential amino acids (Lehninger, 1990)

Amino acid	Quantity (g)
Arginine	0^+
Histidine	Unknown *
Isoleucine	1.30
Leucine	2.02
Lysine	1.50
Methionine	2.02
Phenylalanine	2.02
Threonine	0.91
Tryptophan	0.46
Valine	1.50

+ Required by infants and growing children
* Essential but precise requirement not yet established

Table 1 shows the daily requirement of essential amino acids. Adults require 9 essential amino acids while children require 10, the extra amino acid being arginine. Although the liver normally makes arginine as a step in the synthesis of urea, children cannot make arginine fast enough to support both urea synthesis and the synthesis of body proteins. Dietary protein requirement varies from 30 to 92 g/day. To fulfill this minimum protein requirement, constant research work on providing alternative sources of proteins in the form of microorganisms is in progress worldwide. In this direction, the research activities have given some encouraging results over a decade (Anupma et al., 2000; Ravinder, Venkateshwar Rao and Ravindra, 2003; Ravinder, Chandel, Venkateshwer and Ravindra, 2006; Berge, Baeverfjord, Skrede and Storebakken, 2005).

This review deals with single cell protein (SCP) production technologies, available microorganisms, toxins and their removal, applications and industries involved in production of SCP worldwide.

2. MICROORGANISMS: AN ASSEST IN SCP PRODUCTION

Microorganisms have excelled at producing primary and secondary metabolites from a variety of raw carbohydrates for billions of years under varying cultivation processes for the production of value-added products. In current vogue, the results of studying giant 'microbial libraries' for microbial conversion of alternative and cheap carbohydrates can serve as a raw material for SCP production (Singh and Kumar, 2007). Over the centuries microorganisms have been used by human as biomass in production of various products such as alcoholic drinks, cheese, yogurt and soybean paste (Tuse, 1984). Today, there are many works in food and feed production using microorganisms in the world. Usefulness of SCP is determined by the parameters related to nutritional values and its contents. These parameters are nutrients, vitamins, nitrogen, carbohydrates, fats, cell wall, nucleic acids, and profile of protein and amino acids. Microbial protein is a non-traditional protein; it is not a palatable, desirable food and must be incorporated directly or indirectly into other foods (Anupma et al., 2000). Consumption of microorganisms as a source of protein seems an attractive and sustainable option due to several advantages. The period of doubling time of microorganisms is shorter than the other living beings for SCP production (Table 2).

In recent years, SCP has been produced from many species of microorganisms. These are fungi, bacteria and algae, which are widely used practically. Use of bacteria and fungi for SCP production is more economical when grown on easily available wastes (Pandey, Soccol and Mithchell, 2000a; Pandey, Soccol, Nigam, Soccol, Vandenberghe and Mohan, 2000b; Rudravaram, Chandel, Linga, and Ravindra, 2006). The reason for preferring microorganisms is their fast growth with high protein ratio. The acceptance of a species as food or feed depends on growth rate, substrate utilization and toxicity (Ravinder PhD Thesis, 2003; Ravinder et al., 2003). Filamentous fungi are preferred in SCP production due to easy growth on lignocellulosic wastes and ease of separation from its media (Westhuizen and Pretorius, 1996).

Table 2. The doubling period of mass for some living beings (Bailey and Ollis, 1986)

Organism	Time for doubling mass
Bacteria and yeast	20–120 minutes
Filamentous fungi and algae	2–6 hours
Grass and some plants	1–2 weeks
Chicken	2–4 weeks
Pig	4–6 weeks
Cattle	1–2 months
Human	3–6 months

2.1. Yeast

Yeast is a commonly-used organism in the production of biomass, probably because it is already accepted both in the human food and animal feed industries. Yeast-based processes are the farthest advanced towards commercial production, followed by bacterial processes (Kourkoutas, Sipsas, Papavasiliou and Koutinas, 2007). Yeasts have many convenient characteristics, such as the ability to use a wide variety of substrates like hexoses, pentoses, and hydrocarbons (Moeini, Nahvi and Tavassoli, 2004), susceptibility to induced and genetic variation, ability to flocculate (Labuza, 1975) and high nutritional value (Yanez, Ballester, Fernandez, Gattos and Mönckeberg, 1972). However, attention has often been drawn to the fact that yeast appears to be deficient in essential sulphur amino acids (Worgan, 1973).

2.2. Algae

Unicellular algae are the smallest and simplest photosynthetic apparatus for effective utilization of plentiful solar energy (Venkatratnam, 1978). The advantages and disadvantages of growing algae as protein sources were discussed by several investigators (Chae, Hwang and Shin, 2006). The use of certain algae in foods in certain parts of the world has been well documented. The blue green algae Spirulina is collected and eaten by natives around Lake Chand in Africa traditionally. In countries like Tartary, Mangolia and China it was found that algae was eaten by people and also marketed. The autotrophic, mixotrophic and heterotrophic methods of production of microalgae (Fabregas and Herrero, 1985), Chlorella in Asia has been reviewed by Kawaguchi (1980). Algae processes are still short of full-scale development because they are limited by the requirements for light over a major portion of the year and for a continuous supply of CO_2 or other carbon source (Venkatratnam, 1978). In Taiwan, however, plants for the production of Chlorella feeds using methane generated from manure are now in operation (Srinivasan and Fleenor, 1972).

2.3. Bacteria

A number of different species of bacteria are used in SCP production. Bacteria have a slight advantage over the other microorganisms as a food source, because of their higher growth rates and relatively higher protein content and sulphur-containing amino acids (Kurbanuglu and Algur, 2002). Bacterial SCP is high in protein and certain essential amino acids. The crude protein content is around 80% of the total dry weight. The nucleic acid content, especially RNA, is very high on a dry weight basis and is reported to be 15–16%. Bacterial SCP is rich in methionine, around 2.2-3.0%, which is comparatively higher than that of algae (1.4–2.6%) and fungi (1.8–2.5%) (Singh, 1998). The essential amino acid composition of different *Lactobacilli* group appears comparable to that of the FAO reference protein and SCP from other sources (Erdman, Bergen and Reddy, 1977). The use of bacteria was objected because of their size, which makes harvesting difficult without the use of flocculants or thickeners (Dabbah, 1970). However, a very limited success has been achieved so far with bacterial biomass consumption as human food or animal feed. This is partly because of high nucleic acid content and partly due to the fear of bacterial toxins (Fossum and Almlid, 1977). Production of SCP on SSF by utilizing lignocellulosic waste is not practiced due to less moisture content in lignocellulosic material (Han and Srinivasan, 1968)

2.4. Fungi

The filamentous fungi are extensively used in SCP production. Many strains, such as *Aspergillus, Fusarium, Chaetomium, Rhizopus,* and *Tricoderma*, are being used on a commercial scale for this purpose (Solomans, 1985). Fungi have the ability to provide form and texture (Humphrey, 1975), and hence can be harvested with ease; also, the cost of production may be reduced. Like algae, fungi generally have low nucleic acid content, and accordingly, the dangers of kidney stones and gout are not great even without processing the biomass to lower nucleic acid content (Lipinsky, 1974). Another advantage is that fungi can prosper on a variety of carbohydrates and they are best suited for agroresidues, although growth rates vary considerably with different substrates (Anderson, 1975).

3. SOURCES FOR SCP PRODUCTION AND CULTIVATION METHODS

Bacteria and fungi utilize a variety of substrates (Tannenbaum and Wang, 1975) in SCP production. The main substrate required for algae are carbondioxide and sunlight (Kawaguchi, 1980). Fungal species are cultured on different substrates, mostly cheap wastes, which supply the carbon and nitrogen for growth (Smith, Fermor and Zadrazjil, 1988). Among various substrates, lignocellulosic material such as wood from angiosperms and gymnosperms, grasses, leaves, wastes from paper manufacture, sugarcane bagasse, wheat straw, wheat bran, rice bran, groundnut shell and other agricultural wastes are cheap source for SCP production (Gupte and Madamwar, 1997; Chaudhary and Sharma, 2005).

Lignocellulose is the most abundant organic material on earth, representing 50% of the total plant biomass and presenting an estimated annual production of 5.0×10^{10} tons (Kuhad and Singh, 1993). Plant production from agriculture contributes only a small part of this overall production, but still represents a considerable amount, 1.23×10^8 tons per year (Kuhad et al., 1993). Approximately half of the agricultural biomass produced is not used as food or feed (Smith et al., 1988). The major part of these agricultural residues, if not burned, is fragmented and added to soil as organic fertilizer (Zadrazil, Brunnert and Grabbe, 1983).

Agro-industrial wastes from cereals, wood and fruit processing industry wastes have also been explored for SCP production worldwide. For example the availability of rice bran, which is almost 40 million metric tons rice bran produced every year, has been discarded as unfit for human consumption. After the oil extraction it is called deoiled rice bran (DOB) which contains 9% protein and 36% cellulose and has been thoroughly exploited for protein enrichment using *Aspergillus oryzae* MTCC1846 (Ravinder et al., 2003, 2006).

Waste recycling has been advanced as a method for preventing environmental decay and increasing food supplies. The potential benefits from successful recycling of agricultural wastes are enormous. It may be the only method for large-scale protein production that does not require a concomitant increase in energy consumption. In addition, it may be the most effective method for producing animal and human food from lignocellulose materials apart from their use in bioethanol production (Chandel, Rudravaram, Narasu ML, Rao and Ravindra, 2007a; Chandel et al., 2007b). The lignocellulosic waste materials contain cellulose, hemicellulose and lignin. Cellulose is most abundant of these polysaccharides. These sources are attractive due to their easy availability and low cost (Lin and Tanaka, 2006). These materials however have to be pretreated with some physical or chemical methods since lignin-cellulose complex is tough to unravel and does not allow easy accessibility of enzymes in enzymatic hydrolysis of lignocellulose (Chandel et al., 2007a; Chandel, Kapoor, Singh and Kuhad, 2007c).

4. TOXINS OF MICROFLORA AND THEIR REMOVAL

Toxins are secondary metabolites produced by fungi (Bennett and Keller, 1997) and bacteria (Blancou, Calvet and Riviere, 1978). Algae in general do not produce harmful toxins (Leonard and Compere, 1967). The toxicity of an SCP product must be assessed before marketing, since SCP is considered as animal feed and human food (Anupama et al., 2000).

4.1. Toxins Produced by Bacteria and their Removal

Bacteria produce exotoxins such as enterotoxin, erythrogenic toxin, alpha-toxin, and neurotoxin (Ekenvall, Dolling, Gothe, Ebbinghaus, von Stedingk and Wasserman, 1983). Endotoxins are an integral part of the cell walls of gram-negative bacteria and are liberated upon lysis. Endotoxins are lipopolysaccharides in nature. Endotoxins cause fever in the host and are fatal for laboratory animals at slightly higher doses than exotoxins (Powar and Daginawala, 1995). SCP from *Pseudomonas* species and *Methylomonas methanica*, when used for animal feed purposes, caused febrile reaction and high titres of IgG and IgM due to endotoxins (Ekenvall et al., 1983).

Exotoxins can be easily removed, as these are present in soluble form in the medium (Powar et al, 1995). They are sensitive to temperatures above 60°C. Moreover, 50% alcohol, formaldehyde and dilute acids can denature exotoxins or convert them into non-toxic toxoids (Blancou et al., 1978). Toxoids are beneficial for artificial immunization (Powar et al., 1995).

On the other side, removal of endotoxins is difficult as they are part of cellular components of few gram-negative bacteria. Formation of endotoxins can only be prevented by genetic engineering, where the activity of genes controlling the formation of the unwanted toxins can be modified or suppressed (Anupama et al., 2000). SCP produced by methanotropic and heterotrophic bacteria generally contains a large amount of nucleic acid making the product less suitable for human consumption.

4.2. Toxins Produced by Fungi (Mycotoxins) and Their Removal

Minute amounts of mycotoxins are able to induce allergies, diseases, rashes on skin, neurotoxicity and other disorders. There are many reports on mycotoxins, but aflatoxins are the most important and best understood (Eaton and Groopman, 1994). Prominent fungal toxins include aflatoxins of type B_1, B_2, G_1 and G_2 *Aspergillus flavus,* citrinin from *Penicillium citrinum,* trichothecenes and zearalanone from *Fusarium* species and ergotamine from *Claviceps* species (Bennett et al., 1997). Aflatoxins are produced by *A. flavus, A. parasiticus* and *A. oryzae* (Schlegel, 1996). Aflatoxins cause human liver cancer (Eaton et al., 1994). Apart from aflatoxins, ochratoxins are also important mycotoxins. Ochratoxin are structurally related groups of pentaketides. Ochratoxin A is the most abundant and toxic of the five metabolites in the Ochratoxin group. These metabolites have been found to occur in *Aspergillus* and *Penicillium* species and cause damage to the liver as well as kidneys (Varga, Kevel, Rinyu, Teren and Kozakiewicz, 1996). Trichothecenes are considered the next most important group among mycotoxins. Trichothecenes have a 12, 13 epoxytrichothec-9-ene nucleus which able to induce dermal toxicity and several hematopoietic effects (Joffe, 1986).

Ammoniation is the most successful method to remove mycotoxins as this method can reduce aflatoxin levels by 99% (Park and Liang, 1993). Co-culture of *A. flavus* with other microbes results in reduction of toxins production (Cotty, Bayman and Egel, 1994). The biosynthetic pathways for both aflatoxins (Bennet et al., 1997) and trichothecenes were studied (Desjardins, Hohn and McCormick, 1993). Molecular biology techniques have been exploited for detoxification. Molecular dissection and elimination of genes responsible for mycotoxin synthesis is developing into an exciting area of contemporary mycotoxin research (Keller, Cleveland and Bhatnagar, 1992). The techniques of cloning, probing, sequencing, expression libraries, transcript mapping, gene disruption, and chromosome walking have been employed to isolate the aflatoxin pathway clusters from *A. flavus* and *A. parasiticus* (Bennett et al., 1997). Studies that identify the molecular determinants regulating mycotoxin production are promising rational control strategies. For example, aflR is the regulatory gene controlling the production of both aflatoxins and sterigmatocystin in *Aspergillus*. This forms the target gene to be inhibited for the future control of mycotoxin production in *Aspergillus* (Yu, Butchko and Fernandes, 1996). When *Aspergillus sp* is impaired in asexual reproduction, the production of aflatoxins is inhibited (Kale, Bhatnagar and Bennett, 1994). A report describes how a plasmid vector (pDEL2) was engineered to introduce a deletion within the aflatoxin biosynthesis gene cluster of *A. parasiticus.* Subsequent aflatoxin precursor feeding studies confirmed that the enzyme activities associated with the deleted genes were absent (Cary, Barnaby, Ehrlic, and Bhatnagar, 1999). There are also reports that products of plant defense pathways (i.e., lipoxygenase or the jasmonate pathway) can inhibit aflatoxin production (Zeringue, 1996).

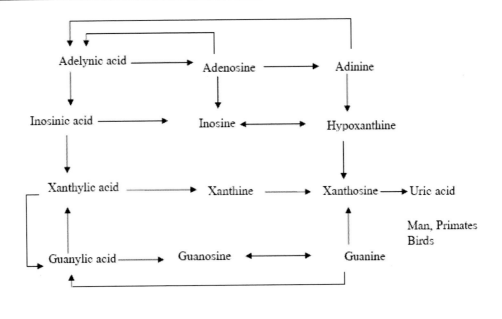

Figure 2. Pathways of nucleic acid metabolism (Calloway, 1974).

5. NUCLEIC ACIDS AND ITS REMOVAL

Bacterial SCP products may have nucleic acids as high as 16% of dry weight. Human consumption greater than 2 g nucleic acid equivalent per day may lead to kidney stone formation and gout (Calloway, 1974). Figure 2 shows the metabolic pathway of nucleic acids. Uric acid, an important metabolite, is slightly soluble at physiologic pH and if the blood uric acid content is elevated, crystals may form in the joints, causing gout or gouty arthritis and urinary problems (Dabbah, 1970).

In rapidly-proliferating microbial cells RNA forms the bulk of the nucleic acids (Singh, 1998). The RNA content of yeast cells is known to be dependent on the culture conditions and carbon/nitrogen (C/N) ratios. The nucleic acid level can be reduced by several means. These include activation of endogenous RNAase by brief heat treatment up to 60–70°C for 20 min, alkaline hydrolysis of nucleic acids, modifications of cultural conditions with respect to nitrogen, carbon, phosphorous and zinc content or chemical extraction and removal of nucleic acid (Worgan, 1973).

Some researchers have extensively discussed safety evaluation of SCP products for human food and animal feed, and specific methods to reduce the nucleic acids are also reported (Taylor, Lucas, Gable and Graber, 1974). Purine content in baker's yeast can be reduced by chemical treatment and autolytic methods (Trevelyan, 1976a, b). Reduction of nucleic acids can be achieved by the endogenous polynucleotide phosphorylase and RNase in *Brevibacterium* (Yang, Thayer and Yang, 1979). Two derivatives of pancreatic RNase and an endonuclease of *Staphylococcus aureus*, immobilized on corncobs, have been used to reduce the percentage of nucleic acids in SCP concentrates of yeasts from 5–15% to 0.5% with a protein loss of only 6% after treatment

(Martinez, Sanchez-Montero, Sinisterra and Ballesteros, 1991). Immobilized nucleases like benzonases on corncobs were also used to reduce the nucleic acid content in protein concentrates. The percentage of DNA was reported to be reduced to 3–6% and RNA to 50% with loss of protein being only 1% (Moreno et al., 1991). An immobilized pancreatic RNase was also investigated for the degradation of yeast ribonucleic acid. The rapid reaction rates obtainable at relatively low temperatures offer a potential alternative method of purifying yeast SCP with minimal loss of derived protein (Dale and White, 1979). Methods for reduction of nucleic acid content in SCP obtained from gas oil are also reported (Abou-Zeid, Khan and Abulnaja, 1995).

6. NEW DEVELOPMENTS FOR IMPROVEMENT IN SCP PRODUCTION

Timely interventions such as strain improvement through mutagenesis, gene cloning and expression, and optimization of potential fermentation parameters enhance the production of selective metabolites (Singh et al., 2007). Mutation has been a successful method for improving the yield of biomass. *Pseudomonas methylotrophus* was isolated by Imperial Chemical Industries (1CI) and then significantly improved through genetic engineering and physiological development. When glutamate dehydrogenase system isolated from *Escherichia coli* was transferred to GOGAT mutant of *P. methylotrophus* by genetic engineering, improved nitrogen assimilation was observed by the mutant. It was used commercially for yield improvement in terms of raw protein, pure protein, and cell dry weight and μ_{max} was significant (Gow, Littlehailes, Smith, and Walter, 1975). Yields of the engineered organisms were about 5% greater than the parent organism (Balasubramanian, Bryce, Dharmalingam, Green and Jayaraman, 1998). The use of genetic engineering is not restricted to designing and production of microbes with high protein yield capability alone. It also finds an application in the improvement of downstream processing in post-fermentation stage. It is used in the production of 'designer' or 'recombinant' proteins of high nutritive value and their subsequent purification. Novel techniques like specifically designed tags or by modification of sequences with target-gene product are used for efficient recovery of native and modified proteins (Murby, Uhlen and Stahl, 1996). Improvement in protein recovery can also be achieved by suppression or elimination of protease enzyme activity in the microbial cells. Mutants with modified protease gene sequence can be isolated and used for the purpose. For example, LDHP1 strain of *S. cerevisiae,* a protease deficient mutant, was produced due to osmotic shock and showed an improved beta-galactosidase yield (Becerra, Cerdan and Gonzalez-Siso, 1997).

Recombinant DNA technology has also been used to isolate a mutant gene that can enhance the overall protein yield or enables the organism to produce high amount of specific amino acids like glutamate, tryptophan, phenylalanine and others along with a high protein yielding capacity. Such genes can be combined together by recombination technique and incorporated in the organism, which has a wide substrate range.

A mutant variety of *S. cerevisiae* produced high concentration of intracellular glutamate (Hill, 1994). Mutants of *S. cerevisiae,* which were resistant to toxic ethionine, showed accumulation of methionine as high as 163 times that of parental strain. Moeini et al., (2004) used mixed culture of the isolated yeast strains with *S. cerevisiae* was used in order to increase the biomass yield and BOD removal. The highest biomass yield (22.38 g/l) and reduction of initial BOD from 30000 to 3450 mg/l were obtained with the mixed culture of *K. lactis* (M2) and *S. cerevisiae*. The property of high accumulation in mutant was gained due to enhanced specific activity and reduced feedback inhibition by threonine of the enzyme aspartokinase and homoserine (Martinez-Force and Benitez, 1993). About 180,000 tons of L-Lysine, is needed annually to improve the nutritive value of food and SCP. Mutant *Corynebacterium glutamicum,* which is auxotrophic for alanine, sensitive to beta-fluoropyruvate and resistant to alpha-chlorocaprolactum, could produce 70 g/l L-Lysine. On improvement of strain by recombinant DNA technique the yield was enhanced to 98 g/l of L-Lysine (Sahm, 1995). The yield of expensive amino acids like L-Threonine can also be improved by use of recombinant *Escherichia coli* (Tsukada and Sugimori, 1971) and that of phenylalanine by genetically manipulated strain *of Rhodotorula rubra (*De Boer and Dijkhuizen, 1990). High cost restricts the use of L-Tryptophan to humans. Enhanced yield can be achieved by classical mutation and genetic engineering. Unification of four mutations by recombination and their incorporation in *S.cerevisiae* helped to produce 12 g/l L-tryptophan (Moller, 1994).

Strain improvement is done by complementing the auxotrophic mutants by wild type genes (Jungehulsing, Arntz, Smit and Tudzynski, 1994). The substrate-range of such strains can be extensively widened when compared to the auxotrophic variety. This would allow the utilization of a variety of unused substrates for the growth of microbes for SCP production.

6.1. Application of New Substrates in Solid State Fermentation (SSF) for SCP Production and Commercialization of SCP

SSF has been traditionally followed for production of fermented foods since ancient times. Some of these foods have worldwide popularity and are used for preparation of delicious curries (Nigam and Vogel, 1990). SSF can be commercially applied for the production of organic acids, mushrooms, cheese and upgradation of lignocellulose, composting, dough fermentation, mycotoxins, starter inoculum, ethanol, cocoa, vanilla, coffee, insecticides and biodegradation of wastes (Pandey et al., 2000a). In recent times, SSF has applications in the field of SCP production in an extensive way due to its advantage of utilization of wastes to produce value-added product (Ravinder et al., 2003; 2006).

Crop residues represent potential source of dietary energy to ruminants if the protein content can be increased (Robinson and Poonam, 2002). Crop residues are renewable and could be a potential raw material for SCP production. They represent a potential solution to feeding animals. Table 4 shows various substrates and micro organisms used for protein enrichment (SCP) by solid state fermentation. Cassava starch an important food for millions of Africa, Asia and South America contains low protein, vitamins and sulphur containing amino acid (Pandey et al., 2000b). Soccol, Leon, Rouses and Raimbault (1993) investigated biotranformation of cassava for nutritional improvement. Apple pomace has been used protein enrichment under SSF by *A. niger* as food and feed for pigs (Nigam and Singh, 1996). The use of protein enrichment of apple pomase by use of co-culture of moulds and yeast has been explored by Bhalla and Joshi (1994). Carob pods contain high amounts of tannins, which have adverse effect on animal growth. Protein enrichment of 20% was obtained in 4 days of SSF of carob pods SSF with 83% of reduction in tannins (Smail, Salhi and Knapp, 1995). Recently, some unconventional sources have been exploited for SCP production. Berge and colleagues (2005) used natural gas for production of bacterial protein meal (BPM) and evaluated growth.

Table 3. Macromolecular composition and general properties of microorganisms (Champagnat, Vernet, Laine and Filosa, 1963)

Parameters	Bacteria	Yeast	Fungi	Algae
Doubling time (hours)	1–3	2–6	5–12	6–24
Crude protein (% dry cell weight)	40–80	40–60	30–45	40–50
Nucleic acids (%)	8–20	5–15	6–13	45–51a
Carbohydrates and fats (%)	10–30	10–40	10–45	34.6–45
Ash content (%)	4–10	4–10	4–10	5–8
Temperature range (°C)	22–55	25–40	25–50	25–32
pH range	5–7	3–5	6–8	6.9–9.6

a Percentage

Table 4. Substrates and microorganisms used for protein enrichment by SSF

Substrate	Microorganism	References
Deoiled rice bran	*A. oryzae*	Ravinder et al., 2003; 2006
Ram horn	*Bacillus cereus* NRRL B-3711, *Bacillus subtilis* NRRL NRS-744, *Escherichia coli*	Kurbanoglu et al., 2002
Rice bran	*Aspergillus niger*	Anupama and Ravindra, 2001
Apple pomace	*C.utilis sp. Kloeckera apicula, A .niger, Tricoderma sp.*	Rahmat, Hodge, Manderson and Yu, 1995 Bhalla et al., 1994
Carob pods	*Agrocybe aegirata, Amellaria mellae*	Nicolini, Volpe, Pezzotti and Carilli, 1993
Canola meal	*Aspegillus carbonarius*	Alasheh and Duvnjak, 1995
Wheat straw	*Chaetomium cellulolyticum Trichoderma reesei*	Viesturs, Apsite, Laukevics, Ose and Bekers, 1981
Coffee pulp	*Aspergillus niger*	Nicolini et al., 1993
Rape-seed meal	*Aspergillus clavatus Fusarium oxysporum*	Viesturs et al., 1981
Sago starch	*Rhizopus oligosporus*	Penbaloza, Molina, Brenes and Bressani, 1985
Sugar beet pulp	*Trichoderma reesei Talaromyces*	Cochrane, 1958
Sugar beet pulp	*Trichoderma viride*	Mitchell, Gumbira-Said, Greenfield and Doelle, 1991
Manure	*Chaetomium cellulolyticum*	Considine, Mehra, Hackett, O'Rorke, Comerford and Coughlan, 1986
Soy beans	*Rhizopus oligosporus*	Reu, 1995

Table 4. (Continued)

Substrate	Microorganism	References
Sawdust	*Chaetomium cellulolyticum*	Durand and Chereau, 1988
Lignocellulosic Substrates	*Chaetomium cellulolyticum*	Chahal, Vlach and Moo-Young, 1980
Cassava starch	*Rhizopus oligosporus*	Mitchell and Lonsane, 1992
Banana wastes	*Aspergillus niger*	Baldensperger, Le Mer, Hannibal and Quinto, 1985
Citrus peel extracts	*Geotrichum candidum*	Ziino, Curto, Salvo, Signorino, Chifal and Giuffrida, 1999
Fodder beets	*Saccharomyces cerevisiae*	Rodrguez, Echevarria, Rodriguez, Sierra, Daniel and Martinez, 1985
Maize	*Aspergillus niger Saccharomyces sp.*	Han and Steinberg, 1986
Bagasse	*Schwanniomyces castelli*	Suacedo-Casteneda, Gutierrez-Rojas, Bacquet, Raimbault and Vinegra-Gonzalez, 1992
Citrus Pulp	*Aspergillus niger, Trichoderma virideae*	De Gregorio, Mandalari, Arena, Nucita, Tripodo and Locurto, 2002

Ghaly and Kamal (2004) used cheese whey as source for SCP production from *Kluyveromyces fragilis* under the optimum conditions (air flow, 3 vvm; agitation, 400 rpm; temperature, 31.6°C). About 99% of lactose utilization was observed after 28 h of growth.

Cheese whey was also used as carbon source for SCP production under fed batch culture using Kefir microflora in anaerobic conditions (Paraskevopaulou, Athanasiadis, Kanellaki, Bekatorou, Blekas and Kiosseoglou, 2003). Deproteinized cheese whey concentrates were investigated for their suitability as substrates for the production of single-cell protein with *Kluyveromyces marxianus* CBS 6556 at 100 L fermenter scale (Schultz, Chang, Hauck, Reuss and Syldatk, 2006). Honda, *Fukushi* and *Yamamoto* (2006) optimized the waste water feeding for SCP production in anaerobic waste water treatment process utilizing purple non sulfur bacteria under mixed culture conditions. Maximum protein content was observed when waste water was fed as organism showed maximum growth.

Tray bioreactors are very simple in design, with no forced aeration or mixing for solid substrate and can be excellent tool to be used in SCP production. These types of reactors are restricted in amount of substrate that can be fermented, as only thin layers can be used, in order to avoid over heating and to maintain aerobic condition (Nigam et al., 1996). Wooden trays were initially used for soy sauce production in Koji fermentations by *A. niger* (Pandey et al., 2000ab). The use of tray fermentors has remained largely unchanged, with only engineering advance being the use of modern materials such as plastic, alluminium and stainless steel. The advantage of modern trays over wooden trays is to make sterilization and cleaning easier and therefore reduce contamination or spoilage (Tengerdy and Szakacs, 2002). However the use of tray fermenters in large-scale production is limited as they require a large operational area and tend to labour intensive with mechanical handling also being difficult (Williams, 2002).

Table 5. Some of the industrial establishments involved in the SCP production

Organization	Microorganisms applied	Used substrates for SCP production
Ammoco, USA	C. utilis	Ethanol
Bellyyeast, France	Kluveromyces	Whey
British petroleum, France	Candida tropicalis	Hydrocarbons
Hoechst, Germany	Methylomonas clara	Methanol
ICI, U.K	Methylotrophus	Methanol
IFP, France	Candida tropicalis	n-alkanes
Liquichemica, Italy	Candida maltosa	n-alkanes
Philips, USA	Pichia sp, Torula sp.	Sugar feed stock
Pekilo, Finland	C. utilis	Starches

There have been increasing demand for protein and recent high prices of fish meal and soya meal have shown that there is a definite market for SCP.

Various companies such as British Petroleum, Ranks, Hovis, and McDougal-Dupont, etc. have shown that pilot plant scale production of SCP is feasible (Table 5).

8. APPLICATIONS OF SCP

8.1. As Feed

The feed industry, through its technology in evaluating, processing and using new ingredients, is offering many opportunities to use substrates that were not used for feed production in the past, in order to minimize feed costs (Caldas, 1999). The acquisition of great volumes of these ingredients by feed industries has been an important factor in prompting vegetable processing industries to invest in the processing of by-products that were previously wasted. A feed is usually produced from agricultural products or by-products such as grains, cereals and residues. However, it is necessary to add micro-ingredients to improve levels of essential amino acids, vitamins and minerals (Edelstein, 1982). New additives such as metabolic modifiers, anti-microbial agents, probiotics and special minerals are also incorporated in order to supply essential nutrients, to enhance growth and to avoid diseases (Wenk, 2000). Even if the raw material is low cost, the addition of these micro-ingredients increases the final prices of feed. Therefore in searching for cheap raw materials to be used in feed production, it is crucial to obtain higher nutritional values than traditional substrates, in order to minimize the need for these high cost additives, and thereby maintain the economic viability of animal production (Villas-Boasa, Granato, Espositob and Mitchellc, 2002).

On the other hand, many of the raw materials currently used in feed production represent products that would be advantageous if used in human nutrition (Brum, Bellaver, Zanotto, and Lima, 1999b). Therefore, it is desirable to use lower quality materials as the basic ingredient of animal feeds. However, most of the agro-industrial by-products and food industry wastes are poor in nutrients such as proteins and vitamins and are rich in fibres with low digestibility.

Such materials are not suitable for non-ruminant animals and, in some cases; the digestibility is so low that they are not even suitable for ruminants (Brum Lima, Zanotto and Klein, 1999a; Brum et al., 1999b). In order to solve the problem, microorganisms (mainly fungi) are utilized to convert agro-industrial wastes in order to obtain products with higher nutritive value, especially in regard to protein and vitamin contents, and with increased digestibility (Kuhad, Singh, Tripathi, Saxena and Eriksson, 1997).

Fungi, bacteria and seaweed, cultivated on a large scale, can be used as feed. These microorganisms are very attractive feed stuffs, because they can be cultivated on agro-industrial wastes, with production of large amounts of cells rich in proteins that commonly contain all the essential amino acids, in addition to favorably high vitamin and mineral levels (Kuhad et al., 1997; Brum et al., 1999 a,b). Further, the growth of microbes in lignocellulosic wastes is able to furnish all the hydrolytic enzymes often added in the preparation of feeds, and also makes the minerals more available for absorption by the animal.

According to Pelczar, Chan and Krieg (1996), there are some advantages in using microorganisms in order to obtain nutritionally-enriched raw materials for feed compared to the traditional way of formulating feeds. Microorganisms grow fast and produce protein in large amounts, because the protein content of microbial cells is very high, typically being around 600 g Kg^{-1}. The substrate contain wastes or industrial by-products such as hydrocarbons from oil refineries, liquid effluents from the cellulose and paper industries, beet and sugarcane molasses and hydrolyzed wood wastes (Pelczar et al., 1996; Kuhad et al., 1997).

Microbial by-products from traditional fermentation industries have already been used in various feed preparations for some time. These products are of major importance to the livestock feed industry. Fermentation of by-products from the brewing and distilling industries has long been utilized for livestock, particularly ruminants, mainly as protein and energy supplements. Numerous by-products of other fermentation industries are also used as sources of nutrients in compounded feed (Shaver and Batajoo, 1995). Nevertheless, some microbial products contain toxic metabolites that could represent serious risks for animal and human health. Therefore, the risk must be evaluated in all bioconversion processes in order to avoid possible toxicity to animals and humans (Brum et al., 1999a, b).

SSF of lignocellulosic by-products has an advantage over submerged fermentation in that it requires only one-tenths of the fermenter size as compared with the latter. The dry matter content in the substrate of solid-state fermentation lies between 250 and 300 g Kg^{-1} and that in submerged fermentation between 20 and 30 g kg^{-1} (Moo-Young, Chahal, Swan and Robinson, 1977). The capacity of lignin degradation by fungi is influenced by the penetration of hyphae into the substrate and the extent of close physical contacts between the substrate and the degrading fungus. Fungal mycelium attaches physically to lignin (Tono, Tani and Ono, 1968) and agitation disrupts this contact (Kirk, Schultz, Conors, Lorenz, and Zeikus, 1978). Therefore, SSF or non-agitated culture in thin layers is always preferred over agitated submerged fermentation as far as lignin degradation is concerned.

8.2. SCP for Human Consumption

Prepared *Aerobacter* aerogenes become unpleasantly slimy and not palatable as food. In human feeding studies, consumption of certain SCP sources lead to wide range of gastrointestinal complaints like bulky stools and flatulence, nausea, vomiting and diarrhea from killed and purified bacteria. Other pathological reactions are peeling of the skin of the palm of the hand, soles of the feet and itching, pain and edema of the toe; these symptoms are due to high levels of nucleic acid. This is the major limitation to the use of SCP as food, unless a safe process for removal of nucleic acid is established.

9. Risk Assessment of SCP in Food and Feed

The purpose of toxicological testing is to define the potential hazard that the microbe-based feed presents, not only to animals but also to humans. As direct testing is impracticable, the evaluation is generally done with laboratory animals. It is recommended that a toxicological evaluation should involve both rodent and target species (Hoogerheid, Yamada, Littlehailes and Ohno, 1979), although the understanding of the basic pathology and biochemistry of farm livestock is less well documented than that of the rodent. The extent of the testing procedure required depends upon the chemical analysis of the product. Materials that have a chemical composition close to that of natural occurring raw materials are unlikely to produce abnormal effects. In contrast, those products with abnormal composition resulting from the biosynthesis of compounds not present in natural feeds may need more careful evaluation. Thus, an extensive chemical analysis of the product is required. The potential for biological hazard is low for the microbially converted feeds so far evaluated (Banerjee, Azad, Vikineswary, Selvaraj and Mukherjee, 2000). Nevertheless, this hazard has to be continually evaluated by various biological studies when a new microbially fermented product is proposed. Tests involving rats are valuable during the development stages of the process. These tests may give some preliminary indication of possible cellular mutagenicity, potential carcinogenicity and teratogenicity (Sinskey and Batt, 1987).

The main problem associated with these types of investigation is the level of inclusion of the test material in the diet, and the subsequent effects that this has on the nutrient balance of the test diet (Hoogerheide et al., 1979). It is possible to induce a "toxicological change" due to simple alterations of dietary minerals or energy. These changes could be wrongly ascribed to microbially fermented feed and hence extreme care must be exercised in the formulation of the test diets. The toxicological evidence gained from the laboratory rodent species will form the principle basis for evaluation of hazard. Nevertheless, it is still necessary to test the converted substrate in the diets of farm animals. Long term life-span studies equivalent to those carried out with the rodent are not necessary. The normal economic life span of a farm animal forms a sensible basis for the evaluation of risk in the actual target species (Hoogerheide et al., 1979).

These experiments with farm livestock should follow the same experimental concepts as those employed for the laboratory animal, namely biochemistry, hematology and terminal pathology (Hoogerheide et al., 1979). Mortality beyond that normally experienced in farm livestock must be carefully evaluated. The experimental designs used in these studies should simulate those used for the laboratory animal, but due regard should be paid to the normal practices followed in animal production. The reproductive potential and the viability of the offspring are important both scientifically and commercially. Multigenerational experiments undertaken with both poultry and pigs need to have a sound statistical basis. Specific teratological investigations should also be made in at least one species apart from the rodent (Hoogerheide et al., 1979).

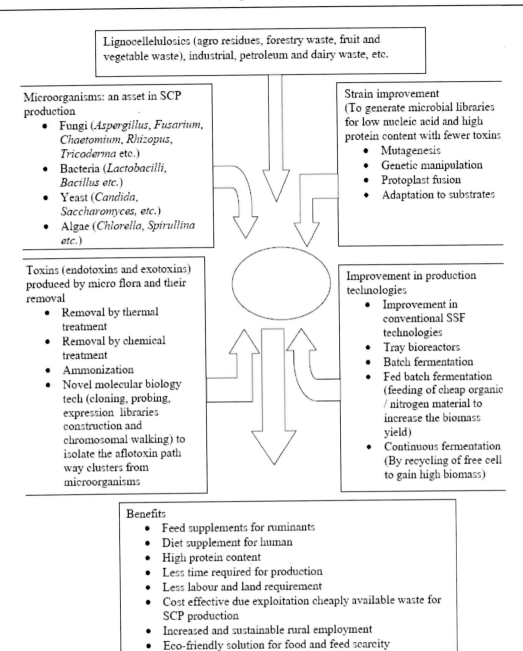

Figure 3. Coordinated action for improvement in SCP production and its long term benefits.

CONCLUSION

The review presents the technological assessments for SCP production, along with the available microorganisms, various lignocellulosic substrates, toxins and their removal, risk assessment, and worldwide industrial scenario of SCP production.

Among the classes of microorganisms used for SCP production, the use of bacteria and fungi is more economical when grown on easily available substrates. The use of cheap and abundantly available sources, viz. lignocellulosic residues, whey and petroleum products are an attractive option over too-costly sources (soymeal and fishmeal, etc.). The successful exploitation of these substrates is beneficial, first, in terms of overall economics of SCP, and, secondly, food production would be less dependent on land. The selection of microorganisms as food or feed depends upon their growth rate with high protein ratio, substrate utilization and toxicity profile. Recent biotechnology-based tools and techniques have been exploited rationally less for SCP production in comparison to other microbial-originated products. The beneficial aspects of SCP over conventional protein sources are its independence from land and climate, its ability to be controlled genetically, and the fact that it causes less pollution, although there are some factors, such as high nucleic acid content, cell wall indigestibility, high color and aroma that impair the use of SCP. Figure 3 presents coordinated actions that are crtically required for the improvement in SCP production and its long-term benefits. However, despite successful laboratory-based research, SCP cannot achieve the desired level of success at the industrial scale. To make the program successful and economically viable, the factors mentioned above need to be addressed for the successful implementation of SCP for food/feed consumption or as a supplement in creating a staple diet.

REFERENCES

Abou-Zeid, A.A., Khan, J.A. and Abulnaja, K.O. (1995). On methods for reduction of nucleic acid content in single cell protein from gas oil. *Biores.Technol.*, 52, 21-24.

Alasheh, S. and Duvnjak, Z. (1995). The effect of phosphate concentration on phytase production and the reduction of phytic acid contents in Canola meal by *Aspergillus carbonarius* during Solid state Fermentation. *Appl. Microbiol. Biotechnol.*, 43, 25–30.

Anderson, J.M, (1975). Succession, diversity and trophic relationships of some soil animals in decomposing leaf litter. *J. Anim. Ecol.*, 44, 475–495

Anupama and Ravindra, P. (2000).Value-added food: Single Cell Protein. *Biotechnol. Advances, 18*, 459-479

Anupama and Ravindra, P. (2001). Studies on production of single cell protein by *Aspergillus niger* in solid state fermentation of rice bran. *Braz. Arch. Biol. Tech.*, 44, 79-88.

Bailey, J.E. and Ollis, D.F. (1986). *Biochemical Engineering Fundamentals*. 2nd Edition. McGraw Hill. Singapore.

Balasubramanian, D., Bryce, C.F.A, Dharmalingam, K., Green, J. and Jayaraman, K. (1998). *Concepts in biotechnology*, Universities Press (India) Limited, COSTED-IBN, pp. 135-174.

Baldensperger, J., Le Mer, J., Hannibal, L. and Quinto, P.J. (1985). Solid-state fermentation of banana wastes. *Biotechnol. Letters, 7*, 743-74

Banerjee, S., Azad, S.A., Vikineswary, S., Selvaraj, O.S. and Mukherjee, T.K. (2000). Phototropic bacteria as fish feed supplement. *Asian Aust. J. Anim. Sci.*, 13, 991–994.

Becerra, M., Cerdan, E. and Gonzalez-Siso, M.I. (1997). Heterologous*Kluyveromyces lacli* beta-galactosidase production and release by *Saccharomyces cerevisiae* osmotic remedial thermosensitive autolytic mutants. *Biochim. Biophys. Acta, 61335* (3), 235-241

Bennett, J. W. and Keller, N. P. (1997). Mycotoxins and their prevention. In: Anke, T [Ed], *Fungal Biotechnology*. An International Thompson Publishing Company, Weinheim, pp. 265-273.

Berge, G.M., Baeverfjord, G., Skrede, A. and Storebakken, T. (2005). Bacterial protein grown on natural gas as protein source in diets for Atlantic salmon, Salmo salar, in saltwater. *Aquaculture, 244* (1-4), 233-240

Bhalla, T.C and Joshi, M. (1994). Protein enrichment of apple pomace by co-culture of cellulolytic moulds and yeasts. *World J. Microbiol. Biotechnol., 10*, 116–117.

Blancou, J., Calvet, H. and Riviere, R. (1978). Single cell protein production from peanut shell. *Rev-Elev-Med-Vet-Pays-Trop., 31* (3), 363-368.

Brum, P.A.R., Bellaver, C., Zanotto, D.L., and Lima G.J.M.M. (1999b). Determinacao de valores de composicao quýmica e da energia metabolize vel em farinhas de carne e ossos para aves. EMBRAPA/CNPSA, *Com. Tec. 239*, 1–2.

Brum, P.A.R., Lima, G.J.M.M., Zanotto, D.L. and Klein, C.H. (1999a). Composicao nutritiva de ingredientes pararacoes de aves. EMBRAPA/CNPSA, *Com. Tec. 241*, 1–4.

Caldas, S.F.N. (1999). Mandioca e resýduos das farinheiras na alimentacao de ruminantes: digestibilidade total e parcial, Master Thesis. Universidade Estadual do Parana, Maringa, Brazil.

Calloway, D.H. (1974). The place of single cell protein in man's diet. In: Davis P, editor. *Single cell protein*. New York: Academic Press, pp. 129–46.

Cary, J.W., Barnaby, N., Ehrlic, K.C. and Bhatnagar, D. (1999). Isolation and characterization of experimentally induced aflatoxin biosynthetic pathway deletion mutants of *Aspergillus parasiticus. Appl. Microbiol. Biotechnol., 51* (6), 587-602

Chae, S.R., Hwang, E.J. and Shin, H.S. (2006). Single cell protein production of Euglena gracilis and carbon dioxide fixation in an innovative photo-bioreactor. Biores Technol., 97, 322-329.

Chahal, D.S., Vlach, D. and Moo-Young, M. (1980). Upgrading the protein feed value of lignocellulosic materials using Chaetomium cellulolyticum in solid-state fermentation.In: Advances in biotechnology. Vol. 2 pp 327-332 Ed: Moo-Young, M, Robinson, C.W, Pergamon press, Torontochahal.

Chamie, J. (2004). World population prospects: The 2002 revision. New York United Nations Department of Economic and Social Affairs Analytical Report III.

Champagnat, A., Vernet, C., Laine, B. and Filosa, J. (1963). Biosynthesis of Protein-Vitamin Concentrates from Petroleum. Nature, 197, 13–14.

Chandel, A.K., Kapoor, R.K., Lakshmi Narasu, M., Viswadevan, V., Saravana Kumaran, S.G., Ravinder Rudravaram., Venkateswar Rao, L., Tripathi, K.K., Lal, B. and Kuhad, R.C. (2007b). Economic evaluation and environmental benefits of biofuel: an Indian perspective, Int. J. Global Energy Issues, (In Press).

Chandel, A.K., Kapoor, R.K., Singh, A.K. and Kuhad, R.C. (2007c). Detoxification of sugarcane bagasse hydrolysate improves ethanol production by Candida shehatae NCIM 3501. Biores Technol., 98, 1947-1950.

Chandel, A.K., Rudravaram, R., Narasu, M.L., Rao, L.V. and Ravindra, P. (2007a). Economics and Environmental Impact of Bioethanol Production Technologies: An Appraisal. Biotechnol Mol Biol Rev, 2, 14-32.

Chaudhary, N. and Sharma Chandra, B. (2005). Production of citric acid and single cell protein from agrowaste. Nat Acad Sc Lett., 28,189-193.

Cochrane, V. W. (1958). Physiology of fungi. John Wiley and Sons, New York.
Considine, P.J., Mehra, R.K., Hackett, T.J., O'Rorke, A., Comerford, F.R and Coughlan, M.P. (1986). Upgrading the value of agricultural residues. Ann. New York Acad. Sci., 469, 304-31.
Cotty, P.J., Bayman, P. and Egel, D.S. (1994). Agricultural, aflatoxins and Aspergillus. The genus Aspergillus. New York: Plenum Press, pp. 1–27.
Dabbah, R. (1970). Protein from Microorganisms. Food Technol, 24 (6), 35.
Dale, B.E. and White, D.H. (1979). Degradation of ribonucleic acids by immobolized ribonuclease. Biotechnol. Bioeng., 21 (9), 1639-1648.
De Boer, L. and Dijkhuizen, L. (1990). Microbial and enzymatic process for L-phenylalanine production. Adv Biochem Eng Biotechnol., 41, 1–27.
De Gregorio, A., Mandalari, G., Arena, N., Nucita, F., Tripodo, M.M. and Locurto, R.B. (2002). SCP and crude pectinase production by slurry-state fermentation of lemon pulps. Biores Technol, 83, 89-94.
Desjardins, A.E., Hohn, T.M. and McCormick, S.P. (1993). Trichothecene biosynthesis in Fusaritim species: Chemical genetics and significance. Microbial Rev., 57, 595-604.
Durand, A. and Chereau, D. (1988). A new pilot reactor for solid-state fermentation: application to the protein enrichment of sugar beet pulp. Biotechnol. Bioeng., 31, 476-486
Eaton, D.L. and Groopman, J.D. (1994). The toxicity of aflatoxins. Human health, veterinary and agricultural significance. Academic Press, San Diego.
Edelstein, H. (1982). Disponibilidades futuras de materias-primas: problemas de quantidade e qualidade. In: Anais do 18 Congresso Brasileiro da Industria de Racoes, Sao Paulo, pp. 145–165.
Ekenvall, L., Dolling, B., Gothe, C.J., Ebbinghaus, L., von Stedingk, L.V. and Wasserman, J. (1983). Single cell protein as an occupational hazard. Br. J. Ind.Med. 40 (2), 212-215.
Erdman, M.D., Bergen, W.G. and Reddy, C.A. (1977). Amino acid profiles and resumptive nutritional assessment of single cell protein from certain lactobacilli. Appl Environ Microbiol., 33 (4), 901–5.
Fabregas, J. and Herrero, C. (1985). Marine microalgae as a potential source of single cell protein (SCP). Appl Microbiol Biotechnol., 23.
Fossum, K. and Almlid, T. (1977). Single-cell protein as a bacterial substrate. Acta Pathol Microbiol Scand (B), 85B (5), 350.
Ghaly, A.E. and Kamal, M.A. (2004). Submerged yeast fermentation of acid cheese whey for protein production and pollution potential reduction. Water Res., 38, 631–644
Gow, J.S., Littlehailes, J.D., Smith, S.R.L. and Walter, R.B. (1975). SCP production from methanol: Bacteria. In: Tannenbaum, S. R. and Wang,D.I.C [Eds], Single cell protein Vol. II, MIT Press, Cambridge, Mass, pp. 370-384.
Gupte, A. and Madamwar, D. (1997). Solid state fermentation of lignocellulosic waste for cellulase and b-glucosidase production by cocultivation of Aspergillus ellipticus and Aspergillus fumigatus. Biotechnol Prog., 13, 166–9.
Han, I.Y and Steinberg, M.P. (1986). Solid-state yeast fermentation of raw corn with simultaneous koji hydrolysis. Biotechnol. Bioeng. Symp., 17, 449-462
Han, Y.W. and Srinivasan, V.R. (1968). Isolation and Characterization of a Cellulose-utilizing Bacterium. Appl Microbiol., 16 (8), 1140–1145.
Hill, F. (1994). Yeast with high natural glutamic acid content. Patent EP592785.

Honda, R., Fukushi, K. and Yamamoto, K. (2006). Optimization of wastewater feeding for single-cell protein production in an anaerobic wastewater treatment process utilizing purple non-sulfur bacteria in mixed culture condition. J. Biotechnol., 27, 565-573.

Hoogerheide, J.C., Yamada, K., Littlehailes, J.D. and Ohno, K. 1979. Guidelines for testing of single cell protein destined as protein source for animal feed (II). IUPAC, 51, 2537–2560.

Humphrey, A.E. "Product Outlook and Technical Feasibility of SCP," in S.R. Tannenbaum and D.l.C. Wang, eds, Single-Cell Protein ll (MIT Press, Cambridge, Mass. USA, and London, 1 975), pp. 1-23.

Joffe, A.Z. (1986). Fusarium species: Their biology and toxicology. John Wiley and Sons, New York.

Jungehulsing, U., Arntz, C., Smit, R. and Tudzynski, P. (1994). The Claviceps purpurea glyceraldehyde-3-phosphate dehydrogenasen gene cloning, characterization and use for improvement of a dominant selection system. Curr. Genet, 25, 101–6.

Kale, S.P., Bhatnagar, D. and Bennett, J.W. (1994). Isolation and characterization of morphological variants of Aspergillus parasiticus deficient in secondary metabolite production. Mycol. Res., 98, 645-652.

Kapitza, S.P. (1996). The phenomenological theory of world population growth. PHYS-USP, 39, 57-71.

Kawaguchi, K. (1980). Microalgae production systems in Asia. In: Algae biomass. Eds. Shelef, G. and Soeder, C.J Elsevier, North Holland Biomedcal Press.

Keller, N.P., Cleveland, T.E. and Bhatnagar, D. (1992). A molecular approach towards understanding aflatoxin production. In: Bhatnagar D, Lillehoj EB, Arora DK, editors. Handbook of applied mycology, vol. 5. Mycotoxins in ecological systems. New York: Marcel Dekker, pp. 287–310.

Kirk, T.K., Schultz, E., Conors, W.J., Lorenz, L.F. and Zeikus, J.G. (1978). Influence of culture parameters on lignin metabolism by Phanerochaete chrysosporium. Arch. Microbiol., 117, 277–285.

Koehnle, T.J., Russell, M.C. and Gietzen, D.W. (2003). Rats rapidly reject diets deficient in essential amino acids. The J Nutr, 133, 2331-2335.

Kourkoutas, Y., Sipsas, V., Papavasiliou, G. and Koutinas, A.A. (2007). An Economic Evaluation of Freeze-Dried Kefir Starter Culture Production Using Whey. J. Dairy Sci., 90, 2175-2180.

Kuhad, R.C. and Singh, A. (1993). Lignocellulose biotechnology: current and future prospects. Crit Rev Biotechnol., 13, 151–172.

Kuhad, R.C., Singh, A., Tripathi, K.K., Saxena, R.K., Eriksson, K.E.L., (1997). Microorganisms as an alternative source of protein. Nutr. Rev., 55, 65–75

Kurbanoglu, E.B. and Algur, O.F. (2002). Single-cell protein production from ram horn hydrolysate by bacteria. Bioresource Technol., 85, 125-129.

Labuza, T. P. (1975). Cell collection, recovery and drying for single cell protein manufacturing. In: Tannenbaum, S. R. and Wang, D. I. C [Eds], Single cell protein, Vol. 2, MIT Press, Cambridge, MA, pp. 69-104.

Lehninger, A.L. (1990). Principles of Biochemistry. CBS Publishers and Distributors Pvt. Ltd. New Delhi, India.

Leonard, J. and Compere, P. (1967). Spirulina plantensis (Gom), Algue Bleue de Grande Valeur Alimentaire par sa Richessse en Proteines. Bull. Jard. Bot. Nat. Belg. 37 (Suppl.), 1.

Levy-Costa, R.B., Sichieri, R., dos Santos, Pontes, N. and Monteiro, C.A. (2005). Household food availability in Brazil: distribution and trends (1974-2003). Rev. Saúde Pública, 39.

Lin, Y. and Tanaka, S. (2006). Ethanol fermentation from biomass resources: Current state and prospects. Appl. Microbiol. Biotechnol., 69, 627-642.

Lipinsky, E.S. and Litchfieid, J.H. (1974). Single-Cell Protein in Perspective. Food Technol., 2815, 16.

Martinez, M.C., Sanchez-Montero, J.M, Sinisterra, J.V. and Ballesteros, A. (1991). New insolubilized derivatives of ribonuclease and endonuclease for elimination of nucleic acids in single cell protein concentrate. Biotechnol. Appl. Biochem., 12 (6), 643-652.

Martinez-Force, E. and Benitez, T. (1993). Regulation of aspartate derived amino acid biosynthesis in yeast Saccharomyces cerevisiae. Curr. Microbiol. 26, 313-322.

Mitchell, D.A. and Lonsane, B.K. (1992). Definition, characteristics and potentials.In: Duelle, H. W, Mitchell, D. A. and Rolz, C. E [Eds], Solid substrate cultivation. Elsevier Science Publishers Ltd. pp. 1-16.

Mitchell, D.A., Gumbira-Said, E., Greenfield, P.F. and Doelle, H.W. (1991). Protein measurement in solid state fermentation. Biotechnology Techniques, 5, 437-442.

Moeini, H., Nahvi, I. and Tavassoli, M. (2004). Improvement of SCP production and BOD removal of whey with mixed yeast culture. Electronic Journal of Biotechnology, 7.

Moller, A. (1994). L-Tryptophan production froms anthranilic acid by amino acid auxotrophic mutants of Candida utilis. Process Biochem., 29, 521-527.

Moo-Young, M., Chahal, D.S., Swan, J.E. and Robinson, C.W. (1977). Single cell protein production by Chaetomium cellitlolyticum, a new thermotolerant cellulolytic fungus. Biotechnol. Bioeng., 19 (4), 527-538

Moreno, J.M., Sanchez-Montero, J.M., Ballesteros, A. and Sinesterra, J.V., 1991. Hydrolysis of nucleic acids in single cell protein concentrates using immobilized benzonases. Biotechnol. Appl. Biochem., 31 (1), 43–51.

Murby, M., Uhlen, M. and Stahl, S. (1996). Upstream strategies to minimize proteolytic degradation upon recombinant production in Escherichia coli. Protein Expr. Purif., 7 (2), 129-136.

Nicolini, L., Volpe, C., Pezzotti, A. and Carilli, A. (1993). Changes in in-vitro digestibility of orange peals and distillery grape stalks after solid state fermentation. Bioresource Technol., 45, 17–20.

Nigam, P. and Singh, D. (1996). Processing of agricultural wastes in solid state fermentation for microbial protein production. J. Sci. Ind. Res., 55, 373–380.

Nigam, P. and Vogel, M. (1990). Process for the production of beet pulp feed by fermentation. European Patent DE 3812612 C2 1.3.

Pandey, A., Soccol, C.R. and Mithchell, D. (2000a). New developments in Solid state Fermentation: I – Bioprocess and products. Proc. Biochem., 35, 1153–1169.

Pandey, A., Soccol, C.R., Nigam, P., Soccol, V.T., Vandenberghe, L.P.S. and Mohan, R.(2000b). Biotechnological potential of agro-industrial residues. II: cassava bagasse. Biores Technol., 4 (7), 81-87.

Paraskevopoulou, A., Athanasiadis, I., Kanellaki, M., Bekatorou, A., Blekas, G. and Kiosseoglou, V. (2003). Functional properties of single cell protein produced by kefir microflora. Food Res Int., 36, 431-438.

Park, D.L. and Liang, B. (1993). Perspective on aflatoxin control for human and animal feed. Trends in Food Sci. Technol., 4, 334-341.

Pelczar Jr, M.J., Chan, E.C.S., Krieg, N.R. (1996). Microbiologia de alimentos: microrganismos como alimentos. In: Pelczar Jr, M.J, Chan, E.C.S, Krieg, N.R. (Eds.), Microbiologia Conceitose Aplicacoes, Vol. II. Makron Books, Sao Paulo, pp. 390–391

Penbaloza, W., Molina, M.R., Brenes, R.G. and Bressani, R. (1985). Solid-state fermentation: an alternative to improve the nutritive value of coffee pulp. Appl. Environ. Microbiol., 49, 388-393.

Powar, C.B. and Daginawala, H.F. (1995). General microbiology, Vol. 2. Himalaya Publishing House, Bombay, pp. 88-131.

Rahmat, R.A. Hodge, G.J. Manderson, P.L. and Yu. (1995). World J. Microbiol. Biotechnol., 11, 168–170.

Ravinder, R. PhD Thesis (2003). Solid state fermentation of de-oiled rice bran for protein enrichmente

Ravinder, R., Chandel, A.K., Venkateshwer, L.V. and Ravindra P. (2006). Optimization of protein enrichment of De-oiled rice bran by solid state fermentation using Aspergillus oryzae MTCC 1846. Int. J. Food Eng., 2 (4).

Ravinder, R., Venkateshwar Rao, L., and Ravindra, P. (2003). Production of SCP from Deoiled Rice Bran. Food Technol. Biotechnol., 41 (3), 243–246.

Reu, J.C. de. (1995). Solid-substrate fermentation of soya beans to tempe. Thesis Agricultural University Wageningen.

Robinson, T. and Poonam, N. (2002). Bioreactor design for protein enrichment of agricultural residue by solid state Fermentation. Biochemical Engineering Journal, 3647, 1–7.

Rodriguez, J.A., Echevarria, J., Rodriguez, F.J., Sierra, N., Daniel, A. and Martinez, O. (1985) Solid-state fermentation of dried citrus peel by Aspergillus niger. Biotechnol. Letters, 7, 577-580

Rudravaram, R., Chandel, A.K., Linga, V.R. and Ravindra, P. (2006). Optimization of Protein Enrichment of Deoiled Rice Bran by Solid State Fermentation Using Aspergillus oryzae MTCC 1846. International Journal of Food Engineering, 2 (4), 1.

Sahm, H. (1995). Metabolic design in the amino acid producing bacterium Corynebacterium glutamicum. Folia Microbiol., 40, 23-30.

Saucedo-Castaneda, G., Gutierrez-Rojas, M., Bacquet, G., Raimbault, M. and Vinegra-Gonzalez, G. (1990). Biotechnol. Bioeng., 35, 802–808.

Schlegel, H.G. (1996). General Microbiology, University Press, Cambridge, pp.357-384.

Schultz, N., Chang, L., Hauck, A., Reuss, M. and Syldatk, C. (2006). Microbial production of single-cell protein from deproteinized whey concentrates. Appl Microbiol Biotechnol., 69, 515-520.

Shaver, R.D. and Batajoo, K.K. (1995). Fermented feeds and feed products. In: Rehm, H.J., Reed, G., Puhler, A. and Stadler, P. (Eds.). Biotechnology, Vol. 9, 2nd Edition. VCH Edition, Weinheim, Germany, pp. 769–793.

Singh, B.D. (1998). Biotechnology, Kalyani Publishers, New Delhi, pp. 498-510

Singh, O.V. and Kumar, R. (2007). Biotechnological production of gluconic acid: future implications. Appl. Microbiol. Biotechnol., 75, 713-722.

Sinskey, A.J. and Batt, C.A. (1987). Fungi as a source of protein. In: Benchat, L.R. (Ed.), Food and Beverage. Von Nostrand Reinhold, New York, pp. 435–471.

Smail, T., Salhi, O. and Knapp, J.S. (1995). Solid state fermentation of Carob pods by Aspergillus niger for protein production: Effect of particle size. World J. Microbiol. Biotechnol., 11, 171–173.

Smil, V. (2005). The next 50 years: Unfolding trends. Population and Development Review, 31, 605-643.

Smith, J.F., Fermor, T.R. and Zadraz¡il, F. (1988). Pretreatment of lignocellulosics for edible fungi. In: Zadraz¡il, F, Reiniger, P. (Eds.), [In] Treatment of Lignocellulosics with White-Rot Fungi. Elsevier, Essex, UK, p. 3–13.

Soccol, C.R., Leon, J.R., Rouses, S. and Raimbault, M. (1993). Growth kinetics of Rhizopus in solid state fermentation of treated cassava. CI. Technol. Letters, 7, 563- 568.

Solomans, G.L (1985). Production of biomass by filamentous fungi. In: Moo-Young, M, Bull, A. T. and Dalton, H [Eds], Comprehensive Biotechnology, Vol.3. Pergamon Press, pp. 483-505.

Srinivasan, V.R. and Fleenor, M.B. (1972). Fermentative and Enzymatic Aspects of Cellulose Degradation. [In] Developments of Industrial Microbiology (American Institute of Biological Sciences, Washington, D.C, USA), pp. 47-53.

Tannenbaum, S.R. and Wang, D.I.C. (1975). Single cell protein, vol. II. MIT Press, Cambridge, MA.

Taylor, J.C, Lucas, E.W, Gable, D.A. and Graber, G. (1974). Evaluation of single cell protein for non-ruminants. In: Davis, P [Ed], Single cell protein. Academic Press, New York, pp. 179-186.

Tengerdy, R.P. and Szakacs, G. (2003). Bioconversion of lignocellulose in solid substrate fermentation. Biochemical Engineering Journal, 13, (2-3), 169-179

Tono, T., Tani, Y., Ono, K. (1968). Microbial treatment of agricultural wastes. Part I. Adsorption of lignin and clarification of lignin containing liquor by moulds. J. Fermen. Technol., 46, 569–576.

Trevelyan, W.E. (1976a). Chemical methods for the reduction of the purine content of the baker's yeast, a form of single cell protein. J. Sci. Food Agric., 27 (3), 225-230.

Trevelyan, W.E. (1976b). Autolytic methods for the reduction of the purine content of baker's yeast, a form of single cell protein. J. Sci. Food Agric., 27 (8), 753–762.

Tsukada, Y. and Sugimori, T. (1971). Induction of auxotrophic mutants from Candida species and their application to L-threoninefermentation. Agr Biol Chem., 35, 1–7.

Tuse, D. (1984). Single- Cell Protein: current status and future prospects. Crit. Rev. Food Science Nutrition, 19, 273-325.

van der Westhuizen, T.H. and Pretorius, W.A. (1996). Production of valuable products from organic waste streams. Water Science Technol., 33 (8), 31–38.

Varga, J, Kevel, E, Rinyu, E, Teren, J. and Kozakiewicz, Z. (1996). Ochratoxin production by Aspergillus species. Appl. Environ. Microbiol., 12, 4461-4464.

Venkatratnam, L.V. (1978). Photosynthetic productivity in mass outdoor cultures of alfae. Proc. Of Int. Symp. On Biological Application of solar Energy, held at Madurai, India.

Viesturs, U.E., Apsite, A.F., Laukevics, J.J., Ose, V.P., Bekers, M.J. and Tengerdy, R.P. (1981). Solid-state fermentation of wheat straw with Chaetomium cellulolyticum and Trichoderma lignorum. Biotechnol. Bioeng. Symp., 11, 359-369.

Villas-Boasa, Granato, S., Espositob, E. and Mitchellc, D.A. (2002). Microbial conversion of lignocellulosic residues for production of animal feeds. Animal Feed Science and Technology, 98e

Wenk, C., (2000). Recent advances in animal feed additives such as metabolic modifiers, antimicrobial agents, probiotics, enzymes and highly available minerals-review. Asian Aust. J. Anim. Sci., 13, 86–95.

Williams, J.A. (2002). EPS: Environmental and Production Solutions, LLC. (www.cepmagazine.org)

Worgan, J.T. (1973). In: Jones, J.G.W. (Eds). The Biological Efficiency of Protein Production. Cambridge University Press, Cambridge, UK, pp. 339-361.

Yanez, E., Ballester, D., Fernandez, N., Gattos, V. and Mönckeberg, F. (1972). Chemical Composition of C utilis and the Biological Quality of the Yeast Protein. J. Sci. Food Agric., 23, 581-586

Yang, H.H., Thayer, D.W. and Yang, S.P. (1979). Reduction of endogenous nucleic acid in single cell protein. Appl. Environ. Microbiol., 38 (1), 143-147.

Yu, J., Butchko, R.A.E. and Fernandes, M. (1996). Conservation of structure and function of aflatoxin regulatory gene aflR from Aspergillus nidulans and Aspergillus flavus. Curr. Gene, 29, 549-555

Zadrazil, F., Brunnert, H. and Grabbe, K. (1983). Edible mushrooms. In: Rehm, H.-J, Reed, G. (Eds.). Biotechnology: A Comprehensive Treatise, Vols. 1–8. Weinheim, Germany, pp. 145–187.

Zeringue, H.J. (Jr.). (1996). Possible involvement of lipoxygenase in a defense response in aflatoxigenic Aspergillus-Cotton plant interactions. Can. J. Bot., 74, 98-102.

Ziino, M., Curto, R.B.L., Salvo, F., Signorino, D., Chifalo, B. and Giuffrida, D. (1999). Lipid composition of Geotrichum candidum. Single cell protein grown in continuous submerged culture. Biores. Technol., 67, 7–11

ZoBell, D.R., Olson, K.C., Wiedmeier, R.D. and Stonecipher, C. (2004). The Effect on Digestibility and Production of Protein and Energy Supplementation of Stocker Cattle on Intensively-Managed Grass Flood-Meadow Pastures. Ag/2004/Beef-02, Utah State University

In: Agricultural Wastes
Eds: Geoffrey S. Ashworth and Pablo Azevedo

ISBN 978-1-60741-305-9
© 2009 Nova Science Publishers, Inc.

Chapter 8

COFFEE PROCESSING SOLID WASTES: CURRENT USES AND FUTURE PERSPECTIVES

Adriana S. Franca and Leandro S. Oliveira*

Departamento de Engenharia Mecânica/UFMG
Av. Antônio Carlos, 6627, 31270-901 Belo Horizonte, MG, Brazil

ABSTRACT

The term "coffee" is applied to a wide range of coffee processing products, starting from the freshly harvested fruit (coffee cherries), to the separated green beans, to the product of consumption (ground roasted coffee or soluble coffee). Coffee processing can be divided into two major stages: primary processing, in which the coffee fruits are de-hulled and submitted to drying, the resulting product being the green coffee beans. This is the main product of international coffee trade, and Brazil is the largest producer in the world with production values ranging from 2 to 3 million tons in the years from 2003 to 2007. During this primary processing stage solid wastes are generated, which include coffee husks and pulps, and low-quality or defective coffee beans. Secondary processing includes the stages that comprise the production of roasted coffee and soluble coffee. The major solid residue generated in this stage corresponds to spent coffee grounds from soluble coffee production. These solid residues (coffee husks, defective coffee beans and spent coffee grounds) pose several problems in terms of adequate disposal, given the high amounts generated, environmental concerns and specific problems associated with each type of residue. Coffee husks, comprised of dry outer skin, pulp and parchment, are probably the major residues from the handling and processing of coffee, since for every kg of coffee beans produced, approximately 1 kg of husks are generated during dry processing. Defective beans correspond to over 50% of the coffee consumed in Brazil, being used by the roasting industries in blends with good-quality coffee. Unfortunately, since to coffee producers they represent an investment in growing, harvesting, and handling, they will continue to be dumped in the internal market in Brazil, unless alternative uses are sought and implemented. Spent coffee grounds are produced at a proportion of 1.5kg (25% moisture) for each kg of soluble coffee. This solid residue presents an additional disposal problem, given that it can be used for adulteration of

---

* E-mail:adriana@demec.ufmg.br. Tel:+55-31-34093512. Fax:+55-31-34433783.

roasted and ground coffee, being practically impossible to detect. In view of the aforementioned, the objective of the present study was to present a review of the works of research that have been developed in order to find alternative uses for coffee processing solid residues. Applications include direct use as fuel in farms, animal feed, fermentation studies, adsorption studies, biodiesel production and others. In conclusion, a discussion on the advantages and disadvantages of each proposed application is presented, together with suggestions for future studies and applications.

I. INTRODUCTION

The most quoted definition for sustainability comes from the Brundtland report: "Sustainable development is a new form of development which integrates the production process with resource conservation and environmental enhancement. It should meet the needs of the present without compromising our ability to meet those of the future" (WECD, 1987). The increasing focus on sustainable agricultural and industrial procedures has resulted recently in an extensive number of new bio-based initiatives. Such interest has been continuously stimulated by environmental pressures and a shift towards the use of agricultural-based raw materials, as well as rapid developments in the science supporting biotechnology (Rogers et al., 2005). In that regard, there are currently several studies being developed on the use of agricultural and industrial solid wastes in compliance with the current need for sustainable development, as extensively discussed in several recent reviews on the subject (Salminen and Rintala, 2002; Laufenberg et al., 2003; Kapdan and Kargi, 2006; Yang, 2007; Tiruta-Barna et al., 2007; Huang et al., 2007; Hargreaves et al., 2008; Oliveira and Franca, 2008; Rebah et al., 2007)

Agricultural and food solid wastes are high volume, low value materials that are highly prone to microbial spoilage, thus limiting their exploitation. Their use is also precluded by legal restrictions and the costs of collection, drying, processing, storage and transportation. Thus, for the most part, these materials are either used as animal feed, combustion feedstock or disposed to landfill, causing major environmental issues (Oliveira and Franca, 2008). However, in recent years, there has been an increasing trend towards more efficient utilization of agri-food residues (Thiagalingam and Sriskandarajah, 1987; Jimenez and Gonzalez, 1991; Pandey et al., 2000ab; Russ and Meyer-Pittroff, 2004; Ioannidou and Zabaniotou, 2007; Matteson and Jenkins, 2007; Prasad et al., 2007). Newer applications include biofuel production, fermentation studies, production of activated carbons and others.

Coffee is deemed a commodity ranking second only to petroleum in terms of currency (usually US dollars) traded worldwide (Illy, 2002). As such, this commodity is quite relevant to the economy of producing coutries, including Brazil, Vietnam, Colombia, Ethiopia, Indonesia, Mexico and India (Franca and Oliveira, 2008). Brazil is the largest coffee producer and exporter in the world, and is the second largest consumer. The production of coffee in Brazil in the last five years ranged from 2.0 to 2.7 million tons (MAPA, 2008). Such production represents an average of over 2.5 million tons of solid residues being generated every year. These solid residues (coffee husks and pulp, defective coffee beans and spent coffee grounds) pose several problems in terms of adequate disposal, given the high amounts generated, environmental concerns and also specific problems associated with each type of residue. In this regard, several studies have been undertaken and are still being developed in terms of alternative uses for such solid residues. A brief review on coffee processing is presented as follows, for a better understanding of solid waste generation during coffee processing.

1.1. Coffee Processing

A schematic view of a coffee cherry is displayed in Figure 1. Two coffee beans are usually found in each fruit, and each bean is covered with a thin closely fitting skin called silverskin. A second yellowish skin, the parchment, loosely covers the silveskin, the whole being encased in a pulp which forms the flesh of the cherry. The green coffee, which is the product of commercial interest, constitutes only 50–55% of the dry matter of the ripe cherry (Vincent, 1987). The remaining material is diverted to various by-products (solid wastes) depending on the processing technique (i.e., dry or wet methods) used.

The general steps involved in the processing of coffee cherries (primary coffee processing) are displayed in Figure 2. There are two major methods: dry and wet processing. Dry processing is the simplest technique for processing coffee cherries. After harvesting, the coffee cherries are dried to about 10–11% moisture content. Thereafter, the coffee beans are separated by removing the material covering the beans (outer skin, pulp, parchment and silverskin) in a de-hulling machine. Generated solid residues are denominated coffee husks (outer skin + pulp + parchment) and silverskin. This processing method is employed for Robusta coffees, but also for the majority of the Arabica coffees processed in Brazil. This is attributed to the simplicity of the method, coupled with the availability of large areas for sun-drying the coffee cherries. Furthermore, this method is convenient where fruits are picked by the stripping method, with ripe, overripe and unripe fruits being simultaneously picked (Vincent, 1987). Drying can be accomplished by either "natural" or "artificial" methods. Natural or sun-drying is the method commonly employed in large farms.

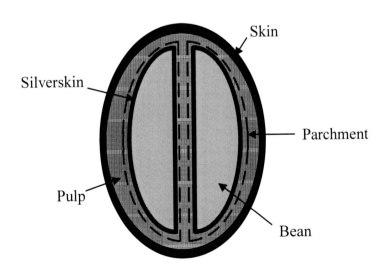

Figure 1. Schematic view of a coffee fruit.

It has the advantages of not requiring equipment investments or energy costs. However, it requires large drying areas (usually concrete surfaces). The process is slow, ranging from three to four weeks, with the berries usually spread out on a thin layer in order to avoid fermentation. Frequent raking is also required to avoid mould proliferation and provide homogeneous drying conditions. Artificial drying can be employed either as substitute or as a complement to natural drying. Several types of equipment are employed, including static, rotary, horizontal and vertical dryers, and the process can be continuous or in batches. A detailed description of the most commonly employed types of dryers can be found in coffee processing textbooks (Sivetz and Desrozier, 1979; Vincent, 1987; Borém, 2008).

Wet processing, on the other hand, does not require drying of the cherries themselves. In this type of processing, first the outer skin and pulp are mechanically removed, thus generating the solid residue, denominated coffee pulp. The beans can be fermented to remove a layer of remaining pulp material, with the processed coffee being called pulped coffee (café despolpado) or can be directly submitted to drying, with the final product being called de-hulled cherry coffee (café cereja descascado). In both cases, after drying to approximately 12% moisture content the beans are again de-hulled to remove the parchment. The resulting solid waste (parchment and silverskin) is collectively termed parchment husks.

The operations that are carried out subsequently to removing the coffee husks/pulp are denominated curing (Vincent, 1987) and include cleaning, size grading, density and colorimetric sorting, and finally storage of the green coffee beans (see Figure 2). Electronic color sorting is the major procedure employed for separation of defective and non-defective coffee beans (Franca and Oliveira, 2008). In the electronic sorters, coffee beans pass, one by one, by an electronic eye or camera system, and depending on wavelength measurements, the bean is either allowed to pass or it is shot with a puff of air into a reject pile. This reject pile will be separated as a mixture of defective (low quality) coffee beans prior to commercialization in external markets, and such mixture is usually dumped on the Brazilian internal market, being employed by the roasting industry in blends with good quality coffee.

After separation from the exportable portion, such beans may be representing more than 50% of the coffee consumed in Brazil (Oliveira et al., 2008a). The presence of defective beans results in a significant decrease in beverage quality, and thus the overall quality of the roasted coffee consumed in Brazil is low (Oliveira et al., 2006).

The next coffee processing step that presents a problem in terms of solid waste generation is termed secondary processing and corresponds to the production of soluble or instant coffee. In this process, roasted and ground coffee beans are treated with pressurized hot water in order to extract the soluble material, which is then submitted to either spray-drying or freeze-drying, and the solid final product (soluble or instant coffee, respectively) is then obtained (Clarke, 1987ab). The insoluble residue (a slurry containing spent coffee grounds) is screw pressed, so the moisture content is reduced from 75-80% to approximately 50%.

A detailed discussion on each specific type of residue (coffee husks and pulp, defective coffee beans and spent coffee grounds, including specifics of their chemical composition and possible solutions for adequate use and disposal is presented throughout the remainder of this chapter.

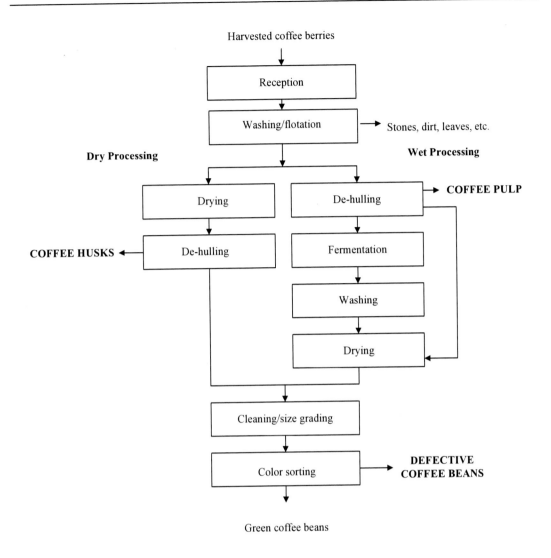

Figure 2. Flow-sheet illustrating primary processing of coffee (adapted from Vincent, 1987).

II. COFFEE HUSKS AND PULP

Coffee husks and pulp are comprised of the outer skin and attached residual pulp, and these solid residues are obtained after de-hulling of the coffee cherries during dry or wet processing, respectively. The moisture content will vary depending on the type of processing. Dry processed coffee husks present moisture contents ranging from 7 to 18%, with such extensive range being attributed to variations in processing and storage conditions (Oliveira et al., 2001; Vilela et al., 2001; Souza et al., 2001, 2003, 2005; Rocha et al., 2005). Wet processed coffee husks (coffee pulp) contain approximately 75% moisture, and are usually let to dry to approximately 13% moisture (Adams and Dougan, 1987; Barcelos et al., 2001). Average values for the chemical composition (dry basis) of coffee husks and pulp are displayed on Table 1.

The high contents of carbohydrates are expected, given the origin of such solid residue, i.e., fruit pulp and outer skin. Some authors also make reference to a specific type of coffee husks known as sticky coffee husks (Gouvea et al., 2008; Vilela et al., 2001; Oliveira et al., 2002; Carvalho, 2008; Parra et al., 2008). Some attributes that differentiate this specific type of coffee husks from the regular ones include its higher density, protein contents and lower fiber contents. However, the major difference relies on its sugar content (see Table 2), which encourages research studies in association with specific applications such as animal feed and fermentation studies.

Even though coffee husks and pulp are rich in organic nature and nutrients, they also contain compounds such as caffeine, tannins, and polyphenols (see Table 1). Due to the presence of the latter compounds, these organic solid residues present toxic nature, which not only adds to the problem of environmental pollution, but also restricts its use as animal feed (Pandey et al., 2000a). Caffeine is an active compound, being one of the nature's most powerful stimulants. It is the major substance to which the stimulation effect of coffee is attributed. It is also present in coffee husks at approximately 1.3% concentration on dry weight basis (Pandey et al., 2000b). Tannins are generally thought to be an anti-nutritional factor and to prevent coffee husks from being used at percentages over 10% in animal feed.

Table 1. Chemical composition of coffee husks and pulp (g/100 g dry basis)[a]

	Coffee Husks (Dry processed)	Coffee Pulp (Wet processed)
Protein	8-11	4-12
Lipids	0.5-3	1-2
Minerals	3-7	6-10
Carbohydrate	58-85	45-89
Caffeine	~1	~1
Tannins	~5	1-9

[a] Compilation of data presented by Adams and Dougan, 1987; Clifford and Ramirez-Martinez, 1991; Pandey et al., 2000; Oliveira et al., 2001; Vilela et al., 2001; Souza et al., 2001, 2003, 2005; Rocha et al., 2005; Gouvea et al., 2008.

Table 2. Sugar contents of coffee husks and pulp (% dry basis)

Coffee pulp	Reducing sugars	14	Adams and Dougan (1987)
	Total sugars	12	
	Sucrose	2	
Coffee husks	Reducing sugars	14	Adams and Dougan (1987)
	Total sugars	12	
	Sucrose	2	
Sticky coffee husks	Reducing sugars	24	Gouvea et al. (2008)
	Total sugars	29	
	Sucrose	4	

Such compounds are found in coffee husks, at approximate levels of 1 and 2.3% for Arabica and Robusta species, respectively (Clifford and Ramirez-Martinez, 1991). According to Teixeira (1992), tannins will react with proteins in the ruminant digestive system, thus affecting the ability to digest cellulose, protein and dry matter.

Barcelos and co-workers (2001) evaluated the levels of caffeine, tannins, lignin and silica in dry and wet processed coffee husks, for Brazilian Arabica coffees, var. Catuaí, Rubi, and Mundo Novo. Wet processed coffee husks were submitted to sun drying down to 13% moisture. After one year storage, there was a 12% increase in caffeine levels, 39% reduction in tannin levels and a slight decrease in lignin content. Dry processed husks presented higher levels of silica in comparison to wet processed, which was attributed to the presence of parchment. It was concluded that 12-month storage improved the quality of coffee husks. However, the increase in caffeine levels was considered a limiting factor in using this residue for animal feed. A detailed description of the specific uses that have been and are currently under evaluation is presented as follows.

2.1. Livestock Feed

Agricultural residues are commonly reused as livestock feed (Nonhebel, 2007). For example, in the Netherlands, approximately 70% of the concentrates fed to pigs, cattle and poultry are based on residues generated by the food processing industry. In the case of coffee husks, it has been pointed out that, its low digestible protein content, in addition to the fact that the starch equivalent is comparable to low quality hay, has prevented its use as animal feed (Adams and Dougan, 1987). However, given the high amounts generated, coupled to its availability at low costs throughout the season, several studies have evaluated its use as a diet supplement for cattle (Jarquin et al., 1973; Barcelos et al., 1999; Filho et al., 2000, 2004; Vilela et al., 2001; Souza et al., 2005, 2006ab; Teixeira et al., 2007), swine (Okai et al., 1985; Oliveira et al., 2001, 2002; Parra et al., 2008; Carvalho, 2008), fish (Bayne et al., 1976; Christensen, 1981; Fagbenro and Arowosoge, 1991; Moreau et al., 2003; Ulloa Rojas and Verreth, 2003) sheep (Demeke, 1991; Furusho-Garcia et al., 2003; Souza et al., 2004) and chicken (Bressani et al., 1977). A summarized description of some of the more recent studies is presented below.

Feeding and digestibility studies were conducted in concrete ponds to evaluate the use of coffee (*Coffea robusta*) pulp as partial and total replacements for yellow maize in low-cost diets for catfish (Fagbenro and Arowosoge, 1991). The percentages of dietary coffee pulp inclusion were 0 (diet 1, control), 10 (diet 2), 20 (diet 3) and 30 (diet 4). All the diets were formulated to contain 37% crude protein using cheap, locally available feed ingredients, and fed daily to experimental fish in replicated dietary treatments at 2–10% of fish biomass for 150 days. Fish survival was high, consequent upon good water quality in all treatments. There were significant differences ($P<0.05$) in growth performance (daily rate of growth, DRG, and specific growth rate, SGR), feed conversion rate (FCR) and apparent nutrient digestibility coefficient (ADC) among fish fed with coffee pulp (CoP) diets and control (yellow maize, YeM) diet. Mean net fish production decreased with progressive increases in percent coffee pulp inclusion in the diets. It is concluded that coffee pulp is potentially useful as replacement for yellow maize in low-cost diets for *C. isheriensis*.

Filho et al. (2000) evaluated the effect of adding coffee husks to animal feed as a substitute for a mixture of corn grain, husks and cobs. Twenty-five Holstein-zebu steers (250 kg average weight) were evaluated for a period of 102 days, with coffee husks added to the feed at the following ratios: 0%, 10%, 20%, 30% and 40% of coffee husks. No significant differences were observed for live weight gain, feed conversion, difference between receipt/cost and total dry matter and crude protein intake and neutral detergent fiber of the roughage and total diet. Neutral detergent fiber intake increased with the increase in the amount of coffee husks. Even though no significant differences in costs were observed, the ratio between receipt and cost was slightly lower for the feed including 40% coffee husks. The authors concluded that employing coffee husks in as a supplement for feeding Holstein-zebu steers was effective up to 30% substitution.

Oliveira et al. (2001) evaluated the technical and economical viability of the inclusion of coffee husks as a corn substitute in isoenergetic diets for growing and finishing pigs. The metabolism essays were conducted with 12 crossbred barrows in growing phase (35 kg average weight) and 12 in finishing phase (35 kg average weight). A total of ninety six crossbred barrows and gilts (35 kg average weight) were employed in the performance essay. Coffee husks were employed as a corn substitute in isoenergetic diets at the following levels: 0, 5, 10 and 15%. An increase in the amount of coffee husks resulted on a linear decrease of the digestibility coefficients of dry matter, crude protein, fiber neutral detergent, nitrogen retention and also of the energetic values of the diets. Both the weight gain and feed intake also showed a linear decrease with the corresponding increase in the percentage of coffee husks in the diet. Results from this study indicated that employment of coffee husks in swine diet is both technically and economically feasible up to 5% levels.

The study by Vilela and co-workers (2001) evaluated the effect of employing of sticky coffee husks (SCH) as a roughage (sugar cane and elephant grass) substitute in cattle diet. Sticky coffee husks were added at the following levels: 0, 15, 30 and 60%. The use of coffee husks affected the intake of dry matter, live weight gain and income:expense ratio but there was no influence on food conversion. The results indicated that it is feasible to substitute roughage by coffee husks for cattle feed up to 40%.

Oliveira et al. (2002) also evaluated the technical viability of the inclusion of sticky coffee husks for swine feed. Metabolism and performance essays were conducted, employing feeds with sticky coffee husks added as a substitute for corn, at the following levels: 0, 5, 10, 15 and 20%. It was observed that an increase in SCH ratio in the feeds provided an increase in feed conversion and a linear reduction in daily weight and carcass fat deposition.

The effect of adding coffee husks (0, 8.7, 17.4, 26.1, and 34.8 kg/ 100 kg of fresh forage) on the proximate composition and in-vitro digestibility of elephant-grass silage was evaluated by Souza et al. (2003). The percentage of dry matter increased linearly with increasing the amount of coffee husks, with an average 0.6% increase. Adding coffee husks resulted on a decrease in pH and increase in crude protein levels, up to 17.4% level. A slight decrease of in-vitro digestibility was observed. The authors concluded that coffee husks is a good additive for ensiling elephant-grass with high moisture content, up to 17.4% added.

Furusho-Garcia and co-workers (2003) evaluated the effect of adding coffee husks to lamb feeds. Twelve crossbred Texel x Bergamácia (TB), 12 crossbred Texel x Santa Inês (TS) and 12 purebred Santa Inês (SI) lambs were individually finished in feedlot for 50 days (from 130 to 180 days of age). Three diets were evaluated: (1) grass (*Pennisetum Purpurium*) silage without coffee husks (control), (2) control diet with 15% added dry coffee husks, and (3) control diet with 15% added urea treated coffee husks. The authors reported that the rumen/reticulum of the animals fed diets 1 and 3 were significantly heavier than those fed diet 1. The abomasums, liver and pancreas of the lambs fed diet 3 were lighter than the animals fed diet 2. The weights of skin, fat, esophagus/trachea, heart and lung were also analyzed and were not affected by the diets, genetic groups and sex. It was concluded that adding coffee husks to the diet provided better development of the rumen/reticulum. The urea treatment provided a reduction of the weight of abomasums, liver and pancreas, organs related to enzymatic digestion.

The study by Moreau et al. (2003) evaluated the effect of adding fresh and ensiled coffee pulps in fish—Nile Tilapia—diet. The inclusion of coffee pulp impaired growth, dietary protein and energy utilization of all food ingredients. Silage process significantly improved growth and feed utilization comparing results obtained with fresh coffee pulp. The general conclusion of this study was that either fresh or ensiled coffee pulp should not be considered as suitable feedstuffs for Nile tilapia. However, the study by Ulloa Rojas and Verreth (2003) showed that, if fish are reared in earthen ponds or pens instead of concrete tanks or raceways, coffee pulp will have a potential as a feed ingredient in fish diets. It was concluded that inclusion of coffee pulp in tilapia diets is feasible up to 130g kg^{-1}, provided that the fish are raised in earthen ponds and natural food is available.

Filho et al. (2004) evaluated the effect of adding coffee husks on the kinetics of ruminal digestion of Holstein cows. Coffee husks at different levels (0, 10, 20, 30 and 40%) were employed for replacing a mixture of ground corn, straws and cobs (GCSC) in iso-energetic and iso-protein feeds. The authors reported that the effective degradabilities of both the feed dry matter and protein increased with the increase in coffee husks levels. However, adding coffee husks to the feed resulted on a reduction of effective degradability of fiber. It was concluded that the replacement of GCSC by coffee husks in cow diet can be satisfactorily accomplished up to the level of 40%.

Evaluation of the use of coffee husks as a substitute for ground corn concentrate in sheep diet was performed by Souza et al. (2004). Coffee husks were added at four different levels (0.0, 6.25, 12.5, 18.75 and 25.0% DM) and the effects on intake and apparent digestibility of diets were observed. Twenty sheep of unknown breed were fed with iso-protein diets, consisting of 10% crude protein (CP), 60% of *coastcross* hay and 40% of corn concentrate. Sheep were maintained in a metabolism cage for 19 days (12 days of adaptation and 7 days of data collection). The intakes of dry matter (1.41), organic matter (1.34), CP (0.15), total carbohydrate (1.17), neutral detergent fiber (0.71) and non-fiber carbohydrate (0.45) and the total digestible nutrients (0.85) kg/day were not affected by the presence of coffee husks in the diet. Coffee husks did not affect apparent digestibility of dry matter (60.1%), organic matter (62.1%), neutral detergent fiber (46.9%), CP (66.3%), total carbohydrate (61.5%) and non-fiber carbohydrate (84.1%). Therefore, it was concluded that coffee husks could be employed as a corn substitute in this specific type of diet up to a level of 25%. In a later study, the same research group studied the effect of adding coffee husks as a substitute for ground corn concentrate in cows diet (Souza et al., 2005). The effects of replacing ground corn with coffee husks on intake, apparent digestibility, milk production and composition were observed, with coffee husks being added at the following levels: 0.0, 8.75, 17.5, and 26.25% of dry mater (DM). Diets were formulated to be iso-proteic (14% protein) and contained a forage:concentrate ratio of 60:40 on DM basis. The inclusion of coffee husks in the diet presented no significant effects on intakes of DM, organic matter, protein levels, and total carbohydrates. However, intake of neutral detergent fiber increased linearly when coffee husks replaced ground corn in the diet. Apparent total tract digestibilities of DM, protein, total carbohydrate and fibers presented a linear decrease with increasing the levels of coffee husks in the diet from 0.0 to 26.25%. Adding coffee husks increased nitrogen excretion in the feces, resulting in a negative nitrogenous balance. The incremental levels of coffee husks did not affect the urinary excretions of allantoin, uric acid, and purine derivatives as well as microbial protein synthesis. Milk yield and contents and yields of fat, protein, and total solids were also not affected by replacing ground corn with coffee husks. The authors concluded that coffee husks can be added to cow feeds in levels up to 10.5% of the total dietary DM.

Souza et al. (2006ab) evaluated the effects of replacing ground corn with coffee husks on nitrogen balance, microbial protein synthesis, nutrient intake, apparent total tract digestibility, and daily weight gain of lactating dairy cows. Twelve crossbred Holstein-Zebu cows yielding on average 23.4 kg/day of milk were used for evaluation of the nitrogen balance and microbial protein synthesis, estimated by excretion of purine derivatives in urine and milk. The animals were fed diets containing coffee husks at the following levels (% DM): 0.0, 3.5, 7.0 or 10.5%. Regression analysis showed no effects of dietary coffee hulls levels on total nitrogen intake (441.3 g/day) and excretion of urine nitrogen (190.8 g/day) and milk nitrogen (114.7 g/day). However, feeding coffee husks to lactating dairy cows increased fecal nitrogen excretion, resulting on a negative nitrogen balance. The increased dietary levels of coffee husks did not affect excretions of milk allantoin (294.6 mmol/day), urinary allantoin (21.3 mmol/day), uric acid (42.3 mmol/day), and purine derivatives (358.2 mmol/day). Microbial protein synthesis estimated by urinary excretion of purine derivatives averaged 266.3 g/day and did not differ across diets. In addition, efficiency of microbial protein synthesis averaged 136.8 g of microbial nitrogen per kg/TDN and also did not differ among diets.

The objective of the second trial was to investigate the effects of partially replacing ground corn with coffee husks (0.0, 8.75, 17.5 and 26.25% of concentrate DM) on nutrient intake, apparent total tract digestibility, and daily weight gain of dairy heifers. Twenty-four Holstein x Zebu heifers were assigned to a randomized complete block design with four treatments and six replicates. Diets were iso-nitrogenous (15.5% CP) and contained 60% of Tifton-85 haylage and 40% of concentrate. Intakes of dry matter (6.75 kg/day), organic matter (6.23 kg/day), CP (1.04 kg/day), total carbohydrates (5.01 kg/day), and neutral detergent fiber (3.11 kg/day) were not changed whereas those of nonfiber carbohydrates (NFC) and total digestible nutrients (TDN) decreased linearly by inclusion of coffee husks in the diet. Apparent total tract digestibilities of dry matter, organic matter, CP, total carbohydrates, neutral detergent fiber, NFC, and TDN all decreased linearly by partially replacing ground corn with coffee husks. Daily weight gain decreased 6.94 g per each percentage unit of coffee husks added to the diet. However, inclusion of up to 17.5% of coffee husks (7.0% of the diet DM) maintained daily weight gain similar to those obtained with 0.0 and 8.75% of coffee husks in the concentrate.

Teixeira et al. (2007) evaluated the effect of replacing corn silage by coffee husks (up to 21%) on performance and total apparent digestibility of dairy heifers. Twenty-four dairy Holstein heifers were employed in the study, being daily fed 2.0 kg concentrate. Dry mass intake increased linearly, while fresh matter intake was not affected by the inclusion of coffee husks in the diets. Digestibilities of DM, OM, CP, CHO, NDF, concentration of NDT in the diet and weight gain presented a linear decrease with the replacement of corn silage by coffee husks. It was concluded that coffee husks could replace corn silage up to 14% in diets for dairy heifers.

Parra and co-workers (2008) presented a comparison of the use of sticky (SCH) and dry coffee husks (DCH) as a feed supplement for growing and finishing pigs. The effects on the performance and carcass quality were evaluated, for inclusion levels of 2.50, 5, 10, 15 and 20%. For pigs in the growing phase, adding either SCH or DCH led to a linear decrease on daily weight gain. SCH presented higher digestibility coefficients in comparison to DCH. For the animals in finishing phase, linear reductions of hot carcass weight and ham weight were observed with the increase in SCH levels in the feeds. Results indicate that the inclusion of SCH can be satisfactorily accomplished at levels of 5 and 9.5% for pigs in growing and finishing phases, respectively. Up to those levels, SCH inclusion is economically viable, presents no effect on performance, and produces leaner meat carcasses.

Carvalho (2008) evaluated the effect of adding sticky coffee husks silage to swine feeds. Experiments were carried out in order to determine the nutritional value and to verify the performance of starting, growing and finishing pigs fed with sticky coffee hull silage (SCHS). The treatments consisted of five diets with increasing levels (0, 4, 8, 12 and 16%) of SCHS. There were no effects of SCHS inclusion on daily fed intake, daily weight gain, feed:gain ratio and plasma urea nitrogen. The study of economic viability indicated the possibility of SCHS inclusion up to 4%. It was concluded that SCHS presents good nutritional value and can be included up to 16% in nursery piglets (15-30 kg) diets without impairing performance. In the growing phase there were no effects of inclusion levels of SCHS on daily fed intake, daily weight gain, feed:gain ratio and plasma urea nitrogen. In the finishing phase, the feed:gain ratio showed a improved with the increasing SCHS inclusion levels. The daily fed intake, daily weight gain and plasma urea nitrogen and loin depth did not differ between SCHS inclusion levels. The results of carcass traits indicated a quadratic effect on back-fat thickness and marbling score of the *longissimus dorsi*. There was a linear increase on weight of empty stomach according to the addition of SCHS on diets. These results suggested that SCHS presented good nutritional value and could be included up to 16% in starting, growing and finishing pigs diets, without impairing performance and carcass traits. However the economic feasibility of using SCHS will depend of feedstuffs prices.

The overall conclusion of the previously discussed studies is that employment of coffee husks as animal feed is limited due to seasonal availability and perishability due to high water content in the case of wet processing, but the major limiting factor is the presence of anti-nutritional compounds, thus restricting the amount that can be used. In that regard, some studies have been conducted with the purpose of removing anti-nutritional substances such as caffeine and tannins from coffee husks. A summarized review of such studies is presented as follows.

1.2.1. Detoxification Studies

The precise nature of the anti-physiological effects of coffee husks and pulp is still unknown, but they are mostly attributed to the combined effect of compounds such as caffeine - stimulatory and diuretic effects - and tannins - decrease in protein availability and inhibition of digestive enzymes (Adams and Dougan, 1987). In that regard, some studies are available on detoxification of coffee husks and pulp by either physical, chemical or biological methods (Molina et al., 1974; Brand et al., 2000; Pandey et al., 2000a; Mazzafera, 2002; Ulloa Rojas et al., 2002, 2003; Gokulakrishnan and Gummadi, 2006; Orozco et al., 2008).

The physiological effects of caffeine in humans and animals have been extensively studied (Bonati et al., 1985; Buckholtz et al., 1987; Landolt et al., 1995; Nehlig et al., 1999; Stavric, 1988, 1992; Meyer and Caston, 2004), and their symptoms have been reported to be similar to those observed in animals fed diets containing coffee husks (Jarquín et al., 1973; Molina et al.,

1974; Bressani et al., 1977). Reviews on the removal of caffeine from coffee husks and pulp are available by Pandey et al. (2000a) and Mazzafera (2002). A general review of enzymatic and microbial methods for caffeine removal is presented by Gokulakrishnan et al. (2005). Comments on some recent studies that are specific for coffee pulp detoxification and were not covered on such reviews are presented below.

Different chemical treatments were evaluated by Ulloa Rojas et al. (2002) with the objective of improving the nutritional value of coffee pulp. The treatments were: (1) alkali (NaOH solutions of 5 and 10% for 24 and 48 h), (2) acid/alkali (treatment with HCl, 1.5 and 3 M for 24 and 48 h, followed by a NaOH solution of 5% for 48 h) and (3) alkali/ensilage (treatment with NaOH solution of 5% for 48 h, followed by ensilage with molasses for up to 3 months). The coffee pulp submitted to both the alkali and acid/alkali treatments presented higher contents of ash, fat and cellulose and lower contents of anti-nutritional factors: polyphenols, tannins and caffeine. The true protein content was reduced in the acid/alkali treatment. The alkali/ensilage treatment resulted in higher true protein, fat and ash contents, no variations in cellulose contents and reductions in polyphenols, tannins and caffeine contents. The reduction of anti-nutritional factors was much higher for the alkali treatment in comparison to alkali/ensilage, with the first providing the best overall results in upgrading the nutritive value of coffee pulp.

The subsequent study by Ulloa Rojas and co-workers (2003) presents a comparison of biological treatments applied to fresh coffee pulp with the purpose of reducing its content of cellulose and anti-nutritional factors (total phenols, tannins and caffeine). The evaluated treatments were: (1) ensiling molasses for 2 and 3 months, (2) aerobic decomposition for up to 42 days, and (3) aerobic bacterial inoculation (*Bacillus* sp.) for up to 28 days. Ensiling treatment resulted in increases in fat, ash and protein contents and decreases in cellulose, tannins and total phenols levels. There was an overall improvement in the nutritional quality of the coffee pulp in association with higher fat and protein contents and lower contents of cellulose, total phenols and tannins. However, the treatment did not present a significant effect on caffeine content. The aerobic decomposition treatment improved the nutritional quality of coffee pulp by increasing true protein and fat contents. Total phenols, tannins, caffeine and cellulose contents were reduced by an increase in treatment time. The bacterial treatment increased the protein content of coffee pulp after 21 days, but decreased it after 28 days. Cellulose, total phenols, tannins and caffeine contents reduced with an increase in time of bacterial degradation. It was concluded that both the aerobic decomposition and the aerobic bacterial degradation seemed to be more suitable to improve the nutritional quality of coffee pulp in comparison to the ensiling procedure, given the reduction in caffeine levels.

Orozco et al. (2008) evaluated the ability of three *Streptomyces* strains to upgrade the nutritional value of coffee pulp residues from Nicaragua by solid-state fermentation. After the growth of the strains, there was a significant decrease in the amount of total polyphenols derived compounds. The analysis of these compounds demonstrated that both monomethoxy- and dimethoxy-phenols were degraded. In addition, the authors observed an increase in the Kjeldahl-based protein of the microbial treated coffee pulp, thus indicating that the treatment was recommended for improving the usefulness of coffee pulp for animal feeding purposes. However, no comments on the effect of the treatment on caffeine content were made.

2.2. Silage

Coffee husks and pulp are rich in potassium (~40 g kg^{-1}) and other mineral nutrients, which has induced some studies on the application of these solid residues as organic fertilizers without any treatment or after composting (Matos, 2008). The use of coffee husks directly as soil coverage is a good option for potassium depleted soils, and can be employed for different types of crops, including coffee. They favor erosion control, decrease temperature fluctuations and water losses by evaporation (Matos, 2008). However, there is always the possibility of phytotoxin production. According to Matos (2008), it is possible to calculate the amount of coffee husks that can be employed, if data on potassium levels in the coffee husks and the recommended levels and recovery efficiency of potassium for the specific crop are available.

Couto Filho et al. (2007) prepared a residue silage based on a mixture of mango production solid residues (husks and residual pulp) mixed with coffee husks at three levels of addition (10, 20 and 30%). It was observed that addition of coffee husks provided an increase in dry matter content and decrease in buffering power. The values of pH and ammonium nitrogen (N-NH$_3$/total N) were within the standard range of a good fermentation process. It was concluded that coffee husks could be added up to 30%, improving the fermentative standard for silages of good quality. The same research group also evaluated a silage based on a mixture of passion fruit solid waste (husks and seeds) with coffee husks added up to 25% (Neiva Júnior et al., 2007). Other additives were also tested including sugar cane bagasse and corn cobs, but only the coffee husks provided an increase in protein contents. All the produced silages were considered to be of average or good quality.

2.2.1. Composting

Composting can be defined as a solid waste management system that accelerates the process of decomposition. In the case of coffee pulp, decomposition occurs spontaneously and if not controlled can result in severe problems including the proliferation of flies and foul odors, soil infiltration and others (Adams and Dougan, 1987). Controlled composting, however, will provide a final product that can be easily handled, stored and applied to the land without the previously mentioned adverse effects.

Composting of coffee pulp have been described in some early studies and is extensively discussed by Adams and Dougan (1987). In the case of coffee husks, Matos (2008) reported that coffee husks have low values for the carbon nitrogen ratio (C/N), indicating that residues with high C/N should be mixed with coffee husks in order to guarantee a good quality final product. It is noteworthy to point out that the produced compost, using either coffee husks or pulp, should be viewed more as a soil conditioner rather than as a fertilizer. Furthermore, it has the physical effect of increasing soil water retention and should improve or at least conserve the long-term quality of the soil (Adams and Dougan, 1987).

2.3. Fuel

Coffee husks have been deemed as a source of cheap fuel with an approximate calorific value of 16 MJ kg^{-1} (Adams and Dougan, 1987). The study by Saenger et al. (2001) investigated the combustion of coffee husks, employing single particle combustion techniques and also combustion in a pilot-scale fluidized bed facility. It was observed that coffee husks present a high content of volatile matter and low contents of fixed carbon and ash. The devolatilization begins at low temperatures (170–200°C), but the increase in particle temperature is quite fast. This means that the coffee husks devolatilize easily upon heating and therefore may require a water cooled feeding system or a very short residence time in the feeder, in order to prevent pyrolysis of the husks taking place in the feeding system leading to blockages and non-uniform fuel flow into the furnace. Because of the tendency to devolatilize rapidly, the feeding systems should be designed such as to achieve uniform distribution of the coffee husks within the cross-section of the furnace to avoid high temperatures near the feeding point and achieve a more or less uniform distribution of heat release in the furnace.

Given that coffee husks are lighter and smaller than wood chips and coals burned in non-pulverized firing systems, there may be a tendency of the particles to be carried out of the furnace with the flue gas. Also, the low density of coffee husks is an indication that such residue is not appropriate for long distance transportation, given the high costs of transportation and storage, so the authors suggest that it should be used in the vicinities of the production area. High values for NO_x emissions (400–500 mg/m^3) were measured, indicating the need for NO_x emission reduction techniques when burning coffee husks. Finally, one of the problems associated with the combustion of coffee husks is the low melting temperature of ash due to the high content of K_2O. Problems of agglomeration, fouling, slagging and corrosion will be expected. Therefore, the authors suggest further research on the use of additives, proper furnace design and co-firing with coal, in order for the direct use of coffee husks as fuel to become viable.

Other studies on the pyrolysis of dry-processed coffee husks were presented by Domíngues et al. (2007) and Menéndez et al. (2007). It was found that the pyrolysis of this solid residue gives rise to a larger yield of the gas fraction compared to the other fractions, even at relatively low temperatures. The gas fraction increased with an increase in pyrolysis temperature. A comparison of microwave-assisted pyrolysis and conventional pyrolysis showed that microwave treatment produces more gas and less oil than conventional pyrolysis. In addition, the gas from the microwave has much higher H_2 and syngas ($H_2 + CO$) contents than those obtained by conventional pyrolysis in an electrical furnace, with CO_2 being the main product. It was also observed that the energy accumulated in the gas increased with the pyrolysis temperature. By contrast, the energy accumulated in the char decreases with the temperature, with this effect being more significant when microwave pyrolysis was employed.

Magalhães et al. (2008) evaluated the use of coffee husks mixed with fire wood as a solid fuel for heating air. Combustion was incomplete, resulting in thermal losses and indicating that the gases could not be employed for direct heating. Nevertheless, combustion efficiency was considered satisfactory for indirect heating, indicating that this solid residue could be employed for heating air to be used for grain drying or other agricultural purposes.

2.4. Fermentation Studies

Several applications of coffee pulp and husks in fermentation studies have been reported, including the production of enzymes (Antier et al., 1993; Minjares-Carranco et al., 1997; Sabu et al., 2006; Niladevi and Prema, 2008), citric acid (Shankaranand and Lonsane, 1994;

Vandenberghe et al., 2000), gibberellic acid (Machado et al., 2004) and flavoring substances (Soares et al., 2000).

Some of the earlier studies on fermentation of coffee pulp and husks have been focused on the production of enzymes. Antier et al. (1993) and Minjares-Carranco et al. (1997) worked with mutant strains of Aspergillus niger to produce pectinases by Solid State Fermentation (SSF) and submerged fermentation (SmF) of coffee pulp. The study by Sabu et al. (2006) presented a comparative evaluation of tamarind seed powder, wheat bran, palm kernel cake, and coffee husks, as substrates for tannase production by *Lactobacillus* sp. ASR S1 under Solid State Fermentation. Maximum tannase production was obtained when SSF was carried out using coffee husks supplemented with 0.6% tannic acid, demonstrating the potential of this solid residue in fermentation studies. Niladevi and Prema (2008) concluded that coffee pulp was the best substrate for laccase production by submerged fermentation employing Streptomyces psammoticus MTCC 7334.

Coffee husks were evaluated by Shankaranand and Lonsane (1994) as a substrate for citric acid production by Aspergillus niger CFTRI under Solid State Fermentation. The authors reported an approximate production of 1.5 g citric acid/10 g dry coffee husks at a conversion of 82% (based on sugar consumed), thus demonstrating the potential of this specific application for coffee husks. A comparative study employing sugar cane bagasse, coffee husks and cassava bagasse as substrates for citric acid production by a culture of Aspergillus niger was presented by Vandenberghe et al. (2000). There was high sugar consumption by A. niger for all substrates (90%, 97% and 87% for sugar cane bagasse, cofee husks and cassava bagasse, respectively), indicating that the strain presented good affinity with all of them. However, citric acid production was highest for cassava bagasse (88 g/kg dry matter) in comparison to sugar cane bagasse (48.7 g/kg dry matter) for sugar cane bagasse and coffee husks (12.7 g/kg g/kg dry matter).

Coffee husks were also tested for flavor production in fermentation studies employing *Ceratocystis fimbriata* (Soares et al., 2000). The coffee husks were steam treated (100°C for 40 min) for removal of inhibitory substances (e.g., caffeine) and supplemented with glucose. It was reported that adding 20 and 35% glucose provided the development of a strong pineapple aroma, with total volatiles production of 6.6 and 5.2 mmol $L^{-1}g^{-1}$, respectively. At 46% glucose, only a weak banana odor was detected, and total volatiles production was reduced. The addition of leucine provided an increase in total volatiles production (ethyl acetate and isoamyl acetate) and resulted on a stronger banana odor. The biosynthesis of volatile compounds was not improved by the addition of soybean oil and was reduced by that of mineral salts. It was concluded that steam treated coffee husks is an adequate substrate for aroma production by C. fimbriata, provided that some glucose is added to increase volatiles production.

Machado et al. (2002, 2004) evaluated the feasibility of employing coffee husks as a substrate for the production of gibberellic acid (GA) in both solid-state fermentation and submerged fermentation tests. The best results were obtained by solid state fermentation with G. fujikuroi LPB-06, employing a mixed substrate of coffee husks and cassava bagasse (7:3, dry wt). GA production was high in comparison to other solid state fermentation studies employing different substrates.

2.4.1. Ethanol Production

Because of its environmental benefits, bioethanol is regarded as a promising biofuel substitute for gasoline in the transportation sector. However, to make it competitive with fossil fuels, it is necessary to reduce production costs by using new, alternative biomass feedstocks. Current industrial processes for bioethanol production still use sugarcane (Southern Hemisphere) or cereal grains (Northern Hemisphere) as feedstocks, but they have to compete directly with the food sector (Wheals et al., 1999). Although these are the predominant feedstocks that are used today, projected fuel demands indicate that new, alternative, low-priced feedstocks are needed to reduce ethanol production costs, since the price of feedstock contributes more than 55% to the production cost (Palmarola-Adrados et al., 2005). Furthermore, it is estimated that ethanol production from agricultural residues could increase in 16 times the current production (Kim and Dale, 2004).

Given the high concentration of carbohydrates in coffee husks, it can be viewed as a potential raw material for bioethanol production. Furthermore, the produced ethanol could be employed for biodiesel production based on coffee oil obtained from defective coffee beans, thus further contributing for the implementation of sustainable development in the coffee and biodiesel production chains (Oliveira et al., 2008a). However, the production of ethanol from coffee husks has not been adopted on a practical scale.

Early studies have indicated that fermented coffee pulp juice contained only 2.5–3.0% w/v of ethanol, which would implicate in high energy costs during the distillation stage (Adams and Dougan, 1987). However, the recent study by Gouvea et al. (2008) demonstrated that fermentation of sticky coffee husks lead to a product containing 14% w/v of ethanol. This was only a preliminary feasibility study, employing baker's yeast as a fermentation agent. However, ethanol production was comparable to other agricultural residues that are being studied for bioethanol production (see Table 3). It can be observed that production of ethanol by fermentation of sticky coffee husks was quite satisfactory in comparison to literature data for other residues, given that most of the other materials were either supplemented with sugar or underwent hydrolysis. The results obtained by Gouvea et al. (2008) indicate that coffee husks present an excellent potential for residue-based ethanol production, given that ethanol production levels can be significantly improved by enzymatic hydrolysis, the use of other microorganisms and nutrient supplementation.

Table 3. Comparison of literature data on residue-based ethanol production

Residue	Ethanol production	Productivity (g / L h)	Reference
Corn stover*	16.8 g/L	—	Ohgrem et al. (2007)
Sweet sorghum*	16.2 g/L	—	Ballesteros et al. (2004)
Sticky coffee husks	14 g/L (9 g/100g)	1.2	Gouvea et al. (2008)
Wheat stillage**	11 g/L	—	Davis et al. (2005)
Sweet sorghum	5 g/100g	—	Sree et al. (1999)
Corn stalks*	5 g/L	1.3	Belkacemi et al. (2002)
Barley straw*	10 g/L	1.3	Belkacemi et al. (2002)

*residue underwent hydrolisis;
** glucose supplemented residue.

2.5. Production of Biogas

The potential of using biogas as a viable alternative source of energy has been widely recognized (Appels et al., 2008). Biogas is the name of a mixture of CO_2 and CH_4, which is produced by bacterial conversion of organic matter, mostly manure and organic wastes, under anaerobic conditions (Raven et al., 2007). Even thought there are biogas plants being currently used, both economic and technical data available indicate that the profitability of many anaerobic digesters is still border line (Pauss and Nyns, 1993; Tafdrup, 1995; Dewil et al., 2006), even though current techniques are being developed to upgrade quality and to enhance energy use (Appels et al., 2008). Some early studies have been developed on the use of coffee husks and pulp for biogas production in anaerobic digestion, with reviews on the subject being presented by Adams and Dougan (1987) and Pandey et al. (2000b). Even though solid coffee residues have been reported to present better performance in terms of CH_4 in comparison to other agricultural residues, the lack of more recent studies in association with coffee processing residues is an indication that this alternative use does not seem to be viable, either due to technical or economical setbacks.

2.6. Production of Mushrooms

Some studies on agri-food solids waste use have focused on the use of such residues as substrates for mushroom growth (Pandey et al., 2000ab). In that regard, there are a few recent studies on the use of coffee husks and pulp for such purposes (Leifa et al., 2001; Salmones et al., 2005; Fan et al. 2006).

Leifa et al. (2001) presented a comparative study employing coffee husks and spent coffee grounds as substrates for the production of edible mushrooms Flammulina. Ideal moisture content for mycelial growth was 60%. First fructification occurred after 25 days of inoculation and the biological efficiency reached about 56% with two flushes after 40 days. Caffeine and tannin contents decreased by 10 and 20%, respectively, after 40 days, with such decrease being attributed to degradation by the culture. The authors concluded that both coffee husks and spent coffee grounds, without any nutritional supplementation, are potential substrates for cultivation of edible fungi.

Salmones et al. (2005) presented a comparative study of cultivation of different strains of Pletorus sp. (P. djamor, P. ostreatus, P. pulmonarius) on coffee pulp and wheat straws. Furthermore, the coffee pulp degraded by mushroom cultures might present further post-production value in the elaboration of forages, vermicomposts, and organic fertilizers.

The study by Fan and co-workers (2006) evaluated the effect of caffeine and tannins on Pleurotus sp. cultivation, in order to ascertain about the feasibility of using coffee husks as substrate for mushroom growth. The increase in caffeine concentration led to a decrease in mycelial growth and biomass production. Furthermore, Pleurotus did not degrade the caffeine, but absorbed it. Tannin under 100 mg/L in the medium stimulated the growth of mycelia, but presented a negative effect above 500 mg/L. It was confirmed that Pleurotus had the capacity of degrading tannic acid. Caffeine content in the husk after cultivation was reduced to 61% and tannins to 79%. The obtained results indicated the feasibility of using coffee husk without any pretreatment for the cultivation of Pleurotus, and also confirmed the results by Salmones et al.

(2005) in the sense that, given the reduction in caffeine and tannins, the coffee husks could afterwards be employed for other uses, including animal feed.

2.7. Adsorption Studies

The association between coffee husks and the production of charcoal has been thought firstly in terms of fuel, given that the calorific value is practically doubled after carbonization (Adams and Dougan, 1987). There was even a development of a commercial product in Kenya based on carbonization of parchment and silverskin, denominated Kahawa coal. However, given the significant amount of published information and increasing research interest on the use of agri-food residues in the preparation of activated carbons (Ioannidou and Zabaniotou, 2007; Oliveira and Franca, 2008), a few recent studies have dealt with the application of coffee husks as either biosorbents (used as adsorbents without the need for carbonization and activation) or as raw materials for the production of adsorbents (Oliveira et al., 2008bcd).

Untreated coffee husks were used by Oliveira et al. (2008b) as potential biosorbents for treatment of dye contaminated waters. Methylene blue was the model dye used in a batch adsorption study. The coffee husks were impregnated with formaldehyde in order to reduce leaching of organic matter and avoid fungal growth on the surface of the biosorbent (Chen and Yang, 2005). The pH of the biosorption system did not present significant effects on the adsorption capacity for values above the determined pH_{PZC} value (~4.5) being, thus, an indication that other mechanisms than ion exchange may be taking place. Evaluation of thermodynamics parameters indicated that adsorption was spontaneous and endothermic. Coffee husks presented excellent adsorption capacity, being more effective than other agricultural residues such as rice husks and wheat shells. Therefore, the major conclusion of that study was that coffee husks presented great potential as an inexpensive and easily available alternative adsorbent for the removal of cationic dyes in wastewater treatments.

Oliveira et al. (2008c) also evaluated the performance of coffee husks as adsorbents for the removal of heavy metal ions from aqueous solutions. The adsorption studies were conducted in batch system using divalent copper, cadmium, zinc and hexavalent chromium as adsorbates. Coffee husks presented better adsorption performance for low concentrations of all metal ions studied. Coffee husks maximum adsorption capacity was compared to the maximum capacity of other biosorbents presented in the literature, being higher than other untreated residues such as sugarcane bagasse, cocoa shell, banana and orange peel, and peanut hulls. The effect of the initial pH in the biosorption efficiency was verified in the pH range of 4–7, and it was demonstrated that the highest adsorption capacity occurred at distinct pH values for each metal ion. Also, the pH of the metal ions solution was monitored before and after sorption tests, and no significant variations were observed, and this fact was attributed to a buffering action by leached potassium. The amount of leached potassium determined after sorption of Cr(VI) indicated that ion exchange may play an important role in the chromium adsorption process. In a later study, thermal and chemical treatments were evaluated in order to improve the adsorption performance of coffee husks (Oliveira et al., 2008d). The activating agent was $CaCl_2$ and carbonization was carried out at 200°C. It was observed that the percent adsorbed of Cr(VI) ions increased after the treatments. The thermal/chemical treatments did not improve adsorption performance in the case of copper, cadmium and zinc adsorption. It was also reported that adsorption efficiency was reduced by the presence of parchment among the husks.

2.8. Other Applications

Coffee husks were employed by Isaac et al. (2007) as mulches for weed management (Commelina diffusa Burm. infestations) in bananas (Musa spp.) cultivated under the Fairtrade system. The non-living mulches evaluated (banana mulch and coffee husks), and a clear plastic mulch, were the best weed management alternatives, providing the highest levels of control. The results obtained in this study indicate that the use of coffee husks mulches is an interesting alternative for sustainable pesticide free production.

Prata and Oliveira (2007) investigated the potential of fresh coffee husks as sources of anthocyanins for applications as natural food colorants. Extraction of pigments was carried out in successive steps, employing an acidified (HCl) methanol solution as the extractant. The pigment extracts were concentrated in a rotary evaporator at 35°C, being also submitted to vacuum treatment in order to minimize pigment degradation by oxidation. The extracted pigments were analyzed by HPLC with photodiode array detection. Cyanidin 3-rutinoside was characterized as the dominant anthocyanin in fresh coffee husks, and its quantification indicated that fresh coffee husks can be considered as a source of this pigment.

3. DEFECTIVE COFFEE BEANS

The removal of defective coffee beans is the last processing step in order to guarantee a good-quality coffee. Such beans are usually associated with specific problems during harvesting and post-harvest processing operations and are known to impart a negative effect on the beverage quality. The most important types of defects are the so-called black, sour, immature and immature-black. Black beans result from dead beans within the coffee cherries or from beans that fall naturally on the ground by action of rain or over-ripening (Clarke, 1987c; Mazzafera, 1999). The presence of sour beans is usually associated with 'overfermentation' during wet processing (Clarke, 1987c) and with improper drying or picking of overripe cherries (Sivetz and Desrosier, 1979; Clarke, 1987c). Immature beans come from immature fruits. Immature-black beans are those that fall on the ground while immature, remaining in contact with the soil and thus being subject to fermentation (Mazzafera, 1999).

There are some studies available on the comparison of physical and chemical characteristics of defective and non-defective coffee beans (Mazzafera, 1999; Franca et al., 2005; Oliveira et al., 2006; Vasconcelos et al., 2007; Mancha Agresti et al., 2008; Mendonça et al., 2008). A detailed review on the subject is presented by Franca and Oliveira (2008). The overall conclusion of these studies indicates that, prior to roasting, it is possible to differentiate defective and non-defective (healthy) beans by color, size, acidity levels, sucrose levels, and the presence of histamine. However, after roasting, only an evaluation of the volatile profile will effectively provide the means for differentiation (Mancha Agresti et al., 2008) and further studies on this subject theme are still needed.

Even though defective beans are mechanically separated from the non-defective ones prior to commercialization in international markets, they are still being commercialized in Brazil and other producing countries. Unfortunately, since to coffee producers they represent an investment in growing, harvesting, and handling, such beans are sold at lower prices to some roasting industries that use them in blends with good quality beans. In order to eliminate these defective beans from the trade market and improve the overall quality of the beverage consumed worldwide, a few recent studies have been developed in terms of alternative uses for such beans (Oliveira et al., 2008a; Nunes et al., 2008, 2009). A description of the major findings from these specific studies is presented as follows.

3.1. Fuel

There are no literature reports on attempts of just burning low quality coffee beans for energy production. This can be attributed to the fact that the separation of defective beans in the farms and cooperatives by color is still an inefficient process, especially for immature beans. Therefore, low quality coffees lots that have been rejected during color sorting still contain a significant amount of non-defective coffees (30–70%), as pointed out in studies employing machine sorted mixtures of defective coffee beans (Franca et al., 2005; Vasconcelos et al., 2007; Mendonça et al., 2008). These low quality coffee mixtures present a reasonable economical value (~US$150 per 60kg coffee bag) in comparison to good quality coffee (~US$220 per 60kg bag). Thus, selling these beans to the coffee roasting industry is viewed by coffee producers as a good alternative from an economical point of view. Therefore, whatever alternative application proposed for this type of residue must be more profitable than selling it to the roasting industry.

The only study that presents any relation between using defective coffee beans for fuel production is the one by Oliveira et al. (2008a), who evaluated the feasibility of producing biodiesel using oil extracted from defective coffee beans. Direct transesterifications of triglycerides from refined soybean oil (reference) and from oils extracted by solvent (hexane) from healthy and defective coffee beans were performed. Oils extracted from healthy and defective coffee beans were successfully converted to alkyl esters of fatty acids (biodiesel) by transesterification with both methanol and ethanol in the presence of sodium methoxide as an alkaline catalyst. The yields for the reactions with the oil of healthy coffee beans were lower than those for the oil of defective beans, indicating the need for correction of the amount of catalyst to be used due to the content of free fatty acids of the oil. Further studies regarding the identification of the factors affecting conversion are needed in order to optimize the production of alkyl esters using the oil from defective coffee beans. Nevertheless, regardless of the ester yields obtained, coffee oil demonstrated potential as a candidate for feedstock in biodiesel production. However, in terms of sustainability, the use of inedible vegetable oils for biodiesel production generates an extensive amount of solid residues (pressed seed cakes), which present an environmental problem in terms of adequate disposal.

3.2. Adsorption Studies

In order for coffee oil-based biodiesel production to become an environmentaly friendly process, alternative proposals for the generated solid residues are necessary. In that regard, the recent study by Nunes et al. (2009) evaluated the potential of defective coffee press cake as raw material for the preparation of activated carbons (DCAC). Batch adsorption tests were performed at room temperature, using methylene blue (MB) as the adsorbate. Preliminary adsorption tests showed that thermal treatment is necessary in order to improve adsorption capacity and that this type of residue cannot be employed as a biosorbent. Equilibrium data indicated favorable and heterogeneous adsorption. The maximum value of uptake capacity obtained for the produced activated carbon (~15mg g^{-1}), was comparable to values encountered in the literature for other similar residue based activated carbons. Actually, the DCAC presented higher adsorption capacity than other residue-based ACs including apricot stones, walnut shells, date pits and almond shells. The results presented in this study indicated that defective coffee press cake presents great potential as an inexpensive and easily available alternative adsorbent for the removal of cationic dyes in wastewater treatments.

In another, Nunes et al. (2008) evaluated the feasibility of employing microwave activation instead of the traditional oven carbonization techniques for the production of DCAC. Batch adsorption tests were performed using methylene blue (MB), with the adsorbent being obtained by carbonization of the defective coffee press cake at 300°C during 6 min in a household microwave oven. No significant variations in terms of adsorption kinetics were observed. However, the activated carbon obtained by microwave activation presented a significantly higher adsorption capacity (3.5 times higher) in comparison to the one obtained by conventional activation of the same type of residue in a muffle furnace (Nunes et al., 2009). These results indicate that microwave activation provides not only a significant reduction in processing time and energy requirements, but also an increase in adsorption capacity and that this solid residue presents excellent potential as a raw material for the production of adsorbents.

4. SPENT COFFEE GROUNDS

Roasted ground beans are treated with high temperature/pressure water in order to extract the soluble material to be used in instant or soluble coffee production. The insoluble residue (spent coffee grounds) is submitted to dewatering in order to reduce moisture content from 80 to approximately 50% (Adams and Dougan, 1987). A summarized description of the chemical composition of this specific residue is displayed in Table 4. The data presented on Table 4 indicate that spent coffee grounds are highly fibrous and present a reasonable amount of oil.

Spent coffee grounds are usually disposed either in landfill sites or by burning as fuel in the boilers of the soluble coffee industry. However, such solid residues are highly pollutant due to high contents of organic substances that demand great quantities of oxygen to decompose. Also, spontaneous combustion has been reported in some storage sites (Silva et al., 1998).

Table 4. Chemical composition of spent coffee grounds (g/100 g)[a]

Crude protein	10–12	Starch equivalent	75 on ruminants
Crude fiber	35–44	Fatty acids	
Lignin	13–16	linoleic	37%
Cellulose	22–28	palmitic	34%
Lipids (ether extract)	22–27	oleic	14%
Minerals	0.3-1	free fatty acids	8%

[a] based on the data reported by Adams and Dougan, 1987.

This solid residue presents an additional disposal problem, given that it can be used for adulteration of roasted and ground coffee, being practically impossible to detect as an adulterant. Thus, the soluble coffee industry has been quite careful with its disposal, and most of the time this residue is simply used as a boiler fuel by the same industry.

There are reports on early attempts to use coffee grounds as fertilizers (Tango, 1971), supplement for animal feed (Sikka et al., 1985; Givens and Barber, 1986; Sikka and Chawla, 1986; Adams and Dougan, 1987) and as fuel (Adams and Dougan, 1987). A detailed discussion on the earlier studies is provided by Adams and Dougan (1987) and therefore the focus of this section will be based on more recent studies on the use of spent coffee grounds. It is noteworthy to mention that we could not find any newer studies on the use of spent coffee grounds on animal feed, so it is our belief that the general conclusions from the earlier studies still hold, i.e., spent coffee grounds can be added to ruminant feed up to 10%.

4.1. Fuel

The calorific value of spent coffee grounds (6930 kcal kg^{-1} dry matter) is reported to be comparable to that of charcoal (Adams and Dougan, 1987). Such characteristic, in association to the fact that this solid residue is practically smoke free during combustion, with low particulate emissions, make burning it for fuel an attractive alternative use. However, its high moisture content presents a problem in terms of fuel use efficiency and thus the spent coffee grounds should be dried to approximately 30% prior to use. According to Adams and Dougan (1987), drying below such level poses a fire risk. A few recent studies have been developed in order to improve spent coffee ground use as a fuel (Xu et al., 2006; Horio et al., 2009; Zhang et al., 2008).

Xu and co-workers (2006) evaluated the feasibility of converting spent coffee grounds into a middle-caloric product gas. The conversion took place in two consecutive steps, the first one being fuel drying/upgrading, followed by pyrolytic gasification of the dried fuel with dual fluidized bed gasification technology (DFBG). Evaluation of the performance of the pilot gasification facility demonstrated that the employed DFBG technology worked stably with the fuel of coffee grounds that was pre-dried to a water content of about 10 wt %. It was possible to convert more than 70% of fuel's C into gas, and the produced gas presented a high heating value (HHV) of over 3500 kcal/m^3. However, the tar load in the product gas was sometimes high. Increasing the steam/fuel mass ratio and decreasing the fuel particle size reduced the tar yield, but the available reduction degree was limited. Inclusion of a small amount of air into steam (gasification reagent) also helped to lower the tar content of the product gas.

Nonetheless, it was concluded that further tar elimination techniques are required for the gasification of coffee grounds via DFBG.

Horio et al. (2009) presented a prototype powdered biomass charcoal fired heater with a heat output of 6 kW. The combustion heater was characterized for charcoal prepared from Japanese oak (*Quercus serrata*) and from several waste biomass sources, including charcoal produced from spent coffee grounds, pure and mixed with soybean fiber. The combustion heater was designed based on the concept of charcoal combustion in a thin bed cross-flow (TBCF) mode, where a very thin uniform bed of charcoal is fixed by air flow on the wall of a cylindrical chamber with an air-penetrable wall. For wood charcoal the heater's thermal efficiency was about 65–86%, and for waste biomass charcoal it was found to be in the range of 60–81%. When the combustion heater was operated at the stable combustion mode, the CO concentration in the exhaust after the flue gas passed through catalyst was less than 5 ppm.

One of the major problems with the use of spent coffee grounds as fuel is related to its high moisture content. In that regard, the recent study by Zhang et al. (2008) describes a novel approach for upgrading biomass fuels with high water content for gasification, using an oil-slurry dewatering process. Wet coffee grounds (CG) were used as a feedstock and dewatered in kerosene with simultaneous addition of calcium. The results obtained show that, for calcium loadings lower than 3 wt%, calcium highly disperses into the CG matrix under dewatering condition, and its catalytic activity for char gasification is comparable to that obtained by impregnation with an aqueous solution of calcium acetate. It was concluded that the dewatering process can provide an effective and practical catalyst loading procedure for biomass fuels with high water content.

4.2. Adsorption Studies

There are a few recent studies on the use of spent coffee grounds in the production of adsorbents (Boonamnuayvitaya et al., 2004, 2005; Namane et al., 2005; Tokimoto et al., 2005; Escudero et al., 2008; Franca et al., 2008). Boonamnuayvitaya and co-workers (2004) investigated the preparation and utilization of spent coffee grounds binding with clay as adsorbent for removal of heavy metal ions in aqueous solutions. Factors affecting the adsorption such as pyrolysis temperature, weight ratio of spent coffee grounds to clay and particle size were investigated. The best results were obtained for the following parameters: pyrolysis temperature of 500°C, spent coffee grounds to clay weight ratio of 80 to 20 and particle size diameter of 4 mm. Langmuir based maximum adsorption capacities were 40, 31, 11, 20 and 13 mg g^{-1} for Cd(II), Cu (II), Pb (II), Zn (II) and Ni (II), respectively. The Cd(II) adsorption increased with increasing pH and temperature, and remained constant at high pH. Evaluation of thermodynamic parameters indicated physical exothermic adsorption. The functional groups studied by FTIR indicated that hydroxyl, carboxyl and amine groups were the main functional groups. Electrical potential study showed that the adsorbent exhibited negative charges that were favorable to attract metal ions. The surface and pore study implied that the high fraction mesopores in the developed adsorbent improved adsorption capacity. In a subsequent study, Boonamnuayvitaya et al. (2005) prepared activated carbons using spent coffee grounds impregnated with zinc chloride and activated at 600°C under nitrogen, carbon dioxide or steam flow. The activated carbon prepared by zinc chloride and nitrogen activation presented the highest adsorption capacity (adsorption of formaldehyde vapor) whereas the one prepared with zinc chloride impregnation coupled with carbon dioxide activation presented the highest total surface area and total pore volume. The authors concluded that, for formaldehyde adsorption, the surface chemistry of the adsorbent plays a major role in increasing adsorption capacity.

Namane et al. (2005) employed spent coffee grounds treated with zinc chloride and phosphoric acid for adsorption of phenol and dyes (acid blue and basic yellow). The produced activated carbons were comparable to a commercial product in terms of adsorption capacity, and the produced activated carbon presented better affinity for basic dyes in comparison to the commercial product.

Escudero et al. (2008) employed grape stalks (solid residue from wine production) and spent coffee grounds as adsorbents for the removal of Cu(II) and Ni(II) from aqueous solutions in presence and in absence of the complexing agent EDTA. Effects of pH and metal–EDTA molar ratio, kinetics as a function of sorbent concentration, and sorption equilibrium for both metals onto both sorbents were evaluated in batch experiments. Metal uptake was dependent of pH, reaching a maximum at pH ~5.5. EDTA was found to dramatically reduce metal adsorption, reaching total uptake inhibition for both metals onto both sorbents at equimolar metal:ligand concentrations. Grape stalks presented higher adsorption capacity (approximately four times) in comparison to spent coffee grounds, for both Cu(II) and Ni(II) adsorption.

The only works that employed spent coffee grounds without further chemical or thermal treatment as adsorbents, were the ones presented by Tokimoto et al. (2005) and Franca et al. (2008), with applications in the removal of lead ions from drinking water and the removal of basic dyes from wastewaters, respectively. The experimental results obtained by Tokimoto et al. (2005) indicated that that proteins present in spent coffee grounds are responsible for the adsorption of lead ions. The rates of adsorption of lead ions by coffee grounds were directly proportional to the amounts of coffee grounds added to the solution. Although activated carbon is widely used in general-purpose water purification, the authors observed that it adsorbed fewer lead ions than either spent coffee grounds or activated clay. When the concentration of lead ions was below 80 µg/L, spent coffee grounds exhibited the highest adsorption capacity. Also, the lead ion adsorption capacity of coffee grounds was not significantly affected by temperature in the temperature range pertaining to tap water, favoring the idea of using coffee grounds to purify tap water.

Franca and co-workers (2008) evaluated the performance of spent coffee grounds as an adsorbent for removal of methylene blue (MB) from aqueous solution. Batch adsorption tests were performed at 25°C and the effects of contact time, adsorbent dosage, and pH were investigated. The typical dependence of dye uptake on kinetic studies indicated the adsorption process to be both chemisorption and diffusion controlled. The maximum value of uptake capacity obtained for the spent coffee grounds/methylene blue system was comparable to values encountered in the literature for other untreated agricultural by-products and wastes. These studies indicate that this residue presents potential to be employed as an adsorbent, even without any further chemical or thermal treatment.

4.3. Other Applications

Nogueira et al. (1999) investigated the applicability of the forced aeration composting process to mixtures of spent coffee grounds and other agricultural wastes. Temperature was used as the main parameter to enable control of the aeration rate. Other parameters used to monitor the process were humidity, pH and carbon/nitrogen ratio. The results obtained were satisfactory and the experiments lead to the production of a high quality compost, with carbon/nitrogen ratios ranging from 13/1 to 15/1.

Leifa and co-workers (2001) evaluated the feasibility of using spent coffee grounds as a substrate for the production of edible mushroom Flammulina under different conditions of moisture and spawn rate. Ideal moisture content for mycelial growth was 55%, with first fructification occurring 21 days after inoculation and the biological efficiency reaching about 78% in 40 days. Tannin content decreased by 28% after 40 days, which was attributed to degradation by the culture. Results showed the feasibility of using both coffee husks and spent coffee grounds as substrates without any nutritional supplementation for cultivation of edible fungus in solid state cultivation, with the later presenting better results.

5. CONCLUSION

Coffee is deemed as a commodity ranking second only to petroleum in terms of currency (usually US dollars) traded worldwide (Illy, 2002). Brazil is the largest coffee producer and exporter in the world and is the second largest consumer. As such this commodity is quite relevant to the country's economy. However, coffee production also represents a yearly average production of over 2.5 million tons of solid residues, including coffee husks and pulp, defective coffee beans, and spent coffee grounds from the soluble coffee industry. Considering the specific characteristics of each residue and the significant amount generated, prevailing difficulties in devising ways for their adequate disposal management are of great relevance, given the increasing awareness and emerging need for properly addressing environmental issues. In this regard, several alternative uses for these solid residues have been proposed in the literature.

Regarding coffee husks and pulp, the majority of the studies have focused on employing this solid residue as a supplement for animal feed. Several in vivo studies have been carried out with a wide variety of livestock (cattle, fish, pigs, sheep and chicken) and the overall conclusion is that its use for animal feed is quite restricted due to the presence of anti-nutritional factors such as caffeine and tannins. Detoxification and fermentation studies have demonstrated that caffeine and tannin levels can be reduced significantly. However, further in vivo studies employing detoxified coffee husks and/or pulp are needed in order to verify if such procedures are sufficient to increase the amount of these solid wastes that can be employed as a supplement or even as a replacement for the currently used animal feed. The use of coffee husks directly as soil coverage seems to be a good option for potassium depleted soils. Composting is also feasible, given that another type of residue be mixed with the husks in order to improve the carbon nitrogen ratio.

Even though coffee husks in their original state present a reasonable calorific value and that this value can be further increased by carbonization, its use as a direct solid fuel still presents several problems. These include the need to reduce NOx emissions, to avoid pyrolysis in the feeding system, dragging of the coffee husks by the flue gas and agglomeration, and fouling and corrosion of the combustion equipment. Recent studies indicated that microwave pyrolysis seems to be an interesting alternative in terms of increasing calorific value, but the aforementioned problems still remain in need of proper addressing.

The lack of recent studies in terms of biogas production can be perceived as an indication that this alternative is not yet technically and economically viable, since the much commented constraints associated with the production techniques, such as large hydraulic retention times (30–50 days) and low gas production in cold weather, must still be overcome (Yadvika et al., 2004; Singh and Prerna, 2009).

Among the most recently proposed uses for coffee husks and pulp, fermentation for the production of a diversity of products, use as a substrate for growth of mushrooms and use as adsorbents or as a precursor for the preparation of adsorbents are worthy mentioning to be promising alternatives. These alternatives generate more value-added products than the aforementioned ones, requiring, in most cases, less processing steps and simpler technologies, thus, being, in theory, relatively more cost effective alternatives. However, all the studies were carried out at laboratory scale and were based on attaining a pre-specified target product using the currently available knowledge of the physical and chemical properties of the residues, which is rather superficial. In the majority of these studies, no attempts to work on the scaling-up and to perform economical analyses of the processes were done and further studies are needed in order to proper address these issues. Also, all of the commented alternative applications still generate residues that must be properly disposed of.

Defective coffee beans are not yet considered residues and are being commercialized in the internal markets of whichever country coffee is produced, thus, depreciating the quality of the beverage consumed locally. Because they represent a fair share of the total amount produced worldwide (~20%), the producers investment in growing, harvesting, handling and processing, and, also, the lack of proposals for profitable alternative uses preclude their withdrawal from the roasting market. Defective coffee beans have been extensively studied as such only recently (Franca and Oliveira, 2008) and proposals for alternative uses are scarce with the production of biodiesel from coffee oil (Oliveira et al., 2008a) and the use of the cold press cake from the oil extraction for the production of adsorbents (Nunes et al., 2009) being the only proposals available in the literature. Regarding spent coffee grounds, the majority of the proposals for alternative use other than energy source are related to the production and use of adsorbents.

Looking at the coffee production solid residues under the concept of a bio-refinery would be a more proper way of addressing the issue of devising adequate management disposal strategies. However, in order to achieve that, a more detailed study of the physical and chemical properties of the residues is necessary. These residues are rich in organic compounds (carbohydrates, proteins, polyphenols, alkaloids, and others) that may be worthy of being recovered or processed into other value-added compounds before they are used as livestock feed, fertilizers or fuels. Even with the currently available proposals for alternative uses, one can envision a sequence of processing steps that could lead to a real application of the bio-refinery concept. One example would be to produce liquid biofuels (such as ethanol) from coffee husks and further use the generated residue (fermented solid) to produce adsorbents for a diversity of applications, with the removal of organic compounds from the water used in the processing of coffee (e.g., in pulping) as an attractive alternative. The organic-loaded adsorbent could then be burned as a solid fuel to produce the necessary energy for the artificial drying of coffee beans. Community-scale bio-refineries could then reduce the costs of coffee production by turning the coffee solid residues into products that can be used by the producers in the several processing steps of coffee, e.g., fuel for the machinery used in harvesting and handling, adsorbents for the treatment of coffee processing wastewater and solid fuel for artificial dryers of coffee beans.

ACKNOWLEDGMENTS

The authors gratefully acknowledge financial support from the following Brazilian Government Agencies: CNPq and FAPEMIG.

REFERENCES

Adams, M.R. and Dougan, J. (1987) Waste Products. In: R.J. Clarke and R. Macrae, (Eds.), *Coffee Technology*, New York: Elsevier Applied Science Publishers Ltd., pp. 257-291.

Antier, P., Minjares, A., Roussos, S., Viniegragonzalez, G. (1993) New approach for selecting pectinase producing mutants of Aspergillus niger well adapted to solid state fermentation, *Biotechnology Advances*, 11, 429-440.

Appels, L., Baeyens, J., Degreve, J. and Dewil, R. (2008) Principles and potential of the anaerobic digestion of waste-activated sludge, *Progress in Energy and Combustion Science*, 34, 755-781.

Ballesteros, M., Oliva, J.M., Negro, M.J., Manzanares, P. and Ballesteros, I. (2004). Ethanol from lignocellulosic materials by a simultaneous saccharification and fermentation process (SFS) with Kluyveromyces marxianus CECT 10875, *Process Biochemistry*, 39, 1843-1848.

Barcelos, A.F.; Paiva, P.C. de A. and Von Tiesenhausen, I.M.E.V. (1999) Desempenho de novilhos de tres grupos genéticos em confinamento recebendo diferentes niveis de casca de café no concentrado (Performance of steers of three genetic groups in a feedlot trial receiving different levels of coffee hulls in the concentrate), *Ciência e Agrotecnologia*, 23, 948-957 (in Portuguese).

Barcelos, A. F., Paiva, P. C., Pérez, J. R. O., Cardoso, R. M. and Snatos, V. B. (2001) Estimativa das frações dos carboidratos, da casca e polpa desidratada de café (Coffea arabica L.) armazenadas em diferentes períodos (Estimate of the carbohydrate fractions of the coffee hulls and dehydrated pulp of coffee (*Coffea arabica* L.) stored for different periods), *Revista Brasileira de Zootecnia*, 30, 1566-1571, (in portuguese).

Bayne, D.R., Dunseth, D. and Garcia R.C. (1976) Suplemental feeds containing coffee pulp for rearing Tilapia in Central América, *Aquaculture*, 7, 133-146.

Belkacemi, K., Turcotte, G. and Savoite, P. (2002) Aqueous/Steam-Fractionated Agricultural Residues as Substrates for Ethanol Production, *Industrial and Engineering Chemistry Research*, 41, 173-179.

Bonati, M., Jiritano, L., Bortolotti, A., Gaspari, F., Filippeschi, S., Puidgemont, A. and Garattini, S. (1985) Caffeine distribution in acute toxic response among inbred mice, *Toxicology Letters*, 29, 25-31.

Boonamnuayvitaya, V., Chaiya, C., Tanthapanichakoon, W., and Jarudilokkul, S. (2004) Removal of heavy metals by adsorbent prepared from pyrolyzed coffee residues and clay, *Separation and Purification Technology*, 35, 11-22.

Boonamnuayvitaya, V., Sae-ung, S. and Tanthapanichakoon, W. (2005) Preparation of activated carbons from coffee residue for the adsorption of formaldehyde, *Separation and Purification Technology*, 42, 159-168.

Borém, F.M. (2008) *Pós-Colheita do Café (Coffee Post Processing)*, Lavras, Brazil: Editora UFLA.

Brand, D., Pandey, A., Roussos, S. and Soccol, C.R. 2000. Biological detoxification of coffee husk by filamentous fungi using a solid state fermentation system, *Enzyme and Microbial Technology*, 27, 127-133.

Bressani, R., Estrada, E., Elias, L.G., Jarquin, R., Valle, L.U. De. (1977) Pulpa y pergamino de café. Efecto de la pulpa de café deshidratada en la dieta de ratas e pollos (Effect of dehydrated coffee pulp on mice and chicken diets), *Revista Cafetalera*, 164:35-44 (in Spanish).

Buckholtz, N.S. and Middaugh, L.D. (1987) Effects of caffeine and L-phenylisopropyladenosine on locomotor activity of mice, *Pharmacology Biochemistry and Behavior*, 28, 179-185.

Carvalho, P.L.O. (2008) Casca de café melosa ensilada na alimentação de suínos (Sticky coffee husks in swine feeds), M.Sc. Dissertation, Universidade Estadual de Maringá, Maringá, PR, Brazil, 77 pp. (in Portuguese).

Chen, J.P. and Yang, L. (2005) Chemical modification of *Sargassum* sp. for prevention of organic leaching and enhancement of uptake during metal biosorption, *Industrial and Engineering Chemistry Research*, 44, 9931–9942.

Christensen, M.S. (1981) Preliminary tests on the suitability of coffee pulp in the diets of common carp /Cyprinus carpio/ L. and catfish /Clarias mossambicus/ Peters, *Aquaculture*, 25, 235-242.

Clarke, R.J. (1987a) Extraction. In: R.J. Clarke and R. Macrae, (Eds.), *Coffee Technology*, New York: Elsevier Applied Science Publishers Ltd., pp. 109-145.

Clarke, R.J. (1987b) Drying. In: R.J. Clarke and R. Macrae, (Eds.), *Coffee Technology*, New York: Elsevier Applied Science Publishers Ltd., pp. 147-199.

Clarke, R.J. (1987c) Grading, Stoarge, Pre-treatment and Blending. In: R.J. Clarke and R. Macrae, (Eds.), *Coffee Technology*, New York: Elsevier Applied Science Publishers Ltd., pp. 35-58.

Clifford, M.N. and Ramirez-Martinez, J. R. (1991) Tannins in wet-processed coffee beans and coffee pulp, *Food Chemistry*, 40, 191-200.

Couto Filho, C.C.C, Silva Filho, J.C., Neiva Júnior, A.P., Freitas, R.T.F., Souza, R.M. and Nunes, J.A.R. (2007) Qualidade da silagem de resíduo de manga com diferentes aditivos (Quality of mango residue silage with different additives), *Ciência e Agrotecnologia*, 31, 1537-1544 (in portuguese).

Davis, L., Jeon, Y., Svenson, C., Rogers, P., Pearce, J. and Peiris, P. (2005) Evaluation of wheat stillage for ethanol production by recombinant Zymomonas mobilis, *Biomass and Bioenergy*, 29, 49-59.

Demeke, S. (1991) Coffee pulp alone and in combination with urea and other feeds for sheep in Ethiopia, *Small Ruminant Research*, 5, 223-231.

Dewil, R., Appels, L., Baeyens, J. (2006) Energy use of biogas hampered by the presence of siloxanes, *Energy Conversion and Management*, 47, 1711-1722.

Domínguez, A., Menéndez, J.A., Fernández, Y., Pis, J.J., Valente Nabais, J.M., Carrott, P.J.M. and Ribeiro Carrott, M.M.L. (2007)Conventional and microwave induced pyrolysis of coffee hulls for the production of a hydrogen rich fuel gas, *Journal of Analytical and Applied Pyrolysis*, 79, 128-135.

Escudero, C., Gabaldon, C., Marzal, P. and Villaescusa, I. (2008) Effect of EDTA on divalent metal adsorption onto grape stalk and exhausted coffee wastes, *Journal of Hazardous Materials*, 152, 476-485.

Fagbenro, O.A. and Arowosoge, I.A. (1991) Growth response and nutrient digestability by Clarias isheriensis (Sydenham, 1980) fed varying levels of dietary coffee pulp as replacement for maize in low-cost diets, *Bioresource Technology*, 37, 253-258.

Fan, L. , Soccol, A.T., Pandey, A., Vandenberghe, L.P.S., Soccol, C.R. (2006) Effect of caffeine and tannins on cultivation and fructification of *Pleurotus* on coffee husks, *Brazilian Journal of Microbiology* (2006) 37:420-424.

Filho, E.R., Paiva, P.C.A., Barcelos, A.F., Rezende, C.A.P., Cardoso, R.M. and Banys, V.L. (2000) Efeito da casca de café (coffea arabica, l.) no desempenho de novilhos mestiços de holandês-zebu na fase de recria (The effect of coffee hulls on the performance of holstein-zebu steers during growing period), *Ciência e Agrotecnologia*, 24, 225-232, (in Portuguese).

Filho, E. R., Paiva, P. C. A., Oliveira, E. R., Barcelos, A. F., Castro, A. L. A. And Santos, J. (2004) Cinética da digestão ruminal da casca de café (*Coffea arabica, L.*) em vacas holandesas (Kinetics of the ruminal digestion of coffee hulls (*Coffea arabica, L.*) in holstein cows), *Ciência e Agrotecnologia*, 28, 627-636 (in Portuguese).

Franca, A.S., Oliveira, L.S., Mendonça, J.C.F. and Silva, X.A. (2005) Physical and chemical attributes of defective crude and roasted coffee beans, *Food Chemistry*, 90, 84–89.

Franca, A.S., Oliveira, L.S., Ferreira, M.E. (2008) Kinetics and equilibrium studies of methylene blue adsorption by spent coffee grounds, *Desalination*, in press.

Franca, A.S. and Oliveira, L.S. (2008) Chemistry of defective coffee beans. In E. N. Koeffer. (Ed.), *Progress in Food Chemistry*, New York: Nova Publishers, pp. 105-138.

Furusho-Garcia, I.F., Perez, J.R.O. and Oliveira, M.V.M. (2003) Componentes Corporais e Órgãos Internos de Cordeiros Texel x Bergamácia, Texel x Santa Inês e Santa Inês Puros, Terminados em Confinamento, com Casca de Café como Parte da Dieta (Body Components and Internal Organs of Texel x Bergamácia, Texel x Santa Inês and Purebred Santa Inês Lambs Finished in Fedlot, with Coffee Hull as Part of the Diet), *Revista Brasileira de Zootecnia*, 32, 1992-1998.

Givens, D. I. and Barber, W. P. (1986) In vivo evaluation of spent coffee grounds as a ruminant feed, *Agricultural Wastes*, 18, 69-72.

Gokulakrishnan, S., Chandraraj, K. and Gummadi,S.N. (2005) Microbial and enzymatic methods for the removal of caffeine, *Enzyme and Microbial Technology*, 37, 225-232.

Gokulakrishnan, S. and Gummadi,S.N. (2006) Kinetics of cell growth and caffeine utilization by *Pseudomonas* sp. GSC 1182, *Process Biochemistry*, 41, 1417-1421.

Gouvea, B.M., Torres, C., Franca, A.S., Oliveira, L.S. and Oliveira, E.S. (2008). Feasibility of ethanol production from coffee husks, *Journal of Biotechnology*, 136, S269.

Hargreaves, J.C., Adl, M.S., Warman, P.R. (2008) A review of the use of composted municipal solid waste in agriculture, *Agriculture, Ecosystems and Environment*, 123, 1-14.

Horio, M., Suri, A., Asahara, J., Sagawa, S., and Aida, C. (2009) Development of Biomass Charcoal Combustion Heater for Household Utilization, *Industrial and Engineering Chemistry Research,* 48, 361-372.

Huang, Y., Bird, R.N. and Heidrich, O. (2007) A review of the use of recycled solid waste materials in asphalt pavements, *Resources, Conservation and Recycling*, 52, 58-73.

Illy, E. (2002) The complexity of coffee, *Scientific American*, June, 86-91.

Ioannidou, O. and Zabaniotou, A. (2007) Agricultural residues as precursors for activated carbon production—A review, *Renewable and Sustainable Energy Reviews*, 11, 1966-2005.

Isaac, W.A.P., Brathwaite, R.A.I., Cohen, J.E. and Bekele, I. (2007) Effects of alternative weed management strategies on *Commelina diffusa Burm.* infestations in Fairtrade banana (*Musa* spp.) in St. Vincent and the Grenadines, *Crop Protection*, 26, 1219-1225.

Jarquín, R., Gonzalez, J.M., Braham, J.E. and Bressani, R. (1973) Pulpa y pergamino de café. II. Utilización de la pulpa de café en la alimentación de ruminates, *Turrialba*, 23, 41-47.

Jimenez, L. and Gonzalez, F. (1991) Study of the physical and chemical properties of lignocellulosic residues with a view to the production of fuels, *Fuel*, 70, 947-950.

Kapdan, I. K. and Kargi, F. (2006) Bio-hydrogen production from waste materials, *Enzyme and Microbial Technology*, 38, pages 569-582.

Kim, S. and Dale, B. E. (2004) Global potential bioethanol production from wasted crops and crop residues, *Biomass and Bioenergy*, 26, 361-375.

Landolt, H.P., Dijk, D.-J., Gauss, S.E. and Borbély, A.A. (1995) Caffeine reduces low-frequency delta activity in the human sleep EEG, *Neuropsychopharmacol*, 12, 229–238.

Laufenberg, G., Kunz, B. and Nystroem, M. (2003) Transformation of vegetable waste into value added products: (A) the upgrading concept; (B) practical implementations, *Bioresource Technology*, 87, 167-198.

Leifa, F., Pandey, A. and Soccol, C.R. (2001) Production of *Flammulina velutipes* on coffee husk and coffee spent-ground, *Brazilian Archives of Biology and Technology*, 44, pp. 205 – 212.

Machado, C.M.M., Soccol, C.R., Oliveira, B.H. and Pandey, A. (2002) Gibberellic Acid Production by Solid-State Fermentation in Coffee Husk, *Applied Biochemistry and Biotechnology*, 102–103, 179-191.

Machado, C.M.M., Oishi, B.O., Pandey, A. and Soccol, C.R. (2004) Kinetics of *Gibberella fujikuroi* Growth and Gibberellic Acid Production by Solid-State Fermentation in a Packed-Bed Column Bioreactor, *Biotechnology Progress*, 20, 1449-1453.

Magalhães, E.A., Silva, J. S., Silva, J.N., Filho, D.O., Donzeles, S.M.L., Martin, S. and Dutra, L. (2008) Casca de café associada à lenha como combustível para aquecimento indireto do ar de secagem (Coffee husks mixed with fire wood as a fuel for indirect heating of air used for drying), *Revista Brasileira de Armazenamento*, Especial Café 10, 66-72 (in Portuguese).

Mancha Agresti, P.D.C., Franca, A.S., Oliveira, L.S., Augusti, R. (2008) Discrimination between defective and non-defective Brazilian coffee beans by their volatile profile, *Food Chemistry*, 106, 787-796.

MAPA—Ministério da Agricultura, Pecuária e Abastecimento–CONAB (Companhia Nacional de Abastecimento, Acompanhamento da Safra Brasileira – Café, 2008, http://www.conab.gov.br/ conabweb/download/safra/2_levantamento_2008.pdf

Matos, A.T. (2008) Tratamento de resíduos na pós-colheita do café (Residues disposal in coffee post-processing). In: F. M. Borém, (Ed.), *Pós-Colheita do Café (Coffee Post Processing)*, Lavras: Editora UFLA, pp. 161-201.

Matteson, G.C. and Jenkins, B.M. (2007) Food and processing residues in California: Resource assessment and potential for power generation, *Bioresource Technology*, 98, 3098-3105.

Mazzafera, P. (1999) Chemical composition of defective coffee beans, *Food Chemistry*, 64, 547–554.

Mazzafera, P. (2002) Degradation of caffeine by microorganisms and potential use of decaffeinated coffee husk and pulp in animal feeding, *Scientia Agricola*, 59, 815-821.

Mendonça, J.C.F., Franca, A.S., Oliveira, L.S. and Nunes, M. (2008) Chemical characterisation of non-defective and defective green arabica and robusta coffees by electrospray ionization-mass spectrometry (ESI-MS), *Food Chemistry*, 111, 490-497.

Menéndez, J. A., Domínguez, A., Fernández, Y. and Pis, J. J. (2007) Evidence of Self-Gasification during the Microwave-Induced Pyrolysis of Coffee Hulls, *Energy and Fuels*, 21, 373-378.

Meyer, L. and Caston, J. (2004) Stress alters caffeine action on investigatory behaviour and behavioural inhibition in the mouse, *Behavioural Brain Research*, 149, 87-93.

Minjares-Carranco, A., Trejo-Aguilar, B.A., Aguilar, G., Viniegra-Gonzalez, G. (1997) Physiological comparison between pectinase-producing mutants of Aspergillus niger adapted either to solid-state fermentation or submerged fermentation, *Enzyme and Microbial Technology*, 21, 25-31.

Molina, M., De La Feunte, G., Batten M. and Bressani, R. (1974) Decaffeination. A process to detoxify coffee pulp, *Journal of Agricultural and Food Chemistry*, 22, 1055–1059.

Moreau, Y., Arredondo, J.L., Perraud-Gaime, I. and Roussos, S. (2003) Dietary Utilisation of Protein and Energy from Fresh and Ensiled Coffee Pulp by the Nile tilapia, Oreochromis niloticus, *Brazilian Archives of Biology and Technology*, 46, 223-231.

Namane, A., Mekarzia, A., Benrachedi, K., Belhaneche-Bensemra, N. and Hellal, A. (2005) Determination of the adsorption capacity of activated carbon made from coffee grounds by chemical activation with $ZnCl_2$ and H_3PO_4, *Journal of Hazardous Materials*, 119, 189-194.

Nehlig, A. (1999) Are we dependent upon coffee and caffeine? A review on human and animal data, *Neuroscience and Biobehavioral Reviews*, 23, 563-576.

Neiva Júnior, A.P., Silva Filho, J.C., Von Tiesenhausen, I.M. E., Rocha, G.P., Capelle, E.R. and Couto Filho, C.C.C (2007) Efeito de diferentes aditivos sobre os teores de proteína bruta, extrato etéreo e digestibilidade da silagem de maracujá (Effects of different additives on the content of crude protein, ether extract and coefficient of digestibility of silage of passion fruit residue), *Ciência e Agrotecnologia*, 31, 871-875 (in Portuguese).

Niladevi, K.N. and Prema, P. (2008) Effect of inducers and process parameters on laccase production by Streptomyces psammoticus and its application in dye decolourization, *Bioresource Technology*, 99, 4583-4589.

Nogueira, W. A., Nogueira, F. N. and Devens, D. C. (1999) Temperature and pH control in composting of coffee and agricultural wastes, *Water Science and Technology*, 40, 113-119.

Nonhebel, S. (2007) Energy from agricultural residues and consequences for land requirements for food production, *Agricultural Systems*, 94, 586-592.

Nunes, A.A., Franca, A.S. and Oliveira, L.S. (2008) Microwave assisted thermal treatment of biodiesel solid residue (coffee beans press cake) for the production of activated carbons, Proceedings of Encit 2008 – 12th Brazilian Congress of Thermal Engineering and Sciences, 6p.

Nunes, A.A., Franca, A.S. and Oliveira, L.S. (2009) Activated carbons from waste biomass: an alternative use for biodiesel production solid residues, *Bioresource Technology*, 100, 1786-1792.

Ohgrem, K., Bura, R., Lesnicki, G., Saddler, J. and Zacchi, G. (2007) A comparison between simultaneous saccharification and fermentation and separate hydrolysis and fermentation using steam-pretreated corn stover, *Process Biochemistry*, 42, 834-839.

Okai, D.B., Bonsi, M.L.K. and Easter, R.A. (1985) Dried coffee pulp (DCP) as an ingredient in the diets of growing pigs, *Tropical Agriculture*, 62(1):62-64.

Oliveira, V. D., Fialho, E. T., Lima, J. A. F., Oliveira, A. I. G. D. and Freitas, R. T. F. D. (2001) Substituição do milho por casca de café em rações isoenergéticas para suínos em crescimento e terminação (Coffee husks as a corn substitute in isoenergetic diets for swines: digestibility and performance), *Ciência e Agrotecnologia*, 25, 424-436 (in Portuguese).

Oliveira, S.L., Fialho, E.T., Murgas, L.D.S., Freitas, R.T.F. and Oliveira, A.I.G. (2002) Utilização de casca de café melosa em rações de suínos em terminação (Use of sticky coffee husks in finishing swine feeds), *Ciência e Agrotecnologia*, 26, 1330-1337 (in Portuguese).

Oliveira, L.S., Franca, A. S., Mendonça, J.C.F. and Barros-Júnior, M.C. (2006). Proximate composition and fatty acids profile of green and roasted defective coffee beans, *LWT - Food Science and Technology*, 39, 235–239.

Oliveira, L.S., Franca, A.S., Camargos, R.R.S. and Ferraz, V.P. (2008a) Coffee oil as a potential feedstock for biodiesel production, *Bioresource Technology*, 99, 3244-3250.

Oliveira, L.S., Franca, A.S., Alves, T.M. and Rocha, S.D. (2008b) Evaluation of untreated coffee husks as potential biosorbents for treatment of dye contaminated waters, *Journal of Hazardous Materials*, 155, 507-512.

Oliveira, W.E., Franca, A.S., Oliveira, LS. and Rocha, S.D. (2008c) Untreated coffee husks as biosorbents for the removal of heavy metals from aqueous solutions, *Journal of Hazardous Materials*, 152, 1073-1081.

Oliveira, L.S., Franca, A.S., Oliveira, W.E. (2008d) Production of Adsorbents from Coffee Husks: Effect of Thermal/Chemical Treatments on the Adsorption of Heavy Metals, 22nd International Conference on Coffee Science, manuscript PF814, 4p.

Oliveira, L.S., Franca, A.S. (2008) Low-Cost Adsorbents from Agri-Food Wastes. In L. V. Greco and Marco N. Bruno. (Eds.), *Food Science and Technology: New Research*, New York: Nova Publishers, pp. 171-209.

Orozco, A.L., Perez, M.I., Guevara, O., Rodriguez, J., Hernandez, M., Gonzalez-Vila, F.J., Polvillo, O. and Arias, M.E. 2008. Biotechnological enhancement of coffee pulp residues by solid-state fermentation with Streptomyces. Py-GC/MS analysis, *Journal of Analytical and Applied Pyrolysis*, 81, 247-252.

Palmarola-Adrados, B., Choteborská, P., Galbe, M. and Zacchi, G. (2005) Ethanol production from non-starch carbohydrates of wheat bran, *Bioresource Technology*, 96, 843–850.

Pandey, A., Soccol, C.R., Nigam, P. and Soccol, V. T. (2000a) Biotechnological potential of agro-industrial residues. I: sugarcane bagasse, *Bioresource Technology*, 74 69-80.

Pandey, A., Soccol, C.R., Nigam, P., Brand, D., Mohan, R. and Roussos, S. (2000b) Biotechnological potential of coffee pulp and coffee husk for bioprocesses, *Biochemical Engineering Journal*, 6, 153-162.

Parra, A.R.P., Moreira, I. Furlan, A.C., Paiano, D., Scherer, C. and Carvalho, P.L.O. (2008) Utilização da casca de café na alimentação de suínos nas fases de crescimento e terminação (Coffee hulls utilization in growing and finishing pigs feeding), *Revista Brasileira de Zootecnia*, 37, 433-442.

Pauss, A. and Nyns, E. -J. (1993) Past, present and future trends in anaerobic digestion applications, *Biomass and Bioenergy*, 4, 263-270.

Prasad, S., Singh, A. and Joshi, H.C. (2007) Ethanol as an alternative fuel from agricultural, industrial and urban residues, *Resources, Conservation and Recycling*, 50, 1-39.

Prata, E.R.B.A. and Oliveira, L.S. (2007) Fresh coffee husks as potential sources of anthocyanins, *LWT - Food Science and Technology*, 40, 1555-1560.

Raven, R.P.J.M., Gregersen, K.H. (2007) Biogas plants in Denmark: successes and setbacks, *Renewable and Sustainable Energy Reviews*, 11, 116-132.

Rebah, F.B., Prevost, D., Yezza, A. and Tyagi, R.D. (2007) Agro-industrial waste materials and wastewater sludge for rhizobial inoculant production: A review, *Bioresource Technology*, 98, 3535-3546.

Rocha, F. C., Garcia, R., Bernardino F. S., Freitas, A.P.W., Valadares, R. F. D., Junqueira, B. A., Rigueira J. P. S. and Rocha, G. C. (2005) Síntese de proteína microbiana em vacas recebendo dietas contendo casca de café (Synthesis of microbial protein in caws fed diets containing coffee husks), *Boletim da Indústria Animal* 62, 149-156 (in Portuguese).

Rogers, P.L, Jeon, Y.J. and Svenson, C.J. (2005) Application of Biotechnology to Industrial Sustainability, *Process Safety and Environmental Protection*, 83, 499-503.

Russ, W. and Meyer-Pittroff, R. (2004) Utilizing Waste Products from the Food Production and Processing Industries, *Critical Reviews in Food Science and Nutrition*, 44, 57-62.

Sabu, A., Augur, C., Swati, C., Pandey, A. (2006) Tannase production by Lactobacillus sp. ASR-S1 under solid-state fermentation, *Process Biochemistry*, 41, 575-580.

Saenger, M., Hartge, E. -U., Werther, J., Ogada, T. and Siagi, Z. (2001) Combustion of coffee husks, *Renewable Energy*, 23, 103-121.

Salminen, E. and Rintala, J. (2002) Anaerobic digestion of organic solid poultry slaughterhouse waste - a review, *Bioresource Technology*, 83, 13-26.

Salmones, D., Mata, G. and Waliszewski, K.N. (2005) Comparative culturing of Pleurotus spp. on coffee pulp and wheat straw: biomass production and substrate biodegradation, *Bioresource Technology*, 96, 537-544.

Shankaranand, V.S. and Lonsane, B.K. (1994) Coffee husk: an inexpensive substrate for production of citric acid by Aspergiilus niger in a solid-state fermentation system, *World Journal of Microbiology and Biotechnology*, 10, 165-168.

Sikka, S. S. and Chawla, J. S. (1986) Effect of feeding spent coffee grounds on the feedlot performance and carcass quality of fattening pigs, *Agricultural Wastes, 18, 305-308.*

Sikka, S. S., Bakshi, M. P. S. and Ichhponani, J. S. (1985) Evaluation in vitro of spent coffee grounds as a livestock feed, *Agricultural Wastes*, 13, 315-317.

Silva, M.A., Nebra, S. A., Machado Silva, M. J. and Sanchez, C. G. (1998) The use of biomass residues in the Brazilian soluble coffee industry, *Biomass and Bioenergy*, 14, 457-467.

Singh, S.P. and Prerna, P. (2009) Review of recent advances in anaerobic packed-bed biogas reactors, *Renewable and Sustainable Energy Reviews,* 13, 1569-1575.

Sivetz, M. and Desrosier, N.W. (1979). *Coffee Technology*. Westport, CO: Avi Publishing Co.

Soares, M., Christen, P., Pandey, A., Soccol, C.R. (2000) Fruity flavour production by Ceratocystis fimbriata grown on coffee husk in solid-state fermentation, *Process Biochemistry*, 35, 857-861.

Souza, A.L., Garcia, R., Pereira, O.G., Cecon, P.R., Filho, S.C.V. and Paulino, M.F. (2001) Composição químico-bromatológica da casca de café tratada com amônia anidra e sulfeto de sódio (Proximate and chemical composition of coffee husks treated with amonium and sodium sulfate), *Revista Brasileira de Zootecnia*, 30, 983-991 (in Portuguese).

Souza, A. L., Bernardino, F. S., Garcia, R., Pereira, O. G., Rocha, F. C. and Pires, A. J. V. (2003) Valor nutritivo de silagem de capim-elefante (Pennisetum purpurem Schum.) com diferentes níveis de casca de café (Nutritive value of Pennisetum purpurem Schum. sillage with different levels of coffee husks), *Revista Brasileira de Zootecnia*, 32, 828-833 (in Portuguese).

Souza, A. L., Garcia, R., Bernardino, F. S., Rocha, F. C., Filho, S. C. V., Pereira, O. G. and Pires, A. J. V. (2004) Casca de café em dietas de carneiros: consumo e digestibilidade (Coffee husks in sheep diet: consumption and digestibility), *Revista Brasileira de Zootecnia*, 33, 2170-2176 (in Portuguese).

Souza, A. L., Garcia, R., Filho, S. C. V., Rocha, F. C., Campos, J. M. S., Cabral, L. S. and Gobbi, K. F. (2005) Casca de Café em Dietas de Vacas em Lactação: Consumo, Digestibilidade e Produção de Leite (Coffee husks in cows diet: consumption, digestibility and milk production), *Revista Brasileira de Zootecnia*, 34, 2496-2504 (in Portuguese).

Souza, A.L., Garcia, R., Bernardino, F.S., Campos, J.M. S., Valadares Filho, S.C., Cabral, L.S. and Gobbi, K.F. (2006a) Casca de café em dietas para novilhas leiteiras: consumo, digestibilidade e desempenho (Coffee hulls in dairy heifers diet: intake, digestibility, and production), *Revista Brasileira de Zootecnia*, 35, 921-927 (in Portuguese).

Souza, A.L., Garcia, R., Valadares, R.F.D., Pereira, M.L.A., Cabral, L.S., Valadares Filho, S.C. (2006b) Casca de café em dietas para vacas em lactação: balanço de compostos nitrogenados e síntese de proteína microbiana (Coffee hulls in diet of lactating dairy cows: nitrogen balance and microbialprotein synthesis), *Revista Brasileira de Zootecnia*, 35, 1860-1865 (in Portuguese).

Sree, N.K., Sridhar, M., Rao, L.V. and Pandey, A. (1999) Ethanol production in solid substrate fermentation using thermotolerant yeast, *Process Biochemistry*, 34, 115-119.

Stavric, B. (1988) Methylxanthines: Toxicity to humans. 2. Caffeine, *Food and Chemical Toxicology* 26, 645-662.

Stavric, B. (1992) An update on research with coffee/caffeine (1989-1990), *Food and Chemical Toxicology* 30, 533-555.

Tafdrup, S. (1995) Viable energy production and waste recycling from anaerobic digestion of manure and other biomass materials, *Biomass and Bioenergy*, 9, 303-314.

Tango, J. S. (1971) Utilização industrial do café e dos seus subprodutos (Industrial utilization of coffee and its byproducts) *Boletim do ITAL*, 28, 48-73 (in Portuguese).

Teixeira, J.C. (1992) *Nutrição de ruminantes* (Ruminant nutrition). Lavras:FAEPE, 267 .p (in Portuguese).

Teixeira, R.M.A., Campos, J.M.S., Valadares Filho, S.C., Oliveira, A.S., Anderson Jorge Assis, A.J., Pina, D.S. (2007) Consumo, digestibilidade e desempenho de novilhas alimentadas com casca de café em substituição à silagem de milho (Intake, digestibility and performance of dairy heifers fed coffee hulls replacing of corn silage), *Revista Brasileira de Zootecnia*, 36,.968-977 (in Portuguese).

Thiagalingam, K. and Sriskandarajah, N. (1987) Utilization of agricultural wastes in Papua New Guinea, *Resources and Conservation*, 13, 135-143.

Tiruta-Barna, L., Benetto, E. and Perrodin, Y. (2007) Environmental impact and risk assessment of mineral wastes reuse strategies: Review and critical analysis of approaches and applications, *Resources, Conservation and Recycling*, 50, 351-379.

Tokimoto, T., Kawasaki, N., Nakamura, T., Akutagawa. J. and Tanada, S. (2005) Removal of lead ions in drinking water by coffee grounds as vegetable biomass, *Journal of Colloid and Interface Science*, 281, 56-61.

Ulloa Rojas, J.B., Verreth, J.A.J., van Weerd, J.H. and Huisman, E.A. (2002) Effect of different chemical treatments on nutritional and antinutritional properties of coffee pulp, *Animal Feed Science and Technology*, 99, 195-204.

Ulloa Rojas, J.B., Verreth, J.A.J., Amato, S., Huisman, E.A. (2003) Biological treatments affect the chemical composition of coffee pulp, *Bioresource Technology*, 89, 267-274.

Ulloa Rojas, J.B., Verreth, J.A.J. (2003) Growth of Oreochromis aureus fed with diets containing graded levels of coffee pulp and reared in two culture systems, *Aquaculture*, 217, 275-283.

Vandenberghe, L.P.S., Soccol, C.R., Pandey, A., Lebeault, J.-M. (2000) Solid-state fermentation for the synthesis of citric acid by Aspergillus niger, *Bioresource Technology*, 74, 175-178.

Vasconcelos, A.L.S., Franca, A.S., Glória, M.B.A. and Mendonça, J.C.F. (2007) A comparative study of chemical attributes and levels of amines in defective green and roasted coffee beans, *Food Chemistry*, 101, 26-32.

Vilela, F. G., Perez, J. R. O., Teixeira, J. C. and Reis, S. T. (2001) Uso da casca de café melosa em diferentes níveis na alimentação de novilhos confinados (Use of "sticky" coffee husks at several levels for feeding confined steers), *Ciênc. Agrotec.*, 25, 198-205 (in Portuguese).

Vincent, J.C. (1987) Green coffee processing. In: R.J. Clarke and R. Macrae, (Eds.), *Coffee Technology*, New York: Elsevier Applied Science Publishers Ltd., pp. 1–33.

Xu, G., Murakami, T., Suda, T., Matsuzawa, Y., and Tani, H. (2006) Gasification of Coffee Grounds in Dual Fluidized Bed: Performance Evaluation and Parametric Investigation, *Energy Fuels*, 20, 2695– 2704.

WECD - World Commission on Environment and Development (1987) *Our common future.* Oxford University Press, Oxford.

Wheals, A.E., Basso, L.C., Alves, D.M.G. and Amorim, H.V (1999) Fuel ethanol after 25 years, *Trends in Biotechnology*, 17, 482–487.

Yadvika, Santosh, Sreekrishnan, T.R., Kohli, S. and Rana, V. (2004) Enhancement of biogas production from solid substrates using different techniques—a review, *Bioresource Technology*, 95, 1–10.

Yang, S.-T. *Bioprocessing for Value-Added Products from Renewable Resources*, 2007, Elsevier Applied Science, 670p.

Zhang, Y., Ashizawa, M. and Kajitani, S. (2008) Calcium loading during the dewatering of wet biomass in kerosene and catalytic activity for subsequent char gasification, *Fuel*, 87, 3024-3030.

In: Agricultural Wastes
Eds: Geoffrey S. Ashworth and Pablo Azevedo

ISBN 978-1-60741-305-9
© 2009 Nova Science Publishers, Inc.

Chapter 9

VERMI-CONVERSION OF INDUSTRIAL SLUDGE IN CONJUNCTION WITH AGRICULTURAL FARM WASTES: A VIABLE OPTION TO MINIMIZE LANDFILL DISPOSAL?

Deepanjan Majumdar[*]
National Environmental Engineering Research Institute
Nehru Marg, Nagpur-440020, India

ABSTRACT

There are environmental concerns associated with industrial sludge disposal, apart from other issues like logistics of disposal, treatment options, cost of disposal, etc. A customary disposal option for many industries is secure landfilling, but more and more industries are now looking at the possibility of recycling and bioconversion of the solid wastes to value-added products. Agro-based industries have often resorted to composting, vermicomposting or biogas generation from their wastes due to their biological substrate value and negligible toxicity. However, this has not been the case with other types of industries like pharmaceuticals, chemicals/petrochemicals, power plants, iron and steel and many others, where the sludge may be unsuitable due to the presence of harmful chemicals, volatiles, persistent organic pollutants (POPs), antibiotics, etc. Sludge generated from water treatment plants in the industrial sector forms a major portion of solid waste requiring disposal, and has been used in some reported cases for culturing earthworms and vermicomposting and could be explored for vermicomposting on case-by-case basis. An acceptable approach would be an initial evaluation of the sludge for screening of known harmful agents and factors to earthworms and then conducting proxy vermicomposting trials on these sludges with prior addition of known substrates of earthworms, such as cured animal manures or crop residues or a combination of both. The quality of the final product—or vermicompost—holds great importance, as the end product may not qualify as good manure. But, it is still not clear as to how one could solve the entire sludge disposal problem only by vermicomposting, as it is time-consuming and industries generate sizeable quantities of sludge every day. It appears that

[*] Ph: +91-712-2249877, Fax: +91-712-2249895, Email: d_majumdar@neeri.res.in/joy_ensc@yahoo.com

vermicomposting could only supplement the normal disposal practices of an industry. This chapter attempts to shortlist the suitable industrial and agricultural wastes for vermiconversion, explores the feasibility of their vermiconversion, and looks at various factors influencing the possible implementation of such a practice in industry.

Key words: earthworm, fertility, nutrient, soil, environment, management

INTRODUCTION

The U.S. Environmental Protection Agency defines industrial sludge as a 'semi-liquid residue or slurry remaining from treatment of industrial water and wastewater' (http://www.epa.gov/owm/mtb/biosolids/). The stockpiling of industrial sludge within an industry's premises and their subsequent disposal in landfills without any end use is a common practice in India and many other countries. Sometimes a specific sludge is sold or distributed to farmers for application as a soil conditioner or manure in agricultural fields due to the presence of appreciable organic matter. As long as the sludge does not contain pathogens, toxins, persistent organic pollutants (POP) and metals and conforms to agricultural land disposal standards, this exercise might be useful and environmentally sustainable. It is fair to demand that common sense prevail with the industries who distribute the sludge for agricultural use, as carelessness and deliberate overlooking of quality of the sludge from their side would end up polluting the environment and endangering human health. Lately, solid waste managers have started considering conversion of sludge into value-added products rather than using conventional disposal methods such as incineration or landfilling (Liang et al., 2003). However, no headway has been made for the hazardous sludges, as they always have been unfriendly to microorganisms and earthworms and thus require special care, which is beyond the scope of this paper.

According to Sinha (1996), India is yet to appreciate full importance of vermiculture, despite a potential production of 400 million tons of vermicompost annually from waste degradation. But, there are efforts from the Indian government to push forward various development measures in this sector. In 1998, the Government of India announced exemption from tax liability to all those institutions, organizations, and individuals in India practicing vermiculture on a commercial scale. Khadi and Village Industries Commission (KVIC), Govt. of India, has been patronizing vermicomposting projects by offering a program called 'Margin Money Scheme' to either (i) individuals (rural artisans/entrepreneurs) for projects up to INR 10 lakh (US$ 20833) or (ii) institutions, co-operative societies and trusts registered with KVIC/Khadi and Village Industries Board (KVIB) for projects up to INR 25 lakh (US$ 52083). For the former type of project, the government offers 25% of project cost as 'margin money' by the way of backend subsidy, while for the latter type, 'margin money' is 25% of INR 10 lakh plus 10% of the remaining cost of the project. For some designated sections of society and people of some designated regions, the 'margin money' goes up to 30% from 25%. Any Nationalized Bank of India provides 90–95% of the project cost as loan (Directorate of Non-conventional Energy and Bio-Technology Cell, 2001).

Vermicomposting plants have been operating in Pune and Bangalore in the past with 100 t day^{-1} capacity. Chennai, Mumbai, Indore, Jaipur and several other Indian cities were also setting up vermiculture at that time (Sinha, 1996). The Bhawalkar Earthworm Research Institute (BERI) is one of the largest non-governmental organisations involved in vermiculture practices at Pune in India, and was operating a vermiculture plant on a commercial scale for the management of municipal wastes (Bhawalkar and Bhawalkar, 1994). Many private firms and non-governmental organizations have understood the market potential of this product and have started producing and marketing vermicompost, though prepared from purely organic wastes of non-industrial origin. Information on these firms and organizations can be found in the World Wide Web.

AGRICULTURAL FARM WASTES: OVERVIEW

Agricultural farm wastes are various materials/by-products left unused after generation by the activities in agricultural farms that may be exclusively crop or horticultural or mixed (crop+animal) farms, while exclusively animal or poultry farms are out of the purview of agricultural farms. In a strict sense, these are not 'wastes' since they can be subsequently utilized for wealth/material/energy generation and vermicomposting remains one of the major options. Agricultural farm waste has a huge stake in the rural development sector and commercial sector as its utility has been recognized by many stakeholders, which include both commercial and non-commercial entities (Figure 1). The use of agricultural solid wastes by recycling can improve soil physical conditions and fertility (Mishra et al., 1989; Bhardwaj, 1994; Sudha and Kapoor, 2000). Some of these wastes can be directly added to soil without any treatment if they do not contain hazardous or toxic pollutants (Lerch et al., 1992).

Until recent times, the majority of research concerning agricultural farm wastes have revolved around organic materials, particularly manure and slurry. Non-natural waste materials, like packaging materials, plastics, tires and oil that have also been generated and found in farms have largely been ignored (Environment Agency, 2001). The outlook has now changed, and since anything used in the farm could end up as wasted materials and add to farm waste, this change in outlook is welcome. But, keeping the scope of this chapter in mind, these wastes—which are by and large unsuitable for vermicomposting—are kept out of the purview of the present discussion. Also, agro-industrial waste is not considered as farm waste but is well within the scope of industrial sludge. It is the organic farm waste that is important and suitable for vermicomposting (Figure 2). These wastes are generated either by field crops, other vegetation, farm animals, especially cattle and buffalo, due to their predominance as farm animals, along with others that are either used in mixed farming or supplementary animal farming or used for other domestic/commercial purpose in farms. Other domestic sources in farms would also generate relevant wastes for vermicomposting, e.g., oilcakes, kitchen waste, papers and boards, etc.

In an average Indian farms and farms in many other countries, major waste products include various crop residues that are not used after harvest. Much of these residues are recycled in fields as organic mulch or organic soil manure or amendments. However, this type of waste holds the key to the raw material supply from the farm sector for composting or vermicomposting and, in fact, these are frequently used for generating farm yard manure (FYM). The potential of these crop residues is huge in terms of nutrients locked in these materials. An estimate done in India projects the nutrient potential of these crop residues generated in India (Table 1). Apart from crop residues, oilseed cakes are also generated on farms, though only on farms having domestic or commercial oil extraction facility. Oilseeds are typical farm products that carry appreciable nutrients (Table 2), especially nitrogen, and in many cases are used in the farm as a source of nutrient in soil or for biopesticide, e.g., neem cake. These cakes can also be used in vermicomposting to improve nutrient status of the final product.

Animal manure has been regarded as the wonder formula among all naturally available agricultural farm wastes, due their unique combination of physical and chemical properties and as a great biological supporting system. Cured animal manure has been considered perhaps the best additive in a vermicomposting mixture due its easy biodegradability and suitability to earthworms. Even without earthworms, manures are best suited for the enhancement of biodegradation of organic waste due to their great substrate value to microflora and microfauna involved in the food web operating in a composting heap, all helping in the quick degradation of waste mixture. Quality of animal manure varies considerably location wise, region wise and country wise, as it is greatly influenced by the animal species and family, diet, health and management (Kimani and Lekasi, 2004). In a country like India and several other similar agrarian countries, animal manure is omnipresent, due to huge animal population (Government of India, Livestock Census Reports) and agriculture being the largest occupation. A summary of animal manure quality is presented in Table 3.

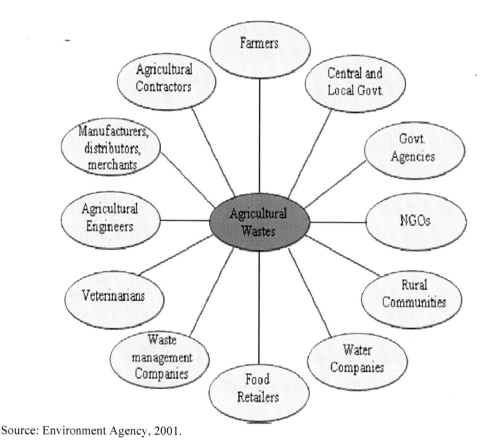

Source: Environment Agency, 2001.

Figure 1. Stakeholders in the management of agricultural waste.

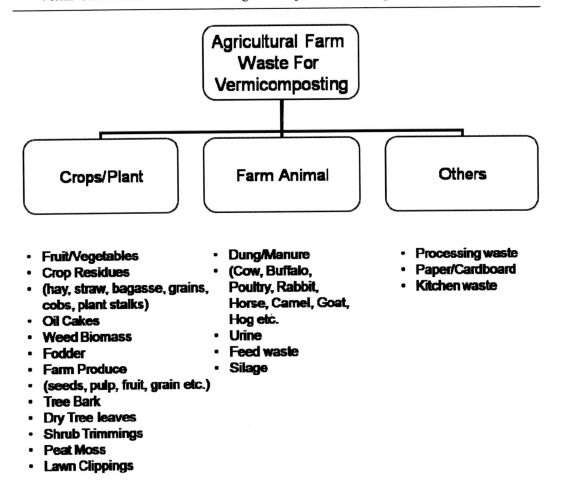

Figure 2. Suitable agricultural farm wastes for vermicomposting.

VERMICOMPOSTING AS A TECHNOLOGY

Vermi is the Latin word for 'worm', and vermicomposting is composting by earthworms. Earthworms for their role in improving soil fertility and health are known as farmers' friends for long. Their role in inverting and mixing soil drew attention of Aristotle who called earthworms the 'intestines of the earth'. Darwin, though popularly known for his theories on evolution, was one of the pioneers who highlighted the role of earthworms in soil health. Darwin's conclusions led to an upsurge of interest in earthworms from the late nineteenth century onwards. Darwin discovered the role of earthworms in the breakdown of dead plant and animal material in soil and forest litter and in the maintenance of soil structure, aeration and fertility (Freeman, 1977). Earthworms not only inhabit soil but by virtue of their activity, contribute to the physical and chemical alterations in the soil, leading to increased soil fertility and plant growth. Some species of earthworms that especially love to feed on pure organics are called as manure worms whereas those who dwell in soil and feed primarily on soil are called as soil dwellers.

Table 1. Estimates of the availability of some crop residues in India and their plant nutrient potential

Residue	Residue: Economic yield	Residue yield* ('000 tonnes)	Nutrient (%) N	P$_2$O$_5$	K$_2$O	Nutrient potential, '000 tonnes Total	Utilizable**	Fertilizer equivalent***
Rice	1.5	110,495	0.61	0.18	1.38	2,398	799	399
Wheat	1.5	82,631	0.48	0.16	1.18	1,504	501	250
Sorghum	1.5	12,535	0.52	0.23	1.34	262	87	43
Maize	1.5	11,974	0.52	0.18	1.35	252	84	42
Pearl millet	1.5	6,967	0.45	0.16	1.14	121	40	20
Barley	1.5	2,475	0.52	0.18	1.30	51	17	8
Finger millet	2.0	5,351	1.00	0.20	1.00	118	39	19
Sugarcane	0.1	22,736	0.40	0.18	1.28	423	423	211
Potato tuber	0.5	7,867	0.52	0.21	1.06	141	141	70
Groundnut (pod)	1.5	10,598	1.60	0.23	1.37	339	339	169
Total	-	273,629	-	-	-	5,609	2.470	1,231

* arrived at by multiplying the economic yield by the given residue: economic yield ratio
**One-third of the total NPK potential assuming that two-thirds of the total residue is used as animal feed on national basis
*** 50% of the utilizable NPK, assuming 50% mineralization of NPK per season
Source: Bhardwaj, 1994.

Table 2. Average nutrient content of common oilcakes

Oilcake sources	N	P	K	Kg N+P$_2$O$_5$+K$_2$O per tonne of cake
Edible Oilseeds	——— percent ———			
Groundnut	7.29	1.65	1.33	103
Mustard	4.52	1.78	1.40	77
Rapeseed	5.21	1.84	1.19	82
Linseed	5.56	1.44	1.28	83
Sesame	6.22	2.09	1.26	96
Cottonseed (decorticated)	6.41	2.89	1.72	110
Cottonseed (undecorticated)	3.99	1.89	1.62	75
Safflower (decorticated)	7.88	2.20	1.92	120
Safflower (undecorticated)	4.92	1.44	1.23	76
Non-edible Oilseeds				
Castor	4.37	1.85	1.39	76
Neem (Azadirachia indica)	5.22	1.08	1.48	59

Oilcake sources	N	P	K	Kg N+P$_2$O$_5$+K$_2$O per tonne of cake
Mahua (Madhuca indica)	3.11	0.89	1.85	59
Karanj (Pongamia glabra)	3.97	0.94	1.27	62
Kusum (Schleichera oleosa)	5.23	2.56	1.37	92
Khakan (Salvadora oleoides)	4.32	2.45	1.24	80

Source: Bhardwaj, 1994.

Table 3. Nutrient content in a few animal manures

Material	N	P$_2$O$_5$	K$_2$O	Ca	Mg
Pig dung	2.27	3.1	1.8	0.21	0.54
Cow dung	1.74	1.7	0.6	0.37	0.53
Horse dung	1.07	2.1	3.6	0.26	0.49
Camel dung	1.51	0.35	1.8	0.7	0.69
Poultry excreta	2.17	2.0	4.2	0.28	1.39
Goat and sheep excreta	0.65	0.5	0.03	-	-
Material	Fe	Zn	Mn	Cu	B
Pig dung	1200	50	70	8.9	-
Poultry excreta	1400	90	210	7.1	5
Goat and sheep excreta	-	2570	150	61	4600

Source: Bhardwaj, 1994.

Vermicomposting is an adequate technology for the transformation of various but not all types of solid wastes into a valuable product like manure (Elvira et al., 1996). The ability of earthworms to consume and break down a wide range of organic residues such as agricultural, animal, industrial and domestic wastes, sewage sludge and crop residues is well known (Delgado et al., 1995; Benitez et al., 1999; Gratelly et al., 1996; Gajalakshmi et al., 2002; Bansal and kapoor, 2000; Hand et al., 1998; Talashilkar et al., 1999; Garg et al, 2006; Mitchell et al. 1980, Edwards et al. 1985, Chan and Griffiths 1988, Hartenstein and Bisesi 1989; Hand et al., 1988; Harris et al., 1990; Logsdon 1994; Ndegwa et al., 2000). Earthworms like to feed slowly on decomposing organic materials like vegetable scraps, plant litter, animal manures or any other organics rich non-toxic materials but also can consume inorganic substrates when interspersed well in organic rich wastes. The "end product" of the vermicomposting is called as "castings", which is actually the excreta of earthworms, which having passed through the gut of earthworms, is full of beneficial microbes and nutrients and other valuable organics such as enzymes, vitamins and organic acids, making it a great fertilizer. Vermicomposting is a better way of waste conversion since it makes quality manure quickly and gives a nutritionally rich and biologically more active product at reasonable cost (Directorate of Non-conventional Energy and Bio-Technology Cell, 2001). Elvira et al. (1997) reported that two factors might limit microbial composting processes: (1) difficult degradation of the structural polysaccharides and (2) low nitrogen content of the waste. Earthworms especially take care of the first limiting factor, i.e., in the process of feeding they fragment the substrates, increasing their surface area for further microbial action (Chan and Griffiths, 1988) and do not depend on nitrogen in particular. But, to obtain stabilized and products appropriate for agricultural purposes, industrial waste/sludge needs to be mixed with other nitrogen rich organic wastes in order provide the much needed nitrogen to the microorganisms who act together with earthworms towards a better output (Butt 1993; Elvira et al., 1997, 1998).

During earthworm mediated bioconversion, major plant nutrients like nitrogen, potassium, phosphorus etc. present in the substrate are converted through microbial action into more soluble forms that are much more available to plants than those in the parent substrate. During vermicomposting, earthworms maintain aerobic condition in the waste pile through burrowing and inverting and biochemical processes are enhanced by microbial decomposition of substrate in the earthworm intestine. Earthworms convert a portion of the organics present in the wastes into worm biomass and excrete undigested and partially digested matter as worm cast (Benitez et al., 1999). Earthworms, by creating aerobic conditions in the waste materials, inhibit the action of anaerobic micro-organisms which release foul-smelling hydrogen sulphide and mercaptans. *Eisenia foetida, Eisenia andrei, Eudrilus euginae, Lumbricusrubellus* and *Perionyx excavatus* are major waste eater and biodegrader earthworm species. They are used worldwide for waste degradation and are found to be very successful functionaries for the ecological management of organic municipal wastes (Edwards, 1988). *E. euginae* and *P. excavatus* are believed to be the more versatile waste managers (Graff, 1981; Kale et al., 1982). Various microflora present in the intestine of earthworm and in the waste are also actively involved in the decomposition of organic carbon, while the gut enzymes play a dominant role in this process (Whiston and Seal 1988; Kavian and Ghatneker 1991). Further, earthworms also enhance soil microbial activity by improving the environment for microbes (Syers et al., 1979; Mulongoy et al., 1989).

Traditional thermophillic composting and vermicomposting have been combined to enhance the overall process and product's quality. Ndegwa and Thompson (2001) used two approaches: (1) pre-composting followed by vermicomposting, and (2) pre-vermicomposting followed by composting. The substrate was biosolid (activated sewage sludge) with mixed paper-mulch as the carbon base. *E. foetida* was used for vermicomposting and results indicated that these processes not only shortened stabilization time, but also improved the product's quality. Combining the two systems resulted in a product that was more stable and consistent (homogenous), had less potential impact on the environment and the product met pathogen reduction requirements.

VERMICOMPOSTING OF INDUSTRIAL SLUDGE

Scientific literature is abound with information on vermicomposting of municipal sewage sludge, evidently due to high organic content and chemically benign nature of sewage sludge. Industrial sludge also has been successfully tried by many, though not as commonly as sewage sludge, due to unfriendly nature of various industrial sludge. To obtain stabilized and products appropriate for agricultural purposes, industrial sludge need to be mixed with other nitrogen rich organic wastes in order provide nutrients and inoculum for microorganisms (Hartenstein, 1978; Elvira et al., 1998). It is very important to know the chemical composition of the wastes and the best mixture ratio to have a good quality and stabilized final compost. Hazardous solid wastes from industries may not prove to be good auxiliary substrate for vermicomposting.

There are reports where ETP sludge and spent mycelium from a pharmaceutical industry or ETP sludge from a CPC green and blue pigment manufacturing industry were not found suitable for vermicomposting (Majumdar et al., 2006; Macwan, 2005). In these cases, probably landfilling or energy generation via incineration (if heat value of waste is found to be suitable) could have been better options.

Wastes like biosolids from industries and sewage treatment plants can produce soil sickness due to presence of contaminants (Ayuso et al., 1996) and pathogens (Hassen et al., 2001). Composting helps stabilizing organic matter and reducing pathogens in sludges to very low levels (Burge et al., 1987; Millner et al., 1987). Composting might be a better option than direct field application since it helps break the complex polymers and help release nutrients, which could be used by the plants directly in fields after application or might even detoxify the wastes and reduce soil and groundwater contamination. Industrial wastes have been used for biogas generation (Jagdeeshan, 2004), direct field applications (Kalra et al., 1998; Mathur, 1995; Yadav, 1995), briquetting (Panwar et al., 2004), energy generation (Reimann, 2000) and landfilling when not recyclable (Hall, 2000).

Decision makers and managers in industries have by and large hesitated to involve industrial sludge in vermicomposting programmes due to concerns of unsuitability of the sludge to earthworms and microorganisms or failure of earthworms to adapt to the waste mixtures containing the sludges in feeding trials. But, there are instances where industrial sludge have been successfully converted into useful compost in conjunction with other organic wastes by different species of earthworms in India and elsewhere, including sludge from various industries and processes. Different sludges which have already been converted into useful compost by different species of earthworms include sewage sludge (Maboeta and van Rensburg, 2003; Benitez et al., 1999); dairy processing plant sludge (Kavian and Ghatneker, 1991; Elvira et al., 1998); paper mill industry sludge (Elvira et al., 1998; Butt, 1993); pig waste (Chan and Griffiths, 1998); vine fruit industry sludge (Atharasopoulous, 1993), etc. Industrial solid wastes, e.g., treatment plant sludge of various industries, have also been utilized in many countries for vermicomposting (Benitez et al., 1999; Elvira et al., 1997; Butt, 1993; Atharasopoulous, 1993). Some other major non-sludge solid waste like fly ash also has been used for vermicomposting (Bhattacharya et al., 2004; Butt, 1993). The apparently non-toxic and friendly sludges have made themselves acceptable to earthworms, or at least non-harmful, if not beneficial, as they are generated from processes where the raw material are benign and the residues of chemicals or additives used or their by products are accepted by microorganisms and earthworms. Various types of industries may produce sludges which could be used for vermicomposting on case to case basis (Figure 3).

Research at Concept Biotech, a turnkey solution provider to industries especially in Gujarat in India, confirms that toxic materials indeed reduces the average life of earthworms from one and half year to 6–8 months, depending upon the concentration of the toxic matter. Even within this shortened span of life, earthworms regenerate 10–20 times. Dead worms too form organic matter. The same quantity of toxic matter gets distributed in multiplied worms leaving manure toxin free, claims the organization. Centipedes and ants in the vermiculture beds provide a proof of a manure's non-toxic nature. Having developed this concept after testing a number of batches with different proportions of industrial sludge and organic substrates (cowdung, bagasse, leaf litter), Concept Biotech has provided services to some medium to large-scale industrial units over last three-four years. An ideal scheme for industrial sludge vermicomposting has been envisioned by Concept Biotech (Figure 4). Some of the executed projects included treatment of textile sludge of Jagdamba Textile at Narol and Samir Textiles at Odhav in Gujarat; Gujarat Refinery waste into non-toxic manure; paper industry sludge of Vepar Pvt. Ltd. at Ahmedabad; dairy waste of Mother Dairy at Gandhinagar; pharma sludge of Aventis Crop Science and Lupin Ltd. at Ankleshwar; and Torrent Pharma at Mehsana in Gujarat. Currently, work on pilot projects at Reliance industries' Hazira, as well as Patalganaga plants, are underway. However, Concept Biotech researchers are of the opinion that no waste from dyes and intermediate plants should be used for vermicomposting, as these have a very high proportion of inorganic compounds in their waste. This statement gets corroboration from the work of Macwan (1995). Only those industrial waste sludge that has good proportion of organic matter, can be converted into non-toxic manure, e.g., sludge from textile industry, dairy, distilleries, pulp and paper industry, food processing, agro industries, etc. The approach should be to carry a detailed characterization of the waste, take up controlled experiments before carrying out large scale experiments and only after establishing the desired results over a sufficiently long period, industrial application should be undertaken (Dabke, Concept Biotech, Personal Communication and http://www.indiatogether.org /2004/sep/env-vermtoxic.htm).

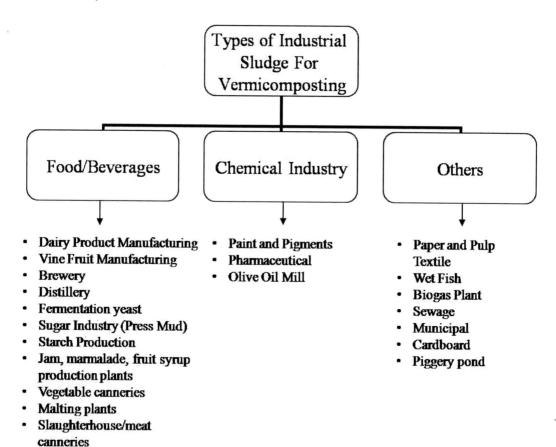

Figure 3. Types of Industrial sludge that have been tested or warrant attention for vermicomposting trials.

University of Agriculture Science in Bangalore has been similarly providing solutions to distilleries in Karnataka. The state's pollution control board (KPCB) has already given the consent to this process. However, the Gujarat Pollution Control Board (GPCB) and Maharashtra Pollution Control Board (MPCB) have not given their consent (http://www.indiatogether.org/2004/sep/env-vermtoxic.htm).

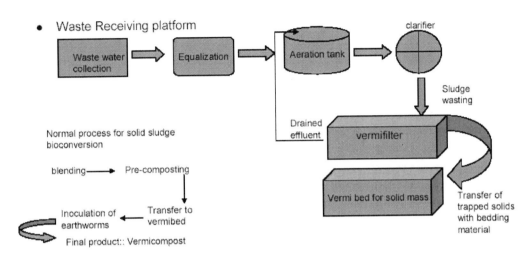

Source: Dabke, Concept Biotech (Personal Communication)

Figure 4. Block diagram of an ideal action plan of sludge vermicomposting.

CASE STUDIES FROM INDIA AND ELSEWHERE

Various reports are available on vermiconversion of industrial sludge in several leading research journals. These papers have reported growth and development of earthworms in waste mixtures containing various types of industrial sludge mixed with various organic wastes as co-substrates and their conversion into nontoxic and fertile manure. Some of these results are discussed.

Textile mill sludge was found to be appropriate for vermicomposting when mixed with cow dung and/or agricultural residues into vermicompost. The maximum growth and reproduction of the worms was obtained in 100% cow dung, as cow dung is one of the best substrates, but worms grew and reproduced favorably in 80% cow dung+20% solid textile mill sludge and 70% cow dung+30% solid textile mill sludge also. Vermicomposting resulted in significant reduction in C:N ratio and increase in Total Kjeldahl Nitrogen (TKN), Total Phosphorus (TP), Total potassium (TK) and Total Calcium after 77 days of worm activity in all the feeds (Kaushik and Garg, 2004). In another study, textile mill sludge spiked with poultry droppings was also found suitable for vermicomposting. Replacement of poultry droppings by cow dung in feed mixtures and vice versa had little or no effect on worm growth rate and reproduction potential. Worms grew and reproduced favourably in 70% poultry droppings (PD)+30% solid textile mill sludge (STMS) and 60% PD+40% STMS feed mixtures. Greater percentage of STMS in the feed mixture significantly affected the biomass gain and cocoon production. Net weight gain by earthworms in 100% CD was 2.9–18.2 fold higher than different STMS containing feed mixtures. As a confirmation of their earlier report, vermicomposting resulted in significant reduction in C:N ratio and increase in nitrogen and phosphorus contents. But, total potassium, total calcium and heavy metals (Fe, Zn, Pb and Cd) contents were lower in the final product than initial feed mixtures (Garg and Kaushik, 2005). Textile sludge when mixed with anaerobically digested biogas plant slurry (BPS) was found effective in aiding vermicomposting. Although *E. foetida* did not survive in fresh sludge but grew and reproduced in sludge spiked with slurry feed mixtures and maximum growth was recorded in 100% BPS. The net weight gain by *E. foetida* in 100% BPS was two-four-fold higher than STMS-containing feed mixtures. After 15 weeks, maximum cocoons (78) were counted in 100% BPS and minimum (26) in 60% BPS+40% STMS feed. Vermicomposting resulted in a pH shift towards acidic, a significant reduction in C:N ratio, and increase in nitrogen, phosphorus, and potassium contents (Garg et al., 2006). Further, vermicomposting with *E. foetida* of solid textile mill sludge mixed with cow dung in different ratios in a 90 days composting experiment resulted in significant reduction in C:N ratio and increase in TKN as reported earlier also. Total K and Ca were lower in the final casts than the initial feed mixture. Total P was higher in the final product than the initial feed mixture. Total heavy metal contents were lower in the final product than initial feed mixture. Microbial activity measured as dehydrogenase activity increased up to 75 days and decreased on further incubation. Solid textile mill sludge could be potentially used as a raw substrate when mixed with up to 30% cow dung on dry weight basis. Increasing proportion of STMS in the feed mixtures promoted a decrease in survival and growth of *E. foetida*. Mortality was recorded in 100% STMS, 90% STMS +10% CD and 80% STMS+ 20% cow dung feed mixtures. Hundred percent cow dung had the highest number of total earthworms and clitellated earthworms but a lower number of cocoons than 60% CD+ 40% STMS, 50% CD+ 50% STMS and 40% CD+ 60% STMS after 90 days. This might have been due to hatching of the cocoons. There was little increase in the total number of earthworms and clitellated earthworms in the feed mixtures having more than 50% STMS, but they had a greater number of cocoons than did 100% CD. This indicated that a greater percentage of STMS in the feed mixture significantly delayed the sexual maturity and reproduction of *E. foetida* (Priya Kaushik and Garg, 2003). In another study, biosolids, ninety percent from varied industrial origins, but mainly from textile industries and the rest from households were vermicomposted with cow manure and oat straw for 2 months at three different moisture contents (60%, 70% and 80% dry weight base). The vermicompost with the best stability and maturity and a weight loss of 18% had the following properties: pH 7.9; organic C content of 163 g kg^{-1}; an electrolytic conductivity of 11 mS cm^{-1}; a humic-to-fulvic acid ratio of 0.5; total N content of 9 g kg^{-1}; water soluble C less than 0.5%; cation exchange capacity of 41 cmolc kg^{-1}; a respiration rate of 188 mg CO$_2$-C kg^{-1} compost-C day^{-1}. The vermicompost gave a germination index of 80% for cress (*Lepidium sativum*) after 2 months while the earthworm production increased 1.2-fold and volatile solids decreased five times. In addition, the vermicompost contained less than 3 CFU g^{-1} *Salmonella spp.*, no fecal coliforms and *Shigella spp.* and no eggs of helminths. Concentration of sodium was 152 mg kg^{-1} dry compost, while concentrations of chromium, copper, zinc and lead were below the limits established by the USEPA (Contreras-Ramos et al., 2005).

Transformation of sugar mill sludge amended with biogas plant slurry (BPS) into vermicompost employing the epigeic earthworm *E. foetida* had also resulted in decrease in pH, TOC, TK and C:N ratio, but increase in TKN and TP, implying an increase in available N and

P. Addition of 30–50% of paper mill sludge with BPS had no adverse effect on the fertilizer value of the vermicompost as well as growth of the worms. Authors noted that vermicomposting can be an alternate technology for the management and nutrient recovery from press mud if mixed with bulking agent in appropriate quantities (Sangwan et al., 2008).

Distillery industry sludge mixed with cow dung in different proportions of 20%, 40%, 60% and 80% resulted in a vermicompost by *Perionyx excavatus* with a significant decrease in pH (10.5–19.5%) and organic C contents (12.8–27.2%) while an increase in total N (128.8–151.9%), available P (19.5–78.3%) as well as exchangeable K (95.4–182.5%), Ca (45.9–115.6%), and Mg contents (13.2–58.6%) was observed. Vermicomposting also caused significant reduction in total concentration of metals: Zn (15.1–39.6%), Fe (5.2–29.8%), Mn (2.6–36.5%) and Cu (8.6–39.6%) in sludge. Bioconcentration factors (BCFs) for metals in different treatments were also calculated and the greater values of BCFs indicate the capability of earthworms to accumulate a considerable amount of metals in their tissues from substrate. Worms showed maximum rate of biomass gain and growth (mg weight worm^{-1} week^{-1}) and cocoon production rate in 40% sludge, while least values of these parameters were in 80% sludge treatment (Suthar and Singh, 2008).

Paper mill sludge was put for vermicomposting using two exotic species (*Eudrilus eugineae* and *E. foetida*) and an indigenous species (*Lampito mauritii*) of earthworm in India, with paper mill sludge added in various proportions of 25%, 50% and 75%. The waste mixtures were subjected to worm treatment for a period of 60 days and results indicated that 25% sludge proportion was ideal and of the three worms used *E. foetida* was the best worm for vermicomposting (Banu et al., 2001). Buch and Patel (2005) and Mehta and Seth (2007), conducted interesting works on the suitability of wastewater sludges generated in paint industries for vermiculturing and found positive results where earthworms adapted easily and produced casting from these sludges mixed with animal manures like cow dung and camel dung.

Apart from Indian subcontinent, few studies have also been undertaken with industrial sludge in other countries. Solid paper mill sludge (SPMS) waste was amended by adding sewage sludge, pig and poultry slurry, characterized by low total solids content and high organic matter and nitrogen. Nine mixtures were established by mixing SPMS with one part of other wastes at three different proportions and moistened to 80%. Immature specimens of *E. andrei* were used for a feeding trial at room temperature and earthworm growth and survival were recorded periodically. Worm mortality occurred in proportions containing higher than 75% sludge. The increasing proportion of sewage sludge in the mixture promoted a decrease in the survival and growth. The 3:2 mixture (sewage sludge: SPMS) produced the highest growth rates and the lowest mortalities. In mixtures of SPMS with pig slurry, earthworms exhibited fast growth and high mortality and their survival after 45 days was always lesser than 25% in all proportions tested. When poultry slurry was added, rapid growth was observed in 3:2 and 2:1 mixtures, although mortality was also high. The maximum weight was achieved when the SPMS was presented in the highest proportion but showed a low growth rate during the entire experiment (Elvira et al., 1997). Subsequently, Elvira et al. (1998) again studied vermicomposting of sludge from a paper mill mixed with cattle manure with *E. andrei* in a six-month pilot-scale experiment. Initially, a small-scale laboratory experiment was carried out to determine the growth and reproduction rates of earthworms in the different substrates tested. In the pilot-scale experiment, the number of earthworms increased between 22- and 36-fold and total biomass increased between 2.2- and 3.9-fold. The vermicomposts were rich in nitrogen and phosphorus and had good structure, low levels of heavy metals, low conductivity, high humic acid contents and good stability and maturity. Researchers opined that these sludges could be potentially useful as raw substrates in larger commercial vermicomposting systems and would reduce the costs related with the exclusive use of different types of farm wastes as feed for earthworms.

In an interesting study, effluent solids from a large recirculating aquaculture facility (Blue Ridge Aquaculture, Martinsville, Virginia) was tested and found suitable for vermicomposting using the earthworm *E. foetida*. In two separate experiments, worms were fed mixtures of sludge removed and shredded from aquaculture effluent. Mixtures containing 0%, 5%, 10%, 15%, 20%, 25%, and 50% aquaculture sludge on dry weight basis were fed to the worms over a 12-week period and their growth (biomass) was measured. Worm mortality, which occurred only in the first experiment, was not influenced by feedstock sludge concentration. In both the experiments, worm growth rates tended to increase with increasing sludge concentration, with the highest growth rate occurring with the mixture containing 50% aquaculture sludge. Effluent solids mixed with shredded cardboard appeared to be a suitable feedstock for vermicomposting (Marsh et al., 2005).

BOTTLENECKS OF IMPLEMENTING SLUDGE VERMICOMPOSTING AT THE INDUSTRIAL LEVEL

Implementation of vermicomposting in industries on regular basis for sludge bioconversion or for that matter any biodegradable solid waste needs a serious look at various relevant issues. These issues may range from the quantity of sludge generated, availability of better disposal options like incineration or landfilling, suitability of the sludge for earthworms and finally, the willingness of the management to adopt vermicomposting, if found applicable. A decision tree is proposed here for streamlining decision making with respect to sludge vermicomposting or other alternative options for that matter (Figure 5). In general, since vermicomposting is a time taking process, it cannot guarantee the recycling of the entire amount of solid waste generated in all types of industrial units. It also demands appreciable land area when load of waste is high, making it difficult to adopt by many industrial units having shortage of land. Having a common vermicomposting facility can also be practically very difficult, due again to lack of land availability, problems in handling huge quantity of solid sludge, slow nature vermiconversion process etc. So, practically vermicomposting could only serve as a supplementary sludge treatment or disposal option and has to be adopted by industrial units individually, or by only few industries together having a common type of sludge. On the operational side, the whole job in a community vermicomposting facility is made extremely difficult by the fact that the sludge keep on changing. And therefore it is impossible to have one method of treatment for all types of sludges.

Use of industrial sludge is possible only after detailed characterization and this entails the industrial units to adopt detailed characterization of the waste on their own or by outsourcing. Also, the sludge quality has to be standardized as frequent characterization is not a practical solution. In India each sludge/solid waste sample could cost anything around or even more than 30,000 rupees (approx. 600 US$) for detailed characterization and that too with skilled personnel. This cost is multiplied as one would have to do many samples and that too over a significant period of time. For some special parameters dioxins, India has only few experts and testing facilities. Further, similar characterization of the vermicompost would have to be done, making the whole affair very expensive. Few beg to differ though. According to few, characterization need not be that costly as knowing a total load estimate, which is not so expensive, will be enough to make a decision. The load estimate is for the level of organic/inorganic compounds in the sludge at the input to the vermicomposting process. However, on estimation of toxicity for post-compost residue, these experts acknowledge that because many batches of sludge need to undergo composting over a considerably long period of time, establishing this level is a time consuming process (http://www.indiatogether.org/2004/sep/env-vermtoxic.htm).

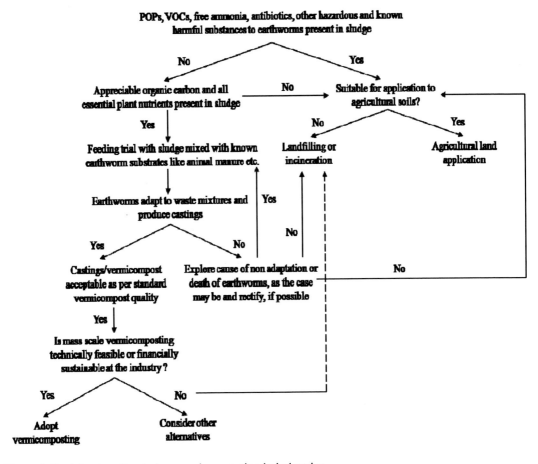

Figure 5. Decision tree for sludge vermicomposting in industries.

Ideally speaking, industries should process the biological waste and send the difficult waste to the secure landfill sites. With vermicomposting, there would be considerable reduction in the volume going to legal landfill or illegal dumping grounds. As per Concept Biotech, the conversion costs are not prohibitive at Rs 3–4 per kg of sludge as against Rs 5–6 per kg for sending it to landfill (http://www.indiatogether.org/2004/sep/env-vermtoxic.htm).

ATTRIBUTES OF AN IDEAL SLUDGE FOR VERMICOMPOSTING

The following are some of the important general attributes of good quality sludge for vermicomposting. Earthworms might also consume materials not having some of these qualities but might not successfully compost them to an enriched product. Unfortunately, there is no acceptable standard for the various elements, inorganic and organic compounds or chemicals, pathogens etc. or other agents for judging suitability or of sludge or solid waste for vermicomposting. But, a general understanding has been developed over the years based on research and field work as to which solid waste can be a suitable for vermicomposting (Table 4). Some of the factors as presented in Table 1 are proven beyond doubt while some are speculated based on their known biological effects in general and needs further confirmation in case of earthworms.

Table 4. Suitability criteria of solid wastes vis a vis industrial sludge for vermicomposting

Avoidable	Preferable
• High conc. of antibiotics (Bauger et al., 2000) • High free NH_3 (Clark, 1998) • High conc. of VOCs (Youn-Joo and Lee Woo-Mi, 2008) • Water saturation or flooding • Anaerobic condition within waste pile (Clark, 1998) • High (>35°C) and low (<10°C) temperature (Clark, 1998) • High acidity (Godbold and Huttermann, 1994; Chan and Mead, 2003) or alkalinity (optimum pH 5.5-8.5) • Addition of fresh manure, as heat generation during its decay may be harmful • High conc. of dioxins, furans (Environment Canada, http:// www.ec.gc.ca/ceqg-rcqe/ English/Html/GAAG_ DioxinsFuransSoil_e.cfm). • High conc. of pesticides (Luo et al., 1999) • Non-biodegradable materials like metal, plastic, glass or acrylic • Lignin rich waste (e.g., wood) as it is difficult to break mechanically • Radioactive substances (Swedish Radiation Protection Authority, 2006) • Chemicals like peroxides, phenols, acids, alkalis • High chloride (>0.5%) and copper salts (>0.1%) (Clark, 1998)	• Rich in soft organic residues containing carbohydrates, sugars, cellulose, hemicellulose, etc. for easy worm penetration and degradation • Nitrogen rich wastes • Waste with small particle size for easy and quick degradation • Waste with good water holding capacity to maintain enough moisture (70-90%) as worms need moisture for maintaining their body fluid and mobility (Clark, 1998).

BASIC MANAGEMENT ISSUES AT A VERMICOMPOSTING FACILITY

The following aspects are important at a vermicomposting facility and could be used as guidelines by industries interested in vermiconversion of the sludge generated at their facility:

- There should be enough earthworms to convert the desired quantity of waste in target time. Large number of worms ensures quicker composting. Steady and emergency supply of earthworms in need should be ensured.
- Industries generating sizeable quantity of wastes on daily basis should only use a limited part of their waste for vermicomposting as this process is slow.
- There should be enough and suitable space available for putting vermibeds, i.e., bedding of waste materials for vermicomposting or vermiculturing which should be protected from vermin and direct rainfall and kept under shade.
- Vermibeds should be sufficiently porous to allow drainage and diffusion of air inside.
- There should be several vermibeds for batch conversion of wastes to take care of regular supply of sludge-.
- Waste quality should be stable and if variable, should be checked periodically.
- There should be assured supply of water for application in the vermibed.
- Organic auxiliary wastes should better be added shredded and decomposed to ensure better and quicker consumption by earthworms.
- Cow dung or any other animal dung should be added after few weeks of curing and decomposition to ward off the thermophillic stage to avoid heat build up in the waste pile which cause discomfort and even death of worms.
- Solid wastes having animal flesh, dairy products, eggs, oily foods, salts, vinegar, butter, bones, cheese, lard, mayonnaise, milk, peanut butter, salad dressing, sour cream, vegetable oils, yogurts, grease and diseased plants should be avoided as they generate foul smell.
- If foul smell is witnessed during vermicomposting, aeration of the waste heap or correction of pH by adding powdered eggshells or lime or letting the excess moisture to evaporate or drain might be necessary.
- In case of mixed type of auxiliary solid wastes, sorting is of utmost importance to ensure addition of only suitable materials for vermicomposting.
- A pre-composting facility may be provided on case to case basis .

VERMICOMPOSTING AS AN INCENTIVE FOR INDUSTRIES

Vermicomposting may gives the following benefits to the industry:
1. It partially removes the waste disposal problem and thus saves money by minimizing transport and labour cost.
2. Vermicompost can be applied on the industry premises for garden, greenbelt and landscape development. In some cases, even a kitchen garden can be maintained on the industry premises. Industry can save money on fertilizers and manures otherwise used for these practices.

3. Vermicompost, if of standard quality (Table 5), and earthworms can be sold to entrepreneurs and farmers.
4. It is one way to test the hazardous nature of the solid waste.
5. Bioconversion of solid wastes by vermicomposting projects 'User of Eco-Friendly Waste Management' image of the industry.

Table 5. Properties of a good-quality bio manure

Sl. No.	Parameter	Value
1.	pH	7-8.55
2.	Organic C	20-25%
3.	Nitrogen	1.5-2%
4.	Phosphorus	1-2%
5.	Potassium	1-2%
6.	Calcium	1-3%
7.	Magnesium	1-2%
8.	Sulphur	<1%
9.	Moisture	15-20%
10.	C:N Ratio	15-20: 1
11.	Copper, zinc, manganese and iron	200 ppm
12.	Pathogens	Absent
13.	Toxic heavy metals	Absent

Source: Directorate of Non-Conventional Energy and Bio-Technology Cell of Khadi and Village Industries Commission.

ENVIRONMENTAL RISKS AND BENEFITS OF SLUDGE VERMICOMPOSTING

There are environmental or ecological concerns associated with solid waste vermicomposting, which may apply in case of industrial sludges also. These issues need to be addressed during vermicomposting or when vermicompost is applied in agricultural fields and whenever possible, apt measures should be undertaken to avoid these environmental footprints.

Water Quality Issues

There was an early belief that vermicomposting may not destroy dangerous pathogens as the process does not reach the high temperatures of conventional composting. In recent years, strong evidence has surfaced that worms indeed destroy pathogens with a success rate equal to conventional composting (Pierre et al., 1982; Eastman, 1999; Eastman et al., 2000), although how is still unknown (Eastman, 1999; Eastman et al., 2000). More recently, it has been found that worms living in pathogen-rich material, when dissected, showed no evidence of pathogens beyond the first five millimetres of their gut. In other words, something inside the worm destroys the pathogens, leaving the castings pathogen-free (Appelhof, 2003). These findings imply vermicompost spread on farm land may not result in pathogen contamination of ground or surface waters. Even this would mean that having land seeded and re-seeded with earthworm cocoons could help to prevent water contamination by pathogens, since fresh manure droppings by grazing animals will be quickly colonized by compost worms. Further, vermicompost, like conventional compost, binds nutrients well, both in the bodies of microorganisms and through their actions, minimizing the risk of nutrient run-off, lessening the chances of eutrophication of surface waters.

Greenhouse Gas Emissions

Climate change is one of the most serious environmental concerns of the present time. Agricultural and animal farms are a significant contributor to climate change, largely through the release of carbon dioxide, methane and nitrous oxide from livestock and their manure. Both composting and vermicomposting address these issues. One of the principal benefits of both composting and vermicomposting occurs through carbon sequestration i.e. locking up carbon in organic matter and organisms. Because composts of all types are stable, more carbon is retained in the soil than would be if raw manure or inorganic fertilizer were applied. Regular application of compost or vermicompost slowly increases soil carbon level.

Composting itself is thought to be neutral with respect to greenhouse gas generation. The United States Environmental Protection Agency (USEPA) found that composting resulted s in same level of GHG emissions as if the materials were allowed to decay naturally. According to other researchers (Paul et al., 2002), GHG benefits from composting do not come from the process itself, but from the avoided processes at both the front and back ends. Front-end savings occur when the organic material, such as manure on farms, is not stored under anaerobic conditions or spread raw on farmers' fields, both of which result in high emissions of methane and or nitrous oxide. The back-end savings result from the replacement of commercial fertilizers by compost, since production and transport of fertilizer over long distances result in high levels of GHG emissions. Unfortunately, these benefits have not as yet been systematically quantified. USEPA also acknowledged potential gains from these factors, but did not include them in their analyses. Whether it is be safe to expect the same potential advantages of composting described above to be valid for vermicomposting also, may be debatable. However, vermicomposting should provide some potentially significant advantages over composting with respect to GHG emissions. First, the vermicomposting process does not require manual or mechanical turning, as the worms aerate the material as they move through it. This should result in fewer anaerobic areas within the piles, reducing methane emissions from the process. Secondly, It also reduces the amount of fuel used by farm equipment or compost turners. Thirdly, vermicompost's increased effectiveness (5 to 7 times) relative to compost in promoting plant growth and increasing yield, implies that five to seven times as much fertilizer could be displaced per unit of vermicompost, decreasing the GHG emissions proportionately. Finally, analysis of vermicompost samples has shown generally higher levels of nitrogen than analysis of compost samples made from similar feedstock. This implies that the process is more efficient at retaining nitrogen, probably because of the greater numbers of microorganisms present in the process. This in turn implies that less nitrous oxide is generated and/or released during the process. Since N_2O is 296 times as potent a GHG as CO_2 (IPCC, 2001), this could be a significant benefit. On the other hand, some preliminary measurement work at the Worm Research Centre in England indicates that, contrary to the above reasoning, large-scale vermicomposting processes may in fact be a significant producer of N_2O. Levels in their process were significantly higher than in comparable windrow processing. They are calling for further research to determine the scope of this potential problem and to assess means of mitigation if it proves to be well founded (Frederickson and Ross-Smith, 2004). It should be noted by the reader that the centre was vermicomposting pre-composted mixed fish and shellfish waste, which are high in nitrogen, so the same results may not be found with manure-based operations. Also, it has not been determined if these emissions are large enough to offset the other gains described above. This is a significant development that should be closely monitored by anyone interested in large-scale vermicomposting.

Below-Ground Biodiversity

This is not an issue that has been discussed much, if at all, in the media or the political arena. Nevertheless, it is a significant issue. Biodiversity is declining rapidly worldwide, so much so that some scientists fear that we are heading for a mass extinction event similar to several that have occurred in Earth's ancient past. These events require millions of years to reverse once they occur, so it is vital to prevent that occurrence.

Earthworms have an extremely important role to play in counteracting the loss of biodiversity. Worms increase the numbers and types of microbes in the soil by creating conditions under which these creatures can thrive and multiply. The earthworm gut has been described as a little "bacteria factory", spewing out many times more microbes than the worm ingests. By adding vermicompost and cocoons to a farm's soil, that soil's microbial community is enriched tremendously. This below-ground biodiversity is the basis for increased biodiversity above ground, as the soil creatures and the plants that they help to grow are the basis of the entire food chain. The United Nations Environment Program (UNEP) has acknowledged the importance of below-ground biodiversity as a key to sustainable agriculture, above-ground biodiversity, and the overall economy (http://www.ciat.cgiar.org /tsbf_institute/ csm_bgbd.htm).

Improvement of Soil Health

Application of vermicompost to soil gives a tremendous boost to soil physical health by improving water-holding capacity, structure formation and also by enhancing fertility (Jeyabal and Kuppuswamy, 2001; Chaoui et al., 2003). Therefore, frequent application of vermicompost would result in increased soil carbon and nutrient levels and better productivity by improvement of soil physical conditions.

ECONOMICS OF SLUDGE MANAGEMENT VIS-A-VIS VERMICOMPOSTING

The cost of sludge management weighs heavily on industries and is set to increase with time as the industries have to manage greater quantities of sludge within tighter quality constraints. Investments in sludge management are usually made with a 20–25 year time horizon, and industries have to make a policy decision as to which option best suits them financially and falls within managerial feasibilities. Both direct and indirect costs are involved that have to be properly calculated and carefully taken into consideration before decision making. Most significant direct cost goes for sludge treatment, if necessary, like pathogen removal or incineration or thermal drying, as strict hygiene standards are imposed in many countries along with sludge disposal regulations if landfill construction cost is involved. Indirect costs could involve handling and disposal of ash and emission control in the case of incineration, and legacy of soil contamination in the case of land disposal. There are benefits such as reduced reliance on fertilizers in the case of composting/vermicomposting, soil conservation and fertility management in the case of land application, reduced dependence on other fuels, and reduced road transport in the case of in situ combustion. Direct benefits are realized if compost/vermicompost are sold directly. An estimate by Hall (2000) of WRc plc, UK, an environmental and water treatment consultant, is presented in Figure 6.

Project Cost of an Industrial Vermicomposting Facility in India

Capital cost and recurring cost of a vermicomposting facility can be appreciable in India, but may be considered fruitful when it helps avert alternative sludge management costs and liabilities. A vermicomposting project for two MT biosludge/day capacity in a petrochemical plant with annual vermicompost output of 432 tons with related cost has been executed by Concept Biotech in Gujarat in India (Table 6).

CONCLUSION

Vermicomposting is an exciting proposition as a method to convert solid waste to manure, but definitely depends on the type and characteristics of the waste sludge. As per the present state of knowledge and trends, industries should try to utilize their sludge to produce manure by vermicomposting after detailed waste characterization and careful planning of the operations on a case-by-case basis. The suitability of other 'waste-to-wealth' options, like composting, briquetting, incineration, producer gas and biogas generation, could also be considered. If the solid waste is devoid of any end-use value, then it could be landfilled. There are financial gains for the waste generator, both direct and indirect, as is evident from various case studies, and this should act as an incentive to push forward vermicomposting initiatives in industries, albeit at a small scale to start with. It cannot be the only solution to sludge management in industries, as it cannot take on the total burden of the sizeable quantity of wastes regularly generated, but can act as one of several options.

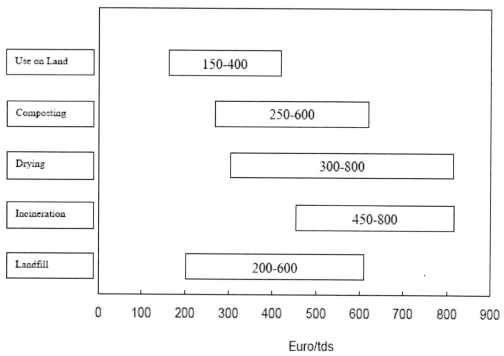

Source: Hall (2000).

Figure 6. Sludge treatment and disposal costs [tds – ton dry sludge].

Table 6. Vermicomposting project cost at a petrochemical plant in India

Item	First year	Second year	Third year
Cost of infrastructure	INR 45 lakh (US$ 93750*)	Nil	Nil
Earthworm culture cost	12 lakh (US$ 25000)	Nil	Nil
Cost of materials (bedding material, composting culture, Vermiaccelerator)	9 lakh (US$ 18750)	9 lakh (US$ 18750)	9 lakh (US$ 18750)
Commissioning cost (labour, supervision, consultancy)	9 lakh (US$ 18750)	9 lakh (US$ 18750)	3 lakh (US$ 6250)
Machinery and transport cost	2 lakh (US$ 4167)	2 lakh (US$ 4167)	Nil
Total Cost	77 lakh (US$ 160417)	20 lakh (US$ 41667)	12 lakh (US$ 25000)

* 1 US$ = INR 48 (approx.) as on 14.10.2008
Source: Dabke, Concept Biotech (personal communication).

Acknowledgments

The author is thankful to Dr. S. Dabke, Concept Biotech of Vadodara, Gujarat, India, for providing some important information on industrial sludge vermicomposting.

References

Albanell, E., Plaixats, J. and Cabrero, T. (1988). Chemical changes during vermicomposting (*Eisenia fetida*) of sheep manure mixed with cotton industrial wastes. *Biol. Fertil. Soils*, 6, 266-269.

Appelhof, M. (2003). Notable Bits. In: *WormEzine*, Vol 2 (5) (http://www.wormwoman.com)

Atharasopoolous, N. (1993). Use of earthworm biotechnology for the management of aerobically stabilized effluents of dried vine fruit industry. *Biotechnol. Lett.*, 15(12), 126-128.

Ayuso, M., Pascual, J.A., Garcia, C. and Hernandez, T. (1996). Evaluation of urban wastes for agricultural use. *Soil Sci. Plant Nutr.*, 42, 105-111.

Baguer, A.J., Jensen, J. and Krogh, P.H. (2000). Effects of the antibiotics oxytetracycline and tylosin on soil fauna. *Chemosphere*, 40, 751-757.

Bansal, S. and Kapoor, K.K. (2000). Vermicomposting of crop residues and cattle dung with *Eisenia fetida*. *Biores. Technol.*, 73, 95-98.

Banu, J.R., Logakanthi, S. and Vijayalakshmi, G.S. (2001). Biomanagement of paper mill sludge using an indegenous (Lampito mauritii) and two exotic (Eudrilus eugineae and Eisenia foetida) earthworms. *J. Environ. Biol.*, 22(3), 181-185.

Benitez, E., Nogales, R., Elvira, C., Masciandaro, G. and Ceccanti, B. (1999). Enzyme activities as indicators of the stabilization of sewage sludges composting with *Eisenia foetida*. *Biores. Technol.*, 67, 297-303.

Bhardwaj, K.K.R. (1995). Recycling of crop residues, oil cakes and other plant products in agriculture. In: H.L.S. Tandon (Ed.), *Recycling of Crop, Animal, Human and Industrial wastes in Agriculture*. pp. 9-30, New Delhi, India, Fertilizer Development and Consultation Organization.

Bhattacharya, S. S. and Chattopadhyay, G. N. (2002). Increasing Bioavailability of Phosphorus from Fly Ash through Vermicomposting. *J. Environ. Qual.*, 31, 2116-2119.

Bhawalkar, V. and Bhawalkar, U. (1994). *Vermiculture Biotechnology*. Pune, India, Bhawalkar Earthworm Research Institute.

Buch, V. and Patel, J. (2005). Characterization and suitability study of paint sludge for vermicomposting. M.Sc. Thesis, Institute of Science and Technology for Advanced Studies and Research, Sardar Patel University, Vallabh Vidyanagar, India.

Burge, W.D., Enkiri, N.K. and Hussong, D. (1987). Salmonella regrowth in compost as influenced by substrate. *Microbial Ecol.*, 14, 243-253.

Butt, K.R. (1993). Utilization of solid paper mill sludge and spent brewery yeast as a feed for soil dwelling earthworms. *Biores. Technol.*, 44, 105-107.

Chan, K.Y. and Mead, J.A. (2003). Soil acidity limits colonisation by *Aporrectodea trapezoides*, an exotic earthworm *Pedobiologia*, 47(3), 225-229.

Chan, P.L.S. and Griffiths, D.A. (1988). The vermicomposting of pretreated pig manure. *Biological Wastes, 24*, 57-69.

Chaoui, H.I., Zibilske, L.M. and Ohno, T. (2003). Effects of earthworm casts and compost on soil microbial activity and plant nutrient availability. *Soil Biology and Biochemistry, 35(2)*, 1 295-302.

Clark. P. (1998) Vermistabilisation is the use of earthworms to treat sewage sludges. In: *Water and Wastewater Treatment. p. 20.* (http://www.edie.net/Library/)

Contreras-Ramos, S. M., Escamilla-Silva, E. M. and Dendooven, L. (2005). Vermicomposting of biosolids with cow manure and oat straw. *Biol. Fertil. Soils, 41(3),* 190-198.

Delgado M., Bigeriego, M., Walter, I. and Calbo, R. (1995). Use of California red worm in sewage sludge transformation. *Turrialba, 45*, 33-41.

Directorate of Non-conventional Energy and Bio-Technology Cell (2001). Bio manure: Rural Industry for Organic farming. Mumbai, India, Khadi and Village Industries Commission.

Eastman, B.R. (1999). Achieving Pathogen Stabilization Using Vermicomposting. BioCycle, p. 62.

Eastman, B.R., Kane, P.N., Edwards, C.A., Trytek, L., Gundadi, B., Stermer, A.L. and Mobley, J.R. (2000). The Effectiveness of Vermiculture in Human Pathogen Reduction for USEPA Biosolids Stabilization. Orange County, Florida, Environmental Protection Division.

Edwards, C.A. (1988). Breakdown of animal, vegetable and industrial organic wastes by earthworms. In: C.A. Edwards and E.P. Neuhauser (Eds.), *Earthworms in Waste and Environmental Management.* The Hague, Netherlands, SPB Academic Publication.

Edwards, C.A., Burrows, I., Fletcher, K.E. and Jones, B.A. (1985). The use of earthworms for composting farm wastes. In: J.K.R. Gasser (Ed.), *Composting Agricultural and Other Wastes,* pp. 229-241, Elsevier, London and New York.

Elvira, C., Goicoechea, M., Sampedro, L., Mato, S. and Nogales, R. (1996). Bioconversion of solid paper-pulp mill sludge earthworms. *Biores. Technol., 57,* 173-177.

Elvira, C., Sampedro, L. , Benitez, E. and Nogales, R. (1998). Vermicomposting of sludges from paper mill and dairy Industries with Eisenia andrei: A Pilot-Scale Study. *Biores. Technol.,* 63, 205-211.

Elvira, C., Sampedro, L., Domonguez, J. and Mato, S. (1997). Vermicomposting of WastewaerSludge from Paper Pulp Industry with Nitrogen Rich Materials. *Soil Biol. Biochem., 29 (3/4),* 759-762.

Environment Agency (2001). Towards sustainable agricultural waste management. Bristol, UK.

Frederickson, J. and Ross-Smith, S. (2004). Vermicomposting of Pre-composted Mixed Fish/Shellish and Green Waste. The Worm Research Centre. SR566. (http://www.wormresearchcentre.co.uk)

Freeman, R. B. (1977). *The Works of Charles Darwin: An Annotated Bibliographical Handlist (Second Ed.), Wm Dawson and Sons Ltd.*

Gajalakshmi, S., Ramasamy, E.V. and Abbasi S.A. (2002). High-rate composting vermicomposting of water hyacinth (*Eichhornia crassipes*, Mart. Solms), *Biores. Technol., 83*, 235-239.

Garg, V.K. and Kaushik, P. (2005). Vermistabilization of textile mill sludge spiked with poultry droppings by an epigeic earthworm *Eisenia foetida*. Bioresourec Technology, 96(9), 1063-1071.

Garg, V.K., Kaushik P. and Dilbaghi, N. (2006). Vermiconversion of wastewater sludge from textile mill mixed with anaerobically digested biogas plant slurry employing *Eisenia foetida. Ecotoxicol. Env. Safety, 65*, 412-419.

Garg, V.K., Priya Kaushik, Dilbaghi, N. (2006). Vermiconversion of wastewater sludge from textile mill mixed with anaerobically digested biogas plant slurry employing Eisenia foetida. *Ecotoxicol. Environ. Safety, 65*, 412–419.

Godbold, D.L. and Hüttermann, A. (1994). Effects of Acid Rain on Forest Processes. Wiley-IEEE, 112 p.

Government of India, Livestock Census Reports.

Graff, O. (1981). Preliminary experiment of vermicomposting of different waste materials using *Eudrilus euginae* Kingberg. In: Proceedings of the workshop on *Role of Earthworms in the Stabilization of Organic Residues.* M. Appelhof (Ed.), pp. 179–191, Michigan, Malanazoo Pub.

Gratelly, P., Benitez, E., Elvira, C., Polo, A. and Nogales, R (1996). Stabilization of sludges from a dairy processing plant using vermicomposting. In: Rodriguez-Barrueco, C. (Ed.), *Fertilizers and Environment.* pp. 341-343, The Netherlands, Kluwer Academic Publishers.

Hall, J. (2000). Ecological and economical balance for different sludge management options. In: Langenkamp, H. and Marmo, L. (Eds.), Proceedings of workshop on *Problems Around Sludge*, pp. 155-172, Stresa, Italy.

Hand, P., Hayes, W.A., Frankland, J.C. and Satchell, J.E. (1988). The vermicomposting of cow slurry. *Pedobiologia, 31*, 199-209.

Harris, G.D., Platt, W.L. and Price, B.C. (1990). Vermicomposting in a rural community, *Biocycle, 31*, 48-51.

Hartenstein, R. and Bisesi, M. S. (1989). Use of earthworm biotechnology for the management of effluents from intensively housed livestock. *Outlook on Agriculture, 18 (2)*, 72-76.

Hassen, A., Belguith, K., Jedidi, N., Cherif, A., Cherif, M. and Boudabous, A. (2001). Microbial characterization during composting of municipal solid waste. *Biores. Technol., 80*, 217-225.

Intergovernmental Panel on Climate Change, 2001. Climate Change 2001: A Scientific Basis, Intergovernmental Panel on Climate Change. Houghton, J.T., Ding, Y., Griggs, D.J., Noguer, M., van de Linden, P.J., Dai, X., Johnson, C.A., Maskell, K. (Eds.), Cambridge University Press, Cambridge, UK.

Jagdeeshan, G. (2004). Briquettes from biomass for household use. In: Pandel, U., Poonia, M.P., Mathur, J. and Mathur, R. (Eds.), *Waste to Energy.* pp. 11-14, New Delhi, India, Prime Publishing House.

Jeyabal, A. and Kuppuswamy, G. (2001). Recycling of organic wastes for the production of vermicompost and its response in rice–legume cropping system and soil fertility. *European Journal of Agronomy, 15(3),* 153-170.

Kale, R., Bano, K. and Krishnamoorthy, R.V. (1982). Potential of *Perionyx excavatus* in utilization of organic wastes. *Pedobiologia, 23,* 419–425.

Kalra, N., Chaudhary, A., Pathak, H., Joshi, H.C., Jain, M.C., Sharma, S.K. and Kumar, V. (1998). Coal-burnt ash characteristics and its incorporation effects on soil properties. In: Kalra, N., Jain, M.C. (Eds.), *Effect of fly ash incorporation on soil properties and crop productivity*. pp. 3-9, New Delhi, India, Division of Environmental Sciences, Indian Agricultural Research Institute.

Kaushik, P. and Garg, V.K. (2004). Dynamics of biological and chemical parameters during vermicomposting of solid textile mill sludge mixed with cow dung and agricultural residues. *Biores. Technol. 94(2)*, 203-9.

Kavian, M.F. and Ghatneker, S.D. (1991). Biomanagement of dairy effluents using culture of red earthworms (*Lumbricus rubellus*), *Ind. J. Env. Prot., 11*, 680-682.

Kimani, S.K. and Lekasi, J.K. (2004). Managing Manures Throughout their Production Cycle Enhances their Usefulness as Fertilisers: A Review. In André Bationo (ed.) Managing Nutrient Cycles to Sustain Soil Fertility in Sub-Saharan Africa. Academy Science Publishers (ASP) in association with the Tropical Soil Biology and Fertility Institute of CIAT. pp. 187-197.

Lerch, R.N., Barbarick, K.A., Sommers, L.E. and Westfall, D.G. (1992). Sewage sludge proteins as labile carbon and nitrogen sources. *Soil Sci. Soc. Am. J., 56*, 1470-1476.

Liang, C., Das, K.C. and McClendon, R.W. (2003). The influence of temperature and moisture contents regimes on the aerobic microbial activity of a biosolids composting blend. *Biores. Technol., 86*, 131-137.

Logsdon, G. (1994). Worldwide progress in vermicomposting. *Biocycle, 35 (10)*, 63-65.

Luo, Y., Zang, Y., Zhong, Y. and Kong, Z. (1999). Toxicological study of two novel pesticides on earthworm *Eisenia foetida*. *Chemosphere, 39(13)*, 2347-2356.

Maboeta, M.S. and van Rensburg, L. (2003). Vermicomposting of industrially produced woodchips and sewage sludge utilizing *Eisenia fetida*. *Ecotoxicol Environ. Safety, 56*, 265-270.

Macwan, P. (2005). Utilization of Paint Industry Sludge by Culturing Earthworms. M.Sc. Thesis, Institute of Science and Technology for Advanced Studies and Research, Sardar Patel University, Vallabh Vidyanagar, India.

Majumdar, D., Patel, J., Bhatt, N. and Desai, P. (2006). CH_4, CO_2 emissions and earthworm survival during composting of pharmaceutical sludge and spent mycelia. *Biores. Technol., 97*, 648–658.

Marsh, L., Subler, S., Mishra, S. and Marini, M. (2005). Suitability of aquaculture e.uent solids mixed with cardboard as a feedstock for vermicomposting. *Biores. Technol., 96*, 413–418.

Mathur, B.S. (1995). Utilization basic slag in agriculture. In: Tandon, H.L.S. (Ed.), *Recycling of crop, animal and industrial wastes in agriculture*. pp. 125-142, New Delhi, India, Fertilizer Development and Consultation Organization.

Mehta, P. and Seth, G. (2007). Vermicomposting of paint industry sludges in combination with cow dung, camel dung and de-oiled Karanja cake. M.Sc. Thesis, Institute of Science and Technology for Advanced Studies and Research, Sardar Patel University, Vallabh Vidyanagar, India.

Millner, P.D., Powers, K.E., Enkiri, N.K. and Burge, W.D. (1987). Microbially mediated growth suppression and death of Salmonella in composted sewage sludge. *Microbial Ecol., 14*, 255-265.

Mishra, M.M., Kukreja, K., Kapoor, K.K. and Bangar, K.C. (1989). Organic recycling for plant nutrients. In: Somani, L.L., Bhandari, C.S. (Eds.), *Soil microorganisms and crop growth*. pp. 195-232, Jodhpur, Divyajyothi Parkashan.

Mishra, R.D. and Ahmed, M. Manual on Irrigation Agronomy. New Delhi, India, Oxford and IBH Publishing Co. Pvt. Ltd., 1987.

Mitchell, M.J., Hornor, S.G. and Abrams, B.I. (1980). Decomposition of sewage sludge in drying beds and the potential role of the earthworm, *Eisenia foetida. J. Env. Qual.*, 9, 373-378.

Mulongoy, K. and Bedoret, A. (1989). Properties of worm casts and surface soil under various plant covers in the humid tropics. *Soil Biol. Biochem.*, 21, 197-203.

Ndegwa, P.M. and Thompson, S.A. (2001). Integrating composting and vermicomposting in the treatment and bioconversion of biosolids. *Biores. Technol.*, 76, 107-112.

Ndegwa, P.M., Thompson, S.A. and Das, K.C. (2000). Effects of stocking density and feeding rate on vermicomposting of biosolids. *Biores. Technol.*, 71, 5-12.

Orozco, F.H., Cegarra, J., Trujillo, L.M. and Roig, A. (1996). Vermicomposting of coffee pulp using the earthworm *Eisenia fetida*: effects on C and N contents and the availability of nutrients. *Biol. Fertil. Soils,* 22, 162-166.

Panwar, S., Endlay, N., Gupta, M.K., Mishra, S., Pandey, A. and Mathur, R.M. (2004). The techno economic feasibility of biomethanation technology for recovery of energy from pulp and paper mill wastes. In: Pandel, U., Poonia, M.P., Mathur, J., Mathur, R. (Eds.), *Waste to energy.* pp. 31-38, New Delhi, India, Prime Publishing House.

Paul, J.W., Wagner-Riddle, C., Thompson, A., Fleming, R. and Alpine, M.M. (2002). Composting as a strategy to reduce greenhouse gas emissions. 14 pp. (http://res2.agr.ca/initiatives/manurenet).

Payal Garg, Asha Gupta, Santosh Satya (2006). Vermicomposting of different types of waste using Eisenia foetida: A comparative study. *Bioresource Technology* 97, 391–395.

Pierre, V., Phillip, R., Margnerite, L. and Pierrette, C. (1982). Anti-bacterial activity of the haemolytic system from the earthworms *Elsinia foetida Andrei.* Invertebrate Pathology 40, 21–27.

Priya Kaushik, Garg, V.K. (2003). Vermicomposting of mixed solid textile mill sludge and cow dung with the epigeic earthworm Eisenia foetida. *Biores. Technol.*, 90, 311–316.

Reimann, D. (2000). Problems about sewage sludge incineration. In: Langenkamp, H. and Marmo, L. (Eds.), Proceedings of workshop on *Problems Around Sludge*, pp. 173-183, Stresa, Italy.

Sangwan, P., Kaushik, C.P. and Garg, V.K. (2008). Vermiconversion of industrial sludge for recycling the nutrients. *Biores. Technol., 99(18)*, 8699-8704.

Shi-wei, Z. and Fu-zhen, H. (1991). The nitrogen uptake efficiency from ^{15}N labeled chemical fertilizer in the presence of earthworm manure (cast). In: G.K. Veeresh, D. Rajagopal and C.A. Viraktamath (Eds.), *Advances in Management and Conservation of Soil Fauna.* pp. 539-542, New Delhi, India, Oxford and IBH publishing Co.

Sinha, R.K. (1996). Vermiculture biotechnology for waste management and sustainable agriculture. In: R.K. Sinha (Ed.), *Environmental Crisis and Humans at Risk*, pp. 233–240. India, INA Shree Publication.

Sudha, B. and Kapoor, K.K. (2000). Vemicomposting of crop residues and cattle dung with Eisenia foetida. *Biores. Technol.*, 73, 95-98.

Suthar, S. and Singh, S. (2008). Feasibility of vermicomposting in biostabilization of sludge from a distillery industry. Sci. Total Environ., 394(2-3):237-243.

Swedish Radiation Protection Authority (2006). Experiments on chronic exposure to radionuclides and induced biological effects on two invertebrates (earthworm and daphnid). Results of experiments carried out within ERICA WP2. DELIVERABLE 5: ANNEX B (*search.ceh.ac.uk*).

Syers, J.K., Sharpley, A.N. and Keeney, D.R. (1979). Cycling of nitrogen by surface-casting earthworms in a pasture ecosystem, *Soil Biol. Biochem.*, *11*, 181-185.

Talashilkar, S.C., Bhangarath, P.P. and Mehta V.P. (1999). Changes in chemical properties during composting of organic residues as influenced by earthworm activity. *J. Ind. Soc. Soil Sci.*, *47*, 50-53.

Whiston, R.A. and Seal, K.J. (1988). The occurrence of cellulases in the earthworm *Eisenia foetida*. *Biol. Wastes*, *25*, 239-242.

Yadav, D.V. (1995). Recycling of sugar factory press mud in agriculture. In: Tandon, H.L.S. (Ed.), *Recycling of crop, animal and industrial wastes in agriculture*. pp. 91-108, New Delhi, India, Fertilizer Development and Consultation Organization.

Youn-Joo, A.N. and Lee, Woo-Mi (2008). Comparative and combined toxicities of toluene and methyl tert-butyl ether to an Asian earthworm *Perionyx excavates*. *Chemosphere, 71(3)*, 407-411.

In: Agricultural Wastes
Eds: Geoffrey S. Ashworth and Pablo Azevedo

ISBN 978-1-60741-305-9
© 2009 Nova Science Publishers, Inc.

Chapter 10

AGRICULTURAL WASTES AS BUILDING MATERIALS: PROPERTIES, PERFORMANCE AND APPLICATIONS[*]

José A. Rabi[*], Sérgio F. Santos[†], Gustavo H. D. Tonoli[‡] and Holmer Savastano Jr.[‡]

Faculty of Animal Science and Food Engineering, University of São Paulo – FZEA / USP
Av. Duque de Caxias Norte, 225, Pirassununga, SP, 13635-900, Brazil

ABSTRACT

While recycling of low added-value residual materials constitutes a present day challenge in many engineering branches, attention has been given to low-cost building materials with similar constructive features as those presented by materials traditionally employed in civil engineering. Bearing in mind their properties and performance, this chapter addresses prospective applications of some elected agroindustrial residues or by-products as non-conventional building materials as means to reduce dwelling costs.

Such is the case concerning blast furnace slag (BFS), a glassy granulated material regarded as a by-product from pig-iron manufacturing. Besides some form of activation, BFS requires grinding to fineness similar to commercial ordinary Portland cement (OPC) in order to be utilized as hydraulic binder. BFS hydration occurs very slowly at ambient temperatures while chemical or thermal activation (singly or in tandem) is required to promote acceptable dissolution rates. Fibrous wastes originated from sisal-banana agroindustry as well from eucalyptus cellulose pulp mills have been evaluated as raw materials for reinforcement of alternative cementitious matrices, based on ground BFS.

Production and appropriation of cellulose pulps from collected residues can considerably increase the reinforcement capacity by means of vegetable fibers. Composite preparation follows a conventional dough mixing method, ordinary vibration, and cure

---

[*] A version of this chapter was also published in *Building Materials: Properties, Performance and Applications*, edited by Donald N. Cornejo and Jason L. Haro, published by Nova Science Publishers, Inc. It was submitted for appropriate modifications in an effort to encourage wider dissemination of research.
[*] jrabi@usp.br
[†] sfsantos1@usp.br
[‡] gustavotonoli@usp.br
[‡] holmersj@usp.br

under saturated-air condition. Exposition of such components under ambient conditions leads to a significant long-term decay of mechanical properties while micro-structural analysis has identified degradation mechanisms of fibers as well as their petrifaction. Nevertheless, these materials can be used indoors and their physical and mechanical properties are discussed aiming at preparing panel products suitable for housing construction whereas results obtained thus far have pointed to their potential as low-cost construction materials.

On its turn, phosphogypsum rejected from phosphate fertilizer industries is another by-product with little economic value up to now. Phosphogypsum may replace ordinary gypsum provided that radiological concerns about its handling have been properly overcome as it exhales ^{222}Rn (a gaseous radionuclide whose indoor concentration should be limited and monitored). Some phosphogypsum properties of interest (e.g., bulk density, consistency, setting time, free and crystallization water content, and modulus of rupture) have suggested its large-scale exploitation as surrogate building material.

INTRODUCTION

Developing countries usually face grave housing deficits and the following hurdles against dwelling construction have been typically pointed: high interest rates, elevated social taxes, high informal labor indexes, and bureaucracy. Lack of loan is an additional problem[1] inasmuch as banks may not be interested on such funding or governmental programs are scarce. As a result, a considerable percentage of population in developing countries still lives at sub-dwelling units. In 2006, estimates suggested that around 7.9 million dwellings were needed in Brazil, most of them (83.3%) located in urban areas, particularly in the so-called Metropolitan Regions[2] surrounding Brazilian state capitals (2.2 million dwellings) [1]. São Paulo and Rio de Janeiro Metropolitan Regions present the greatest housing deficits, adding together to almost 1.2 million dwellings [1].

Aiming at lowering costs, scientific attention has been given to non-conventional building materials with similar features as those presented by construction materials traditionally used in civil engineering. Quest for such surrogate materials can be two-fold interesting as (*i*) it may help to reduce dwelling deficits (particularly in developing countries) inasmuch as cheaper houses become economically feasible and (*ii*) it can be environmentally friendly as low-value wastes can be recycled or exploited. Accordingly, this chapter is particularly interested in agroindustrial residues or by-products as prospective non-conventional construction materials.

VEGETABLE FIBERS AS NON-CONVENTIONAL BUILDING MATERIAL

Vegetable fibers are widely available in most developing countries. They are suitable brittle matrix reinforcement materials even though they present relatively poor durability performance. Accounting for the mechanical properties fibers as well as their broad variation

---

[1] This is particularly during economic and financial crisis as the one worldwide experienced in 2008.

[2] Metropolitan Regions of major interest in Brazil refer to the following capitals (corresponding state in brackets): Belém (Pará), Fortaleza (Ceará), Recife (Pernambuco), Salvador (Bahia), Belo Horizonte (Minas Gerias), São Paulo (São Paulo), Rio de Janeiro (Rio de Janeiro), Curitiba (Paraná), and Porto Alegre (Rio Grande do Sul).

range, one may develop building materials with suitable properties by means of adequate mix design [2], [3].

The purpose of fiber reinforcement is to improve mechanical properties of a given building material, which would be otherwise unsuitable for practical applications [4]. A major advantage concerning fiber reinforcement of a brittle material (e.g., cement paste, mortar, or concrete) is the composite behavior after cracking has started. Post-cracking toughness produced by fibers in the material may allow large-scale construction use of such composites [4].

There are two approaches for the development of new composites in fiber-cement [5]. The first one is based on the production of thin sheets and other non-asbestos components. The later are very similar to asbestos-cement ones and they are produced by well-known industrial-scale processes such as Hatschek and Magnani methods commercially used with high acceptance for building purposes [6]. The second one consists of producing composites for different types of building components such as load-bearing hollowed wall, roofing tiles, and ceiling plates, which are not similar to components commercially produced with asbestos-cement.

Estimated to be approximately several million of tons per year [7], consumption of fiber-reinforced cement building components is rapidly increasing, especially in developed countries. This is because such type of material allows the production of lightweight building components with good mechanical performance (mainly regarding impact energy absorption) and suitable thermal-acoustic insulation, while being economically attractive [4]. Fibers naturally occur in tropical countries (like Brazil), where they have been essentially targeted to cordage, textile, and papermaking sectors. Their heterogeneity and perishing allied to restricted market for their use have lead to intense generation of residues with high pollution potential. For example, each ton of commercially used sisal fibers yields three tons of residual fibers, whose dumping has originated environmental hazards [8].

Table 1. Physical or mechanical properties of some fibers.

Fiber	Density (g/cm^3)	Tensile strength* (MPa)	Modulus of elasticity (GPa)	Elongation at failure (%)	Water absorption (%)
Jute (*Corchorus capsularis*) [a]	1.36	400 - 500	17.4	1.1	250
Coir (*Cocos nucifera*) [b]	1.17	95 - 118	2.8 [d]	15 - 51	93.8
Sisal (*Agave sisalana*) [b]	1.27	458	15.2	4	239
Banana (*Musa cavendishii*) [a]	1.30	110 - 130	---	1.8 - 3.5	400
Bamboo (*Bambusa vulgaris*) [b]	1.16	575	28.8	3.2	145
E-glass [c]	2.50	2500	74	2 - 5	---
Polypropylene [c]	0.91	350 - 500	5 - 8	8 - 20	---

Tensile strength strongly depends on type of fiber, being a bundle or a single filament; a [11], b [5], c [12], d [13].

As reported in [9], vegetable fibers contain cellulose (which is a natural polymer) as the main reinforcement material. The chains of cellulose form microfibrils, which are held together by amorphous hemicellulose and lignin in order to form fibrils. The later are then assembled in various layers to build up the fiber structure. Fibers or cells are cemented together in the plant by lignin, which can be dissolved by the cement matrix alkalinity [10]. The usual denomination for fibers is indeed a reference to strands with significant consequences on durability studies.

Banana fibers cut from the plant pseudo-stem and sisal by-products from cordage industries are examples of widely available fibers. Eighteen types of potential fibers have been identified, including cellulose pulp recovered from newspaper, malva, coir, and sisal [5]. However, if costs and availability issues are accounted for, the number of suitable fiber types reduces drastically. Coir, sisal chopped strand fibers, and eucalyptus residual pulp have already been identified as fibrous waste materials suitable for cement reinforcement [8] and Table 1 presents some of their mechanical or physical of interest.

The present chapter is particularly interested in three different types of fibrous residues typically found in Brazil, namely:

- Sisal (*Agave sisalana*) field by-product: This material is readily available (e.g., 30,000 tons per year from a given producers' association) and it has presently no commercial value. Its use offers an interesting additional income for rural producers and simple manual cleaning via a rotary sieve provides a suitable starting-point material.
- Banana (*Musa cavendishii*) pseudo-stem fibers. This by-product has high potential availability from fruit production (e.g., 95,000 ton per year, based on Brazil's main producing area). This material has no market value and a simple low-cost fiber extraction process is only required.
- *Eucalyptus grandis* waste pulp. Accumulating from several Kraft and bleaching stages, this resource has low commercial value (USD 15/ton) and is readily available (e.g., 17,000 ton per year from one pulp industry in Brazil's southeast). Disadvantages of this material include short fiber length and high moisture content (about 60% of dry mass).

Chopped fibrous residues can be introduced directly into the cement for reinforcement as reported elsewhere but the utilization of further chemical processing of these residual fibers has proved to improve the performance of the building products [14]. Pulped fibers are preferred for composites production using slurry vacuum de-watering technique, which is a laboratory-scale crude simulation of Hatschek process. During the de-watering stage, pulp forms a net responsible for retaining cement grains. Small fibers remain homogeneously distributed in two directions (2-D) into the matrix [15] and this fact suggests some advantages of using sisal pulp (individualized fibers) in relation to sisal strand fibers [14]. Reinforcement is more distributed into the composite leading to effective capacity of reinforcing and bridging cracks during bending tests. Cellulosic pulps can be produced from residual crops (non-wood) fibers or wood species, reaction with alkalis liquors (e.g. sodium hydroxide, i.e., Kraft process), or organic solvents (e.g., ethanol).

Low performance of natural fiber reinforced composites (NFRC) has been associated to the use of chopped strand fibers as reinforcement for ordinary brittle cement matrices produced by conventional dough mixing methods [5]. This has been identified as the main reason for the low acceptance of these products by industry. Consequently, in several developing countries asbestos-cement remains the major composite in use although health hazards have become a major concern [16]. In view of that, agricultural residues fibers discussed in the present chapter were further prepared in order to fit their use and to achieve improved performance in composites.

PHOSPHOGYPSUM AS NON-CONVENTIONAL BUILDING MATERIAL

Another research line has pointed to the replacement of ordinary gypsum (calcium sulphate hemihydrate). Basically composed by calcium sulphate dihydrate ($CaSO_4 \cdot 2H_2O$), gypsite is raw material for ordinary gypsum production through thermal dehydration. When this way produced, gypsum achieves rigidity and hardness after being mixed-up with water. Nevertheless, its large-scale utilization in developing countries can become restricted or economically unfeasible due to transport costs from production sites (mines) to potential consumption places (e.g. large cities), which is scenario observed in Brazil.

According to the Brazilian National Department of Mineral Production (subordinated to the Ministry of Mining and Energy) [17], around 98% Brazil's gypsite mines adding up to 1.3×10^9 tons are located in northern or north-eastern states, namely, Pará (30.3%), Bahia (42.7%), and Pernambuco (25,1%). Altogether in 2007, their corresponding gypsite production comprised 1.9×10^6 tons so that 89% (1.7×10^6 tons) was due to Pernambuco alone, which is a state more than 2400 km away from Rio de Janeiro and São Paulo Metropolitan Regions (both located in Brazil's southeast). Gypsum is mainly consumed as constructive boards (panels) and on minor extent as wall covering, but compared to US (world's largest gypsum consumer and producer), Brazilian consumption is very low, not only due to the aforesaid transportation costs and logistics from gypsite mines but also due to related energy costs and lack of necessary facilities.

Conversely, growing demands for phosphate fertilizers have yielded enormous amounts of phosphogypsum for years worldwide [18]. Despite being essentially $CaSO_4 \cdot 2H_2O$, such sort of gypsum has currently little economic value (if any) due to environmental issues notably regarding radon-222 (^{222}Rn) exhalation. Together with its decay products, ^{222}Rn is a radioactive noble gas responsible for most human natural exposure to radiation and assessment of its exhalation rate is crucial for radiological protection design [19].

Half-life of ^{222}Rn (3.824 days) is long enough to allow its transport through porous media or open air. If inhaled, its progeny is relatively short-lived allowing its eventual decay to ^{210}Pb (half-life = 22.3 years) before removal by physiological clearance mechanisms. Lung cancer risk related to dangerous exposure to the radiation released from ^{222}Rn and from its short-lived decay products was addressed in the 1950s among uranium miners while indoor-air ^{222}Rn concentration issues were claimed in the 1970s. Since then, scientific attention to ^{222}Rn exposure has increased [20]. Initial evidence had pointed to soil as a major natural source for high indoor concentrations [21] but building materials can also play an important role [22].

Such by-product has been simply piled up nearby phosphate fertilizers industrial units [23], this way requiring considerable open-air space. For inactive phosphogypsum stacks in the US territory, US-EPA (Environmental Protection Agency) has once restricted ^{222}Rn exhalation rates to 0.74 $Bq \cdot m^{-2} \cdot s^{-1}$ [24]. Nonetheless, large-scale application of phosphogypsum is an motivating dilemma and research has been conducted so as to overcome difficulties related to the disposal and handling of such industrial waste. Feasible solutions have suggested its exploitation as soil amendment in agriculture, mine recovery, road base, and embankment filling.

Of particular interest, encouraging (i.e., radioactively safe) evidence has already become available for phosphogypsum use as a non-conventional building material with similar features as ordinary gypsum [25]. Bearing in mind indoor air ^{222}Rn concentration, a major concern refers to the maximum amount of phosphogypsum (pure or blended to building materials) to be present in constructive elements such as pre-fabricated blocks or panels for low-cost dwellings. Relying on protection procedures as air renewal and/or building openings (doors and windows), limits must be set so as to avoid hazardous radiological impacts, thus enabling occupants to be exposed to tolerable levels[3], i.e., without considerably increasing naturally occurring doses.

In line with its agroindustrial expansion, the Brazilian phosphogypsum scenario could not be different so that millions of tons have been similarly piled-up close to fertilizers plants [23]. Yet, the later are located in Brazil's southeast (basically in Minas Gerais and São Paulo states), therefore close to the biggest Metropolitan Regions. Aiming at low-cost dwellings, such strategic coincidence might be convenient while motivating for prospective application of phosphogypsum as a surrogate building material.

Engineering precaution should then be exercised in terms of constructive aspects as well as in view of protection against ionizing radiation whose level should be as low as those recognized by federal regulations and/or international standards. For instance, the International Commission on Radiological Protection (ICRP) has recommended 200 to 600 Bq·m^{-3} for indoor-air ^{222}Rn concentrations at residences and workplaces [26] whereas the National Commission for Nuclear Energy (CNEN) is the Brazilian organization responsible for setting upper limits for indoor-air ^{222}Rn concentrations at these two aforesaid human environments.

VEGETABLE FIBERS: COMPONENTS AND PERFORMANCE

A main drawback of using vegetable fibers is their durability in a cementitious matrix and the compatibility between both phases. Alkaline media weaken most natural fibers, especially the vegetable ones, which are actually strands of individual filaments liable to get separated from one another. The mineralization phenomenon proposed elsewhere [27], [28] can be associated to the long-term loss of composite tenacity. The severe degradation of exposed composites can also be attributed to the interfacial damages due to continuous volume changes exhibited by the porous vegetable fibers inside the cement matrix [29].

At the Construction and Thermal Comfort Laboratory (Faculty of Animal Science and Food Engineering, University of São Paulo, Brazil), the Research Group on Rural Construction has adopted two approaches to improve the durability of vegetable fibers. One is based on fibers protection by coating to avoid water effect, mainly alkalinity. The other approach aims at free alkalis reduction within the matrix by developing low alkaline binders based on

---

[3] Those more acquainted to human comfort may realize that, as opposing to thermal environment, a "radiological comfort zone" concept is not applicable here as the ideal lower level will be always zero rather than a minimum value (below which there would be discomfort). Accordingly, one may think of a "radiological stress zone" whose lower threshold would correspond to a maximum acceptable ^{222}Rn concentration.

industrial and/or agricultural by-products [30]. Similar reduced alkalinity effect can be reached by fast carbonation process as studied in [31], [32]. Studies and strategies to improve durability of vegetable fibers have been basically carried out on two types of building components, namely, flat sheets for wall panels and roofing tiles. In what follows, methodologies and results obtained for each type of component are presented and discussed.

FLAT SHEETS FOR WALL PANELS

In order to evaluate the performance of the fiber-reinforced cement-based composites under various conditions, flat sheets were produced. Production method followed the slurry vacuum de-watering process aiming at the viable use of such materials in civil construction. Matrix materials were added to an appropriate amount of moist fibers pre-dispersed in water so as to form slurries within a 20-30% range (solid mass basis). After homogeneously stirred, slurry was immediately transferred to an evacuable box. Water was drawn off under vacuum until the pad appeared dry on the surface, whereupon it was flattened carefully with a tamper. Pads were then pressed at 3.2 MPa for 5 min and sealed inside a plastic bag for cure under saturated-air condition or water immersion for future mechanical tests at a total age of 28 days. In some cases, further samples were prepared to evaluate their performance after natural or accelerated ageing experiments. Tests were carried out so as to evaluate prospective effects of matrix modification by less alkali blends, different fiber contents, fiber mineralization reduction by chemical modification, fiber improvement to cement bonding, and distance reduction between fibers. In what follows, related results are presented and discussed.

Matrix Modification By Less Alkali Blends

In order to improve composites durability, matrix alkalinity reduction was attempted. The main component for paste matrix production was Brazilian alkaline granulated iron blast-furnace slag (BFS) ground to 500 $m^2 \cdot kg^{-1}$ Blaine fineness, presenting the following oxide composition (mass-basis) as provided in [33]: SiO_2 - 33.78%, Al_2O_3 - 13.11%, Fe_2O_3 - 0.51%, CaO - 42.47%, MgO - 7.46%, SO_3 - 0.15%, Na_2O - 0.16%, K_2O - 0.32%, free CaO - 0.10%, and CO_2 - 1.18%.

In Brazil, more than 6 million tons of basic ground-BFS (GBFS) are available every year and about 1/3 of such amount is stocked without any use, bringing about concerns to the steel industry as well as to the environment [34]. Consequently, GBFS costs can be as low as USD 10.00 per ton. For cement production, slag must be ground to fineness at least similar to that of ordinary cement, which adds a further cost of USD 15.00 per ton, and it must be activated with chemical procedures. Gypsum for agricultural purposes and lime (calcium hydroxide for civil construction) were elected as chemical alkali-activators for BFS in proportions of 10% and 2% (binder mass basis), respectively, as previously discussed in [35]. Standard commercial ordinary Portland cement (OPC), Adelaide Brighton brand type "general purpose" (GP), minimum compressive strength of 40 MPa at 28 days (Australian Standards AS 3972 and AS 2350.11) was also adopted as matrix for comparison with the alternative BFS cement.

As detailed in [14], sisal (*Agave sisalana*) and banana (*Musa cavendishii* - nanicao variety) waste strand fibers generated during crop stages were initially cut to around 30 mm in length. About 0.5% (mass basis) of commercial production from a cellulose mill, waste *Eucalyptus grandis* pulp from Kraft and bleaching stages was collected by filtration from drainage lines prior to effluent biological treatment. Some pulp and fiber properties are summarized in Table 2.

Aiming at cheaper price and economical viability at small-scale production (if compared with chemical pulps), strand fibers were submitted to low-temperature chemithermomechanical pulping (CTMP) in line with [37], [38]. Additionally mechanical beating provides important internal and external fibrillation of filaments, leading to conformable fibers and thus to fiber-matrix bonding improvement. Initial preparation comprised soaking in cold water overnight, followed by simple and low-pollutant chemical pre-treatment based on 1-hour cooking in boiling saturated lime liquor. Such step effectively attacked residual slivers, which could be easily broken by fingers, thus showing adequate preparation for subsequent mechanical treatments. Appropriate chemical attack is of utmost importance for reducing energy consumption, which represents one of the major concerns related to mechanical pulps [6].

Table 2. Some pulp and fiber properties of interest.

Fiber	By-produced sisal CTMP [6]	Banana CTMP [6]	*E. grandis* Kraft waste
Screened yield (%)	43.38	35.57	N/A
Kappa number [1]	50.5	86.5	6.1
Freeness (mL) [2]	500	465	685
Length-weighted average length (mm) [3]	1.53	2.09	0.66
Fines (%) [4]	2.14	1.55	7.01
Fiber width (μm) [5]	9.4	11.8	10.9
Aspect ratio	163	177	61

[1] Appita P201 m-86 [36]; [2] AS 1301.206s-88; [3] Kajaani FS-200; [4] Arithmetic basis; [5] Average from 20 determinations by SEM; N/A – not available data; [6] Chemithermomechanical pulping (CTMP).

Asplund laboratory defibrator provided 103 kPa steam gauge pressure corresponding to 121°C in presence of the pre-treatment solution with pre-steaming by 120 s and defibration by additional 90 s. At those conditions [6], well fibrillated softwood fibers could be obtained from low temperature (i.e., 125-135°C) CTMP instead of yield smooth, lignin encased and un-fibrillated fibers from high temperature (i.e., 150-175°C) pulping. The pulp in preparation passed through a Bauer 20 cm laboratory disc refiner equipped firstly with straight open periphery plates (1 pass at 254 μm clearance) and subsequently with straight closed periphery plates (1 pass at 127 μm and 1 pass at 76 μm). Partial de-watering before each refinement process provided pulp with adequate consistency so as to improve the defibration process.

Sisal and banana pulps passed through a 0.229 mm slotted Packer screen for separation of shaves and then through a supplementary Somerville 0.180 mm mesh screening and subsequent washing to reduce particles with length below 0.2 mm. Fibers shortening and fines generation are expected although undesirable results from beating procedures can be controlled via appropriate energy applied to the stock during the mechanical treatment [37]. Finally, produced pulps were vacuum de-watered, pressed, crumbed, and stored in sealed plastic bags under refrigeration. *Eucalyptus grandis* waste pulp was employed just as received after a 2 min disintegration and washing in hot (nearly boiling) water using a closed-circuit pump system. Figure 1 shows results from mechanical and physical tests on composites reinforced with 8% of produced sisal and banana pulps as well with waste *Eucalyptus grandis* pulp from Kraft and bleaching stages.

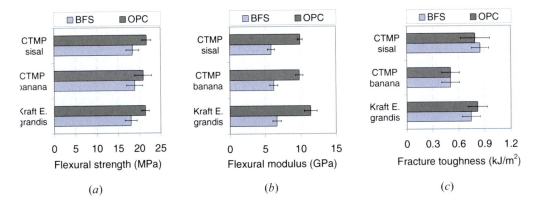

Figure 1. Average values and standard deviations for mechanical properties of composites as a function of the type of pulp fibers and matrix.

Flexural strength around 18 MPa for 8% residual fiber-BFS composites is considered an acceptable achievement when using mechanically pulped fibrous raw materials in comparison with similar results [38] for sisal Kraft pulp as reinforcement of air-cured BFS based matrices in 17 MPa and 1.4 kJ·m^{-2} ranges for flexural strength and toughness, respectively. Current results may also be considered a significant advance over a previous work [39] using disintegrated paper reinforced OPC with flexural strength up to 7 MPa at least 30% less than the corresponding control matrix under dough mixing method for fabrication.

OPC-based composites performed superior mechanical strength around 21 MPa at 28 days of age due to lower water absorption in comparison to BFS composites with all tested fibers in 8% content. BFS matrix seemed to lack hydration improvement, which could be achieved by high-temperature cure up to 60°C [40] or by adopting another alkaline activator (e.g., sodium silicate) as proposed in [41].

The high standard deviation associated with the flexural strength could be justified by the heterogeneity of the reinforcement fibers based on the following facts:

- Fibrous wastes generally present high moisture content and are thus expected to undergo fast biological decay [42], [43], leading to weak fibers in the pulp;
- Mechanical refinement often originates bunches of individual fibers mutually linked by non-cellulose compounds (e.g., lignin), as indicated in Table 2 by high Kappa numbers for sisal and banana CTMP. Strand fibers thus tend to perform poor distribution in the cement matrix.

Figure 2. Sequence of scanning electron microscope (SEM) images (hydration age between brackets): (a) 4% sisal CTMP in BFS matrix (73 days); (b) 4% banana CTMP in BFS matrix (32 days); (c) 4% *Eucalyptus grandis* waste Kraft in BFS matrix (51 days); (d) 4% sisal CTMP in OPC matrix (67 days); (e) 4% sisal CTMP in OPC matrix - higher magnification (67 days).

BFS composites reinforced with banana fiber presented lower values for fracture toughness compared to both sisal and *Eucalyptus grandis* composites. It can be understood in view of high length and aspect ratio of banana fiber, leading to stronger anchorage in the matrix and to the predominance of fiber fracture during mechanical test before any further fiber displacement could occur, as depicted in Figure 2(b). Figure 2(d) and 2(e) show images of by-product sisal CTMP in OPC with the desirable coexistence between fractured and pulled out filaments, pointing to the proximity of critical length [44] of that fiber in the specified matrix under ambient moisture condition. Twisted and bent fibers reinforce the idea of optimum interaction between both phases as well as of high energy dissipation during fiber pullout. Such favorable microscopic behavior explains the suitable compromise between flexural strength (21.7 MPa) and toughness (0.792 kJ·m^{-2}) for 8% sisal CTMP in OPC (Figure 1).

Freeness values within 465- 685 mL range (Table 2) provide adequate water drainage and prevent cement particles loss during suction stage of industrial systems based on Hatschek model for panels fabrication as pointed in [45], [46]. Low Kappa number value (Table 2) for *Eucalyptus grandis* Kraft pulp is an indication of bleached fiber with low lignin content. On the other hand, high values for mechanical pulps suggest damaged and fibrillated filaments or even remaining slivers expected to present exposed lignin to undesirable alkaline attack inside cement matrices.

As also concluded in similar research [47], [38], density, water absorption and porosity are interrelated properties. Low densities are preferable to reduce transportation costs and carriage effort while they are commonly connected to higher water absorption values with the inconvenient load increase during utilization and to risk of excessive permeability. As a reference, Brazilian standards (NBR 12800 and NBR 12825) limit water absorption to 37% (mass basis) for undulated roofing fiber cement sheets.

Elevated presence of fines in waste *Eucalyptus grandis* Kraft (Table 2) could contribute to pore-filling effect inside cement matrices, thus leading to denser materials with lower water absorption and apparent porosity, if compared to other OPC and BFS composites for the same fiber content (Figure 3). Such observation is consistent with results from wastepaper fiber-cement composites research reported in [48].

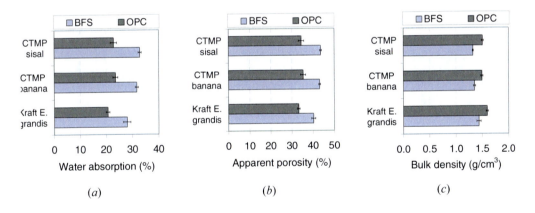

Figure 3. Average values and standard deviations of the physical properties of the composites in function of the type of pulp fibers and matrix.

Effects of Fiber Content

Figure 4 and Figure 5 present the effect of the fiber content respectively on mechanical and on physical performance of composites. The formulations follow those for BSF composites as presented in the previous section, except for variable fiber content. All BFS composites showed a considerable increase (at least 20%) in flexural strength within the 8-

12% fiber loading interval if compared to the corresponding 4% fiber content composites (Figure 4).

The short length of *Eucalyptus grandis* (Table 2) allowed the inclusion of up to 16% of fiber in binder mass although losing flexural strength, which could be associated to the high volume of permeable voids. As *Eucalyptus grandis* fiber content increased from 4% to 16%, the elastic modulus in bending of BFS composites decreased from 9 GPa to 4 GPa, a behavior that was equally observed for the other similar fiber-cements. Materials with 8% fiber OPC showed significantly higher modulus than corresponding BFS ones, likely due to insufficient hydration attributed to the alternative binder as previously commented.

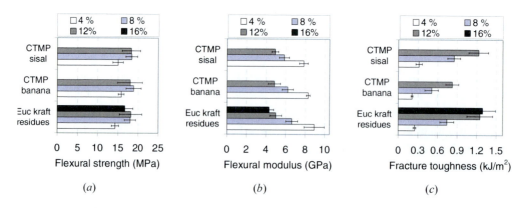

Figure 4. Variation of mechanical properties as a function of fiber content (%, mass basis) for composites with different pulp fibers.

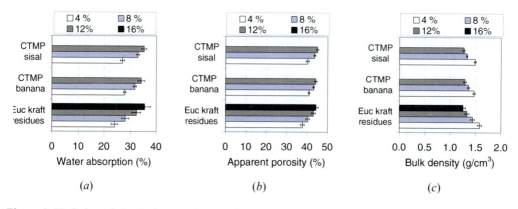

Figure 5. Variation of physical properties as a function of fiber content (%, mass basis) for composites with different pulp fibers.

By fiber reinforcement fracture toughness was greatly enhanced up to 5-fold improvement from 4 to 12% load interval. As observed in Figure 4(c) for *Eucalyptus grandis* BFS composites, best energy absorption seems to fall between 12 and 16% of fiber content with toughness in the vicinity of 1.3 kJ·m^{-2}. Predominance of fiber pullout for sisal and *Eucalyptus grandis* composites (see Figure 2(a) and (c)) is related to the high frictional energy absorption. In the specific case of *Eucalyptus grandis* fibers, the short length is compensated by larger number of filaments for a fixed fiber content and, hence, by higher probability of a matrix micro-crack interception in its initial stage of propagation. Increase of fiber content leads to poor packing during composites making, especially at the pressing production stage. Cellulose fibers are highly conformable and play a spring effect inside cement matrix immediately after press release. As a consequence of compaction reduction in conjunction with fibers low bulk density (~ 1.5 kg·dm^{-3}) and high water absorption (often over 100% mass basis), permeable voids volume and water absorption increase linearly with fiber content.

Reducing Mineralization of the Fibers By Their Chemical Modification

Eucalyptus Kraft pulp fibers were submitted to chemical treatment in order to reduce their hydrophilic character. Procedure for surface treatment of Eucalyptus pulp fibers and option for silanes were based in studies as developed in [49]. Silanes used were methacryloxypropyltri-methoxysilane (MPTS) and aminopropyltri-ethoxysilane (APTS), in a proportion of 6% mass basis of cellulose pulp. Silanes were pre-hydrolyzed during 2 hours under stirring in 80/20 volume basis ethanol-distilled water solution. Cellulose pulp was used (5% mass basis relative to the solution) followed by further 2 hours stirring (after cellulose addition). Composites were prepared by the same procedure as described in the previous section, using 5% mass basis of (untreated or treated) pulp in a matrix composed by 85% of OPC and 15% of carbonate filler.

Effect of treating pulp fibers on their mineralization was analyzed using scanning electron microscope (SEM) and back-scattered electron (BSE) detector to view cut and polished surfaces (BSE images allow easy identification of composite phases via atomic number contrast). Figure 6 presents SEM micrographs of composites whose pulp fibers are either impregnated with or free from silane coupling agents, where black areas are cross-sections of pulp fibers. In composites with untreated pulp fibers, Figure 6(*a*), one observes that the majority of lumens were filled up with reprecipitated products from cement, while in composites with treated pulp fibers, Figure 6(*b*), fiber lumens are free from hydration products.

Table 3 shows the effect of pulp treatment on the mechanical performance of composites. At 28-days cure, composites with APTS silaned pulp presented significantly higher modulus of rupture (MOR) than composites with untreated pulp or MPTS silaned pulp whereas toughness of composites was not influenced by silane treatment. Similar results in composites reinforced with silane-treated fibers were found in [50], [51].

Average MOR values decreased notably after 200 ageing cycles for composites with either treated or untreated fibers compared to those after 28 days cure. MOR differences after ageing were not observed between composites with treated or untreated fibers. As suggested in Table 3, the fact that MPTS-treated pulp did not present fibers filled up with products from cement seems to influence the higher toughness of composites after 200 ageing cycles. Yet, for untreated and APTS-treated pulps, composite capacity to absorb energy was markedly decreased most likely due to reprecipitation of hydration products into fibers with consequent composite embrittlement.

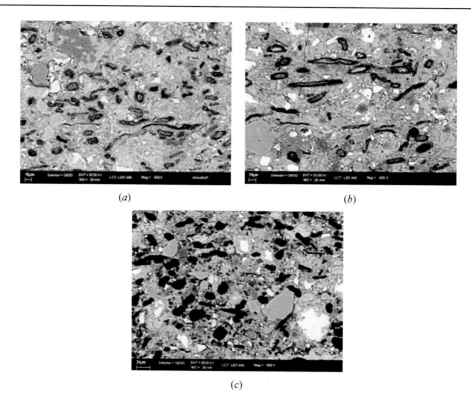

Figure 6. SEM-BSE images of composites reinforced with Eucalyptus bleached pulps: (*a*) untreated, (*b*) APTS treated, and (*c*) MPTS treated (after 200 ageing cycles).

Table 3. Average values and standard errors for composites properties: limit of proportionality (LOP), modulus of rupture (MOR), modulus of elasticity (MOE), |and toughness (TE)

Fiber treatment	Condition	LOP (MPa)	MOR (MPa)	MOE (GPa)	TE (kJ/m^2)
Untreated		6.9 ± 1.1	9.9 ± 1.4	13.3 ± 1.2	0.86 ± 0.25
MPTS	28 days	6.5 ± 1.0	10.7 ± 1.3	16.3 ± 1.7	0.83 ± 0.46
APTS		7.8 ± 1.3	12.1 ± 1.4	16.3 ± 2.5	0.82 ± 0.29
Untreated		6.3 ± 0.9	7.5 ± 0.5	17.7 ± 1.1	0.13 ± 0.07
MPTS	200 cycles	7.2 ± 0.9	8.0 ± 1.0	18.6 ± 4.6	0.30 ± 0.12
APTS		6.9 ± 1.7	8.3 ± 1.0	18.4 ± 3.8	0.13 ± 0.07

The effect of silane treatment on composites physical properties is presented in Table 4. Significant differences were not observed between composites with treated or untreated pulp at 28 days. However, after 200 ageing cycles, composites with APTS-treated pulp presented lower water absorption and apparent porosity in relation to composites with untreated pulp and MPTS-treated pulp. Bulk density of composites with APTS-treated pulp was significantly higher. This behavior suggests that chemical treatment increased the interaction with cement, which then influenced physical properties of the composite. Lower porosity of composites reinforced with silaned carbon fibers was reported in [52]. In such study, authors attributed this behavior to the hydrophilic character of the silane used, which improved fiber-matrix bond. The fact that fibers are filled up with cement hydration products also explains the porosity decrease of composites.

Table 4. Average values and standard error for composites properties: water absorption (WA), apparent void volume (AVV), and bulk density (BD)

Fiber treatment	Condition	WA (%)	AVV (%)	BD (g/cm^3)
Untreated		16.4 ± 0.9	29.0 ± 1.0	1.77 ± 0.04
MPTS	28 days	17.7 ± 1.3	30.8 ± 1.5	1.75 ± 0.04
APTS		16.7 ± 0.8	29.9 ± 1.0	1.79 ± 0.03
Untreated		15.2 ± 1.2	26.5 ± 1.9	1.75 ± 0.03
MPTS	200 cycles	16.2 ± 1.7	27.9 ± 2.4	1.72 ± 0.08
APTS		13.5 ± 0.5	24.6 ± 0.7	1.83 ± 0.03

The effect of chemical composition of pulp fibers seems to exert significant influence on composites durability as well. Lignin is an amorphous chemical species with high solubility in alkaline medium and its removal is essential part of pulping process [4]. Further lignin extraction from pulps is normally referred to as bleaching treatment. Figure 7 presents the influence of bleaching Eucalyptus pulp fibers with oxygen, chlorine dioxide, and ozone in order to improve adhesion between fibers and matrix. One observes that bleaching process makes fiber more susceptible to mineralization as it extracts compounds from fiber cell wall structure.

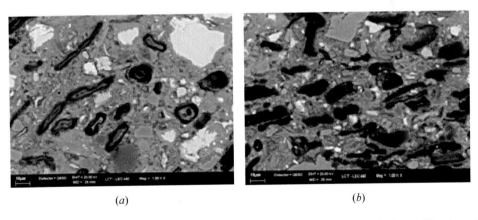

(a) (b)

Figure 7. SEM-BSE images of composites reinforced with Eucalyptus: (*a*) bleached and (*b*) unbleached (28 days).

Improving Fiber to Cement Bonding

One possible treatment to enhance mechanical performance of composites reinforced with cellulosic pulp is the refining process, which is carried out in the presence of water, usually by passing the suspension of pulp fibers through a disc refiner composed by a relatively narrow gap between rotor and stator [53], [54]. Cellulosic fibers are intrinsically strong and refinement improves their ability to be processed, which is necessary if the composite is manufactured using Hatschek production method [55]. The main effect of refinement in cellulosic fiber structure as a result of mechanical action is the fibrillation of fibers surface [56]. These fibrillated and shorter fibers are responsible for the formation of a net inside the composite mixture with the consequent retention of cement matrix particles during de-watering stage of manufacturing process. Better fiber / matrix interface adhesion and mechanical performance can be achieved by increasing the fiber aspect ratio (i.e., specific surface area), by reducing fiber diameter and producing a rough surface proportioning better mechanical anchorage in the matrix [55].

Figure 8(a) presents the poor adhesion of unrefined sisal pulp fibers and depicts void sizes up to 3 μm at the interface between fiber and matrix. In composites with refined pulp fibers, external layers were partially pulled out from cell wall after refining and these external layers then improve fibers anchorage into the cementitious matrix. In Figure 8(b), one may see external layers of refined fibers largely bonded to the cementitious matrix. The refined fiber-cement paste bond seems to be stronger than the unrefined fiber-cement paste one, as asserted in [57]. Hence, the large superficial contact performed by refined cellulosic pulp has enhanced the mechanical performance and has improved the load transfer from matrix to fibers.

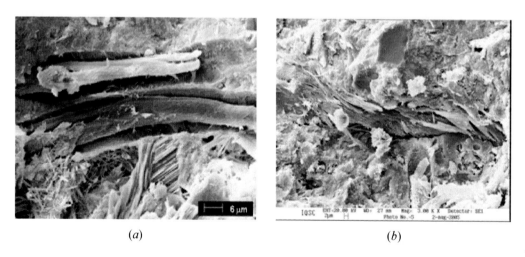

(a) (b)

Figure 8. SEM micrographs of fractured surfaces of sisal fiber-cement composites with (a) unrefined pulp (CSF 680 mL) - voids at the fiber-matrix interface after 100 soak / dry ageing cycles; (b) pulp refined at CSF 20 mL and after 100 soak / dry ageing cycles.

The state of surface structure of vegetable pulp fibers may vary either as a function of their natural source or as function of pulping process. Roughness of Eucalyptus and Pinus pulps were evaluated via atomic force microscopy (AFM). The surface of Eucalyptus fibers presented some fibrillar structure in the majority of samples, Figure 9(a). In Pinus fibers, typical surface structure was granular, Figure 9(b), possibly related to amorphous non-carbohydrates (lignin and extracts) in fiber surface. Fibrillar surface structures of Eucalyptus fibers provide a higher rough mean square (RMS = 74 ± 18 nm) than Pinus fibers (RMS = 52 ± 10 nm), which suggests a higher potential for Eucalyptus fibers to anchorage in the cement matrix.

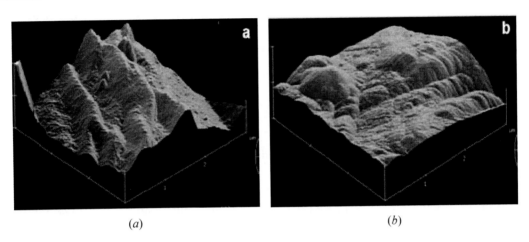

Figure 9. AFM topography images of (a) unbleached Eucalyptus fibers and (b) unbleached Pinus fibers. Image sizes are 3 μm × 3 μm.

Similar to procedure previously cited, interface between pulp fibers and cement matrix was analyzed utilizing SEM-BSE. Arrows show improved interface of Eucalyptus fibers, Figure 10(a), compared with Pinus ones, Figure 10(b), in composites cross-sections after accelerated ageing cycles. Figure 10(c) presents larger pores (within the 1-10 μm range) found by mercury intrusion porosimetry (MIP) in composites reinforced with Pinus, which might be attributed to voids in the fiber-matrix interface (arrows in Figure 10(b)).

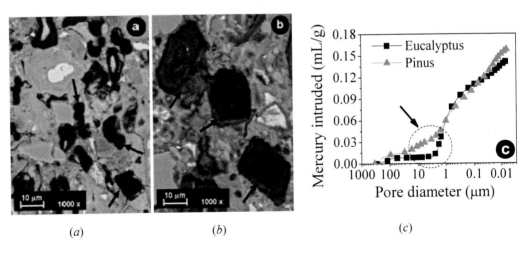

Figure 10. SEM-BSE image of composites reinforced with (a) Eucalyptus pulps and (b) Pinus pulps, after accelerated ageing, and (c) related cumulative mercury intrusion porosimetry (MIP).

Effect of Decreasing the Distance Between Fibers

Mechanical properties of fiber-cement composites are very sensitive to the uniformity of fibers volume distribution (dispersion) whereas the distance (spacing) between fibers is a critical geometrical parameter for composites performance [58]. As a rule, cracks initiate and advance from a composite section that has larger fiber-free matrix regions and fiber clumping [59]. Crack initiation requires less energy if it increases the size and the number of matrix regions that are not reinforced by fibers and such phenomenon becomes more pronounced in view of the progressive cement matrix embrittlement throughout its ageing.

Likely to be more homogeneous in length, Eucalyptus pulp presents shorter fibers (0.83 ± 0.01 mm) than Pinus pulp (2.40 ± 0.09 mm). As the use of short fibers might lead to higher number of fibers per volume or weight in relation to long fibers, the former may reduce fiber-free areas, i.e., the distance between fibers. Additionally, the smaller the fiber length is (which usually refers to lower aspect ratio), the easier the fiber dispersion becomes [60].

Comparing Figure 11(*a*) to Figure 11(*b*), short Eucalyptus fibers are better distributed than Pinus fibers. Bridging fibers share and transfer the load to other parts of the composite, this way increasing composite MOR and toughness. Calculated fiber spacing is at least two times higher for Pinus fibers relative to Eucalyptus ones [58]. Furthermore, due to their longer length, fibers in Pinus pulp are prone to cling to one another, thus jeopardizing the reinforcement.

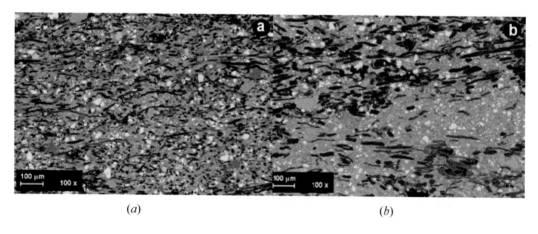

(*a*) (*b*)

Figure 11. SEM-BSE image of composites reinforced with (*a*) Eucalyptus pulp and (*b*) Pinus pulp.

ROOFING TILES

The Research Group on Rural Construction (at the Faculty of Animal Science and Food Engineering, University of São Paulo, Brazil) has also studied other cement-based composites containing vegetable fibers or particles for rural constructions [61], [62], [63]. Better results for fiber-cement materials should be expected by using refined pulp and slurry dewatering process, followed by pressing [64]. The energy consumption increase during these procedures may justify the improved composites performance. Such production model is similar to Hatschek industrial method still employed for asbestos-cement-based products in Brazil and it can be worthy for corrugated sheet fabrication in the near future by using of natural fibers or agricultural residues.

Improving Tiles Performance By Accelerated Carbonation

The present study was carried out as an attempt to produce durable fiber-cement roofing tiles by slurry dewatering technique and using sisal (*Agave sisalana*) Kraft pulp as reinforcement. Effects of accelerated carbonation on physical and mechanical performances of vegetable fiber-reinforced cementitious tiles were evaluated along with their consequent behaviors after ageing. Cement raw materials mixture was prepared with approximately 40% of solids (comprising 4.7% sisal pulp, 78.8% cement, and 16.5% ground carbonate material). With approximate dimensions as 500 mm long, 275 mm wide, and 8 mm thick, roofing tiles were produced by means of the device sketched in Figure 12.

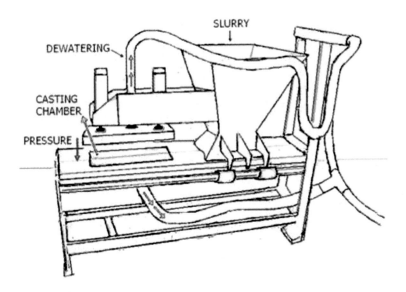

Figure 12. Schematic view of the semi-automated device used to mould roofing tiles.

Initial cure was carried out in controlled environment (i.e., temperature: 25 ± 2°C, relative humidity - RH: 70 ± 5%) so that roofing tiles remained in moulds protected with plastic bags for two days. Afterwards, roofing tiles were removed from moulds and immersed in water for further 26 days. After total curing period (28 days), tiles were submitted to both physical and mechanical tests. Remaining tiles series were intended to soak and dry-accelerated ageing tests as well as to accelerated carbonation so that roofing tiles were tested in saturated condition after immersion in water for at least 24 hours. Accelerated carbonation of roofing tiles was carried out in a climatic chamber providing environment saturated with carbon dioxide (CO_2) and controlled temperature (20°C) and humidity (75% RH). Roofing tiles were submitted to climatic chamber environment during one week until complete carbonation of samples. Carbonation degree was evaluated via exposure to 2% phenolphthalein solution diluted in anhydrous ethanol as described in [4].

Figure 13 shows cross-sections of tiles after the application of phenolphthalein solution so that violet regions refer to non-carbonated areas. Non-carbonated tiles underwent carbonation in peripheral regions only (tile 1 in Figure 13), which probably occurred during exposure inside the laboratory environment itself. As cross-section of tile 2 (Figure 13) was not violet colored, such tile was fully carbonated after being exposed to accelerated carbonation.

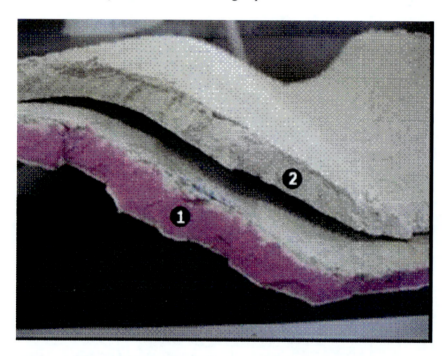

Figure 13. Cross-sections of tiles after the application of phenolphthalein solution: (1) non-carbonated areas and (2) areas exposed to fast carbonation.

Results for physical properties of undulate roofing tiles are presented in Table 5. Values for WA, AVV, and BD of roofing tiles for 28 days ageing (condition A) were similar to those found in [63] concerning the evaluation of roofing tiles produced with binder based on GBFS (ground blast furnace slag) reinforced with 3% (mass basis) of sisal pulp and processed by vibration. Results in [63] for WA, AVV, and BD were 31.0%, 42.3%, and 1.35 g/cm^3, respectively.

Table 5. Average values and standard deviations for water absorption (WA), apparent void volume (AVV), and bulk density (BD) of roofing tiles evaluated under different conditions

Condition	WA (%)	AVV (%)	BD (g/cm^3)
A (28 days)	32.8 ± 1.1	44.2 ± 0.7	1.35 ± 0.03
B (100 cycles)	33.3 ± 0.9	44.0 ± 0.9	1.32 ± 0.01
C (100 cycles, fast carbonation)	23.3 ± 0.7	35.8 ± 1.8	1.56 ± 0.01

In general, ageing cycles contributed to mitigate leaching and to reduce porosity of roofing tiles. Accelerated carbonation followed by 100 cycles (condition C) was the treatment that most affected physical properties of roofing tiles. Porosity reduction provided by carbonation can be responsible for mechanical properties improvement while accelerated carbonation reduced tiles apparent void volume (AVV) by approximately 20%. Significant water absorption reduction and carbonated roofing tiles densification suggested the effective carbon dioxide adsorption as well as the formation of new hydration products in the cement matrix. An estimated 15% reduction of cellulose fiber-cement porosity after its accelerated carbonation was also reported elsewhere [65].

Figure 14 presents typical load-deflection curves of roofing tiles. The maximum load (ML) supported by roofing tiles did not experience a significant reduction after ageing cycles. These results are considerably above the 425 N limit as recommended in [66] for 8 mm thick tiles. Ageing (condition B) did not cause significant decrease in ML and toughness (TE) of roofing tiles in relation to non-aged tiles tested with 28 days (condition A). Moreover, ML and TE were superior to those found in preceding works with roofing tiles produced by vibration (Figure 14). ML and TE values around 550 N and 1.6 kN·mm, respectively, for roofing tiles reinforced with 2% (volume basis) of unrefined coir, sisal macro-fibers and eucalyptus waste pulp at 28 days of age were obtained in [8].

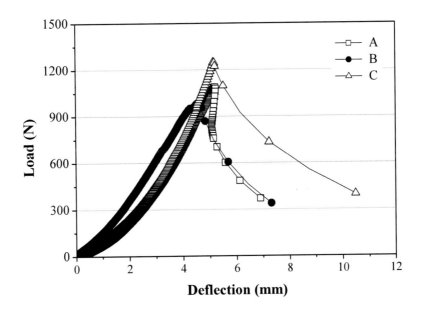

Figure 14. Load-deflection curves for tiles at different treatment conditions.

It seems that pulp refinement and its dispersion in the composite, as adopted in the present research, contributed to homogeneous fibers distribution during roofing tiles molding, leading to better fibers anchorage in the matrix, thus improving product strength. Among other variables, fibers net was more efficient at retaining cement particles during vacuum dewatering process, hence providing suitable packing during the pressing stage as well as more effective fiber-matrix bonds. Figure 15 presents a BSE image of carbonated tile with refined sisal Kraft fibers, showing the advantage of refined fibers in generating a high contact area with the matrix.

The cementitious phases of samples at each condition were analyzed by X-ray diffraction so that related patterns in Figure 16(a) and (b) show the presence of calcium hydroxide, $Ca(OH)_2$, in samples not submitted to accelerated carbonation (conditions A and B). Conversely, $Ca(OH)_2$ was not identified in fast carbonated samples (condition C). The formation of greater amount of calcium carbonate instead of other carbonates can be associated to high percentage of calcium (about 60%) in the ordinary Portland cement used in the present work [67]. Such result suggests a successful CO_2 adsorption in the cement based matrix while high carbonation can be associated to the consumption of hydroxyls (OH) present within the cement matrix due to CO_2 adsorption after its diffusion into composite pores. High apparent porosity (AVV) of roofing tiles (around 44% at 28 days) contributed to such fast diffusion [68].

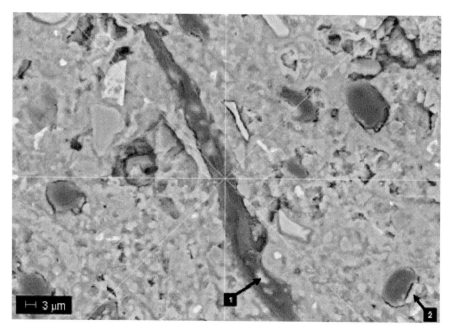

Figure 15. SEM-BSE image of fast carbonated and aged tile (condition C in Table 5). Fibers are well adhered to matrix, suggesting good composite packaging. Arrows point to reprecipitated calcium hydroxide (1) inside fiber lumen and (2) around fiber surface.

Series of accelerated carbonated roofing tiles after 100 ageing cycles (condition C) showed better mechanical performance in comparison to other series, including notably higher toughness (5.9 ± 1.9 kN·mm) and deflection at toughness (DTE = 9.1 ± 2.5 mm) in relation to non-aged (condition A) and fast-aged (condition B) series. Cement based specimens reinforced with sisal and coir chopped fibers with similar behavior have been studied elsewhere. While a strength increase of aged material achieved in [69] was attributed to calcium hydroxide elimination due carbonation treatment (109 days in a CO_2 incubator), mechanical performance improvement of flat sheets reinforced with 12% (mass basis) of Eucalyptus pulp after accelerated carbonation and ageing cycles was reported in [65].

Agricultural Wastes as Building Materials 241

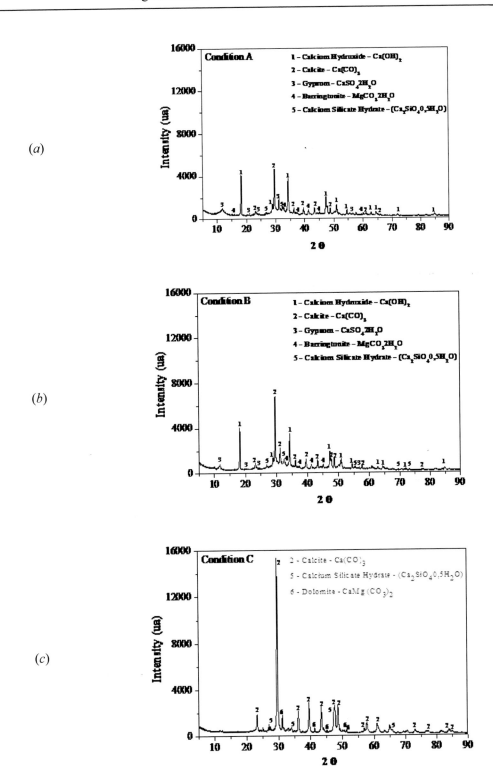

Figure 16. X-ray diffraction pattern of tiles: (a) non-aged (condition A), (b) after 100 ageing cycles (condition B), and (c) fast carbonated after 100 ageing cycles (condition C).

In addition to chemical analysis, calcium carbonate (CaCO$_3$) microstructure in tile fracture surfaces was visualized via SEM - secondary electron (SE) images. CaCO$_3$ formation in crystals as well packed in layers was observed in fast carbonated tiles (Figure 17). As reported in [70], morphology of CaCO$_3$ crystalline state plays an important role in determining binder strength so that improved strength was then attributed to the layered morphology of CaCO$_3$.

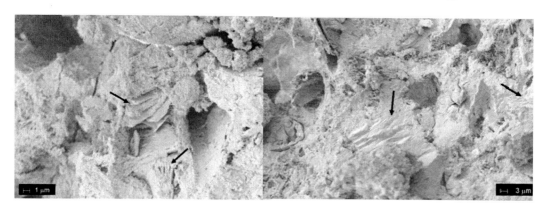

Figure 17. SEM of fracture surfaces of fast carbonated and aged tiles (condition C). Arrows indicate layered CaCO$_3$.

Last but not least, Figure 18 shows roofing tiles exposed to natural ageing. Such undulated shape is typically employed in Brazil.

Figure 18. Roofing tiles exposed to natural ageing in Brazil.

Thermal Properties of Produced Tiles

Table 6 lists values for thermal conductivity k, specific heat c, and thermal diffusivity $\alpha = k/(\rho \cdot c)$ of composites, obtained by parallel hot wire method as described in [71].

Aforementioned thermal properties were determined at room temperature (i.e., 25°C) as well at 60°C.

Table 6. Values for thermal conductivity k, specific heat c, and thermal diffusivity α = k/(ρ·c) of composites, obtained by parallel hot wire method

Mix design	Test temperature (°C)	k [W/(m·K)]	c [J/(kg·K)]	α (10^{-7} m^2/s)
Fiber-cement sisal (S4_7)	25	0.716	1278.7	3.060
	60	1.214	1071.5	6.191
Fiber-cement sisal + PP (S3PP1_7)	25	0.816	1201.1	3.951
	60	0.823	1294.4	3.694
Fiber-cement asbestos	25	0.837	1167.6	4.048
	60	0.967	1315.8	4.150

One notes that composites have different thermal behavior at room temperature (~ 25°C) and at 60°C. At the former, S4_7 formulation exhibited more favorable thermal properties for thermal comfort, namely, lower thermal conductivity k and higher specific heat c, thus leading to lower thermal diffusivity α. At 60°C, formulation with PP fibers (S3PP1_7) presented more adequate values for thermal comfort. At both test temperatures, the mix-design with asbestos presented less appropriate values in comparison to the formulations without asbestos.

Raising test temperature to 60°C increased thermal conductivity and thermal diffusivity of composites with sisal (S4_7) up to 45% and 100%, respectively, compared to the other mix-design. One may point to the presence of moisture within prismatic specimens used to influence measurements, which may increase thermal conductivity for higher temperatures [72]. The fact that non-asbestos mix-design (S4_7) shows larger cellulose fiber content (thus presenting great water absorption) may explain such thermal conductivity increase. Thermal behavior was also evaluated with respect to downside surface temperatures of non-asbestos roofing tiles produced (fiber-cement sisal – S4_7) compared with ceramic and asbestos-cement corrugated roofing tiles currently available in Brazilian market. Thermal behavior of covering roofing tiles might also be influenced by product dimensions, shape, and color; however, a comprehensive discussion on those aspects is beyond the scope of the present chapter.

Thermal performance analysis of roofing tiles considered prospective influence of thermal conductivity k, tiles cross-sectional area A, tiles thickness Δx, and temperature difference $\Delta T = T_{air} - T_{down}$, where T_{air} is outside air temperature while T_{down} is downside surface temperature of tiles. Absolute values for heat transfer rate, $dQ/dt = |\dot{Q}|$, across tiles were calculated according to one-dimensional Fourier's law [73]:

$$\dot{Q} = -Ak\frac{dT}{dx} \Rightarrow \frac{dQ}{dt} = \left|Ak\frac{\Delta T}{\Delta x}\right| \quad (1)$$

At hottest hours of the day, ceramics roofing tiles presented downside surface temperatures up to 10°C lower than those of asbestos tiles and 6°C lower in relation to those of sisal fiber-cement tiles. Among fiber-cement roofing tiles, non-asbestos ones presented downside surface temperatures 3°C lower. Heat transfer resistance is directly related to material microstructure (which can be assessed by its thermal diffusivity) and dimensions (mainly thickness) so that observed differences might be explained. Ceramic, non-asbestos, and asbestos roofing tiles presented thicknesses around 12 mm, 8 mm, and 4 mm respectively. Heat transfer rates dQ/dt were calculated employing thermal conductivity data at 25°C as presented in Table 6 for fiber-cement tiles while literature values were assigned to ceramic tiles, namely, 1.05 W/(m·K) [74].

Solar radiation flux and heat transfer rate along different hours of the day are shown in Figure 19 for the two hottest days over the evaluated period in summer. As one may see, heat transfer rates through asbestos-cement tiles were higher than rates related to ceramic and non-asbestos tiles. As a consequence, such higher rates lead to a hotter indoor environment.

Related to thermal diffusivity, thermal inertia is an important attribute for roofing tiles as it tends to diminish thermal variation amplitude inside buildings. For thermal inertia analysis, one may divide results into two periods, namely, morning and afternoon. In Figure 19(*a*), one observes that the highest radiation flux peak (about 1100 W/m^2) occurred at around 9:30 am, whose effect was observed 1.5 h later with the increase of heat transfer rates through tiles. Such phenomenon can be attributed to thermal inertia of tiles. In the second period, the highest radiation peak happened at 1:00 pm whereas the maximum heat transfer rate through asbestos roofing tiles was reached at 2:00 pm. Maximum heat transfer rate was reached at 2:15 pm for non-asbestos roofing tiles and at 2:30 pm for ceramic ones. Such 15 min difference among tiles might be equally attributed to thermal inertia. Furthermore, bearing in mind a barrier for thermal radiation, results suggested a better performance of ceramic and non-asbestos fiber-cement tiles.

Similar behavior is observed in Figure 19(*b*) so that highest solar radiation peaks (around 1100 W/m^2) occurred at 10:00 am and at 11:30 am, which increased heat transfer rates through roofing tiles one hour later. Ceramic tiles presented maximum heat transfer rate 15 min later than non-asbestos tiles and 30 min later than asbestos ones.

In view of their lower heat transfer rates, sisal fiber-cement tiles rendered good capacity of lowering outdoor temperature peaks (more efficiently than asbestos fiber-cement ones). One may then claim resultant advantages concerning the reduction of thermal radiation penetration, which is a result in accordance to previous studies. Differences were found for sisal-reinforced fiber-cement roofing tiles (produced by vibration) around 11.5°C lower (at 12:00 pm) compared with asbestos corrugated tiles [63] while significant differences were also observed for animal thermal comfort as non-asbestos fiber-cement roofing tiles presented superior performance in relation to simple asbestos fiber-cement counterparts [75]. Using fiber-cement roofing components based on blast furnace slag and vegetable pulps (Eucalyptus), differences of 7°C between internal and external superficial temperatures were found in [76].

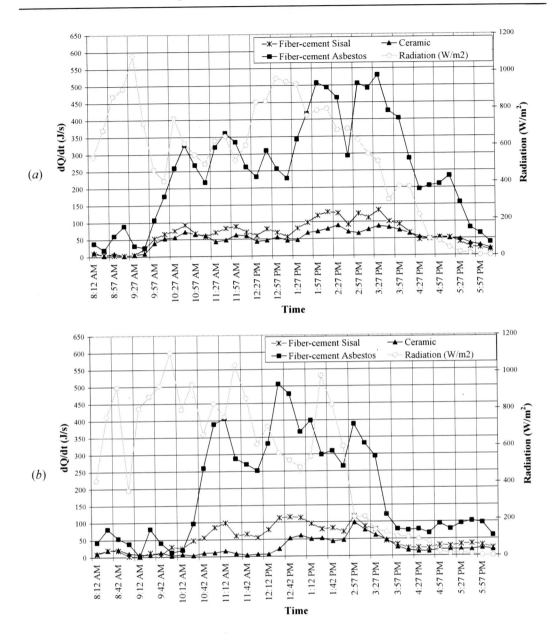

Figure 19. Variation of solar radiation and heat transfer rate through distinct roofing tiles as a function of time. Data collected on (a) December 20, 2005 and (b) December 22, 2005.

Ceramic tiles showed slightly lower heat transfer rates compared to sisal fiber-cement ones, indicating a possible influence of tiles surface reflectance [77]. Although darker coloration of red ceramics tile suggests larger absorption (66%) in the visible spectrum, infrared reflectance is high enough (78%) to provide a total reflectance of about 67%. The opposite was observed for light gray non-asbestos fiber-cement tiles, with a total reflectance of approximately 40% [77].

PHOSPHOGYPSUM: RADIOLOGICAL CONCERNS AND PERFORMANCE

Most of total radiation received by humans has to do with ^{222}Rn, which is a decay product from uranium, actinium, and thorium. As such radioactive noble gas can naturally occur in soil formations, it can penetrate into dwellings through microscopic cracks in the building structure. Exposure risks to inhabitants can equally be attributed to ^{222}Rn exhaled from building materials themselves, in which case phosphogypsum-bearing materials are likely to fall. Belonging to ^{238}U decay chain, ^{222}Rn results from α-decay of ^{226}Ra, which is an impurity frequently observed in phosphogypsum. Hence, ^{226}Ra trapped in phosphogypsum-bearing materials eventually decays to ^{222}Rn and such gaseous radionuclide may percolate the porous matrix, reach up open air (local atmosphere or indoor air), and be eventually inhaled by nearby humans or animals.

Radon exposure belongs to the scope of environmental toxicology and its exhalation rate closely depends on prevailing distribution in the porous medium. Research has been conducted in order to measure and correlate those rates to known physical parameters regarding the porous medium such as emanation rates from ^{226}Ra, moisture content, grain size, porosity, permeability, species (^{222}Rn) diffusivity, and temperature [20].

Though environmental (radiological) issues concerning phosphogypsum management still remain, there is a broad interest in finding its large-scale exploitation. Accordingly, besides the constructive performance viewpoint, use of such agroindustrial waste as a substitute for ordinary gypsum deals with ^{222}Rn exhalation so that a comprehensive understanding of ^{222}Rn generation and transport can be useful to assess related radiation exposure, to set up acceptable radiological standards, and to devise radiological protection based on human health threats.

MODELING AND SIMULATION OF RADON-222 EXHALATION

Like several other applications [78], [79], ^{222}Rn exhalation from phosphogypsum-bearing materials involve transport phenomena in porous media. Besides diffusion and convection, ^{222}Rn accumulation in dwellings can be concurrently affected by emanation (from ^{226}Ra), adsorption, absorption, and self-decay [80], [81]. Initial model frameworks have considered diffusion and convection with interstitial air flow driven by prescribed pressure differences in line with Darcy's law [82], [83], [84], [85]. Including pressure fluctuations [86], such approach has been adopted until lately [87] as it yields a Poisson equation (for transient transport) or a Laplace equation (for steady-state transport) to be solved for pressure, which simplifies the numerical implementation of governing equations into a computational code[4]. Including entry rates from water and building materials, a steady-state balance for indoor ^{222}Rn concentration was proposed in [88] whereas a transient model for exposure to phosphogypsum panels was recently presented in [89]. Further theoretical contributions comprise transient models for ^{222}Rn diffusion and decay in activated charcoal [90] and for ^{222}Rn and ^{210}Pb transport through the atmosphere [91].

Models have assessed ^{222}Rn entry rates and accumulation in building or enclosures via bulk values. Though accounting for time variation and sources (e.g. from distinct building

---

[4] Otherwise, one should additionally evoke and solve complete (and stiffer) momentum equations.

materials [22]), models have been zero-order concerning spatial coordinates so that ^{222}Rn concentration is allegedly uniform throughout (it is a representative value for the entire domain). Whenever point-to-point variation should be analyzed, failure of zero-order approaches is expected as they solely provide volume-averaged ^{222}Rn concentrations so that detailed ^{222}Rn indoor distributions can only be achieved by means of higher order model frameworks in space [92].

Comprehensive models for indoor-air ^{222}Rn exhalation and accumulation become stiffer as more and more influencing processes are entailed. Experimental data do help to asses and depict real and complex behavior but field or laboratory data acquisition might be risky or should be avoided due to safety concerns, technological constraints, or economic issues (costly human or material resources). Since such situation is applicable to the alternative use of phosphogypsum, numerical simulation may then play an important role by rapidly (and safely!) investigating any prospective scenario evoking as few simplifying assumptions as possible while accounting for effects sometimes conveniently neglected or simply ignored.

Nuclear physicists or engineers may rely on simulation if, for instance, non-linear and/or transient behavior, three-dimensional domains, or irregular geometry should be analyzed. Such real-world problems (for which computational time was prohibitive years ago) may help one to define proper radiological protection standards or design. A finite-difference method to predict ^{222}Rn entries into house basements from underneath soil gas was implemented following Darcy's law in [83] while a simulator for ^{222}Rn transport in soil was developed in [85]. Assuming air flow driven by user-defined pressure differences, a finite-volume method code was developed in [87] including time variation, three-dimensional domains, and several controlling parameters whereas fluctuations in such driving pressure differences were numerically investigated in [86].

Modeling ^{222}Rn Exhalation from Phosphogypsum Materials: Primitive Variables

A porous medium is a solid matrix with an interconnected void (interstices) through which fluids may flow [78]. If only one fluid saturates the void space, single-phase flow occurs; in a two-phase flow liquid and gaseous phases share the interstices [79]. Macroscopic measurements are normally achieved on regions or samples comprising several pores so that space-averaged physical quantities are assumed to continuously vary on time and on spatial coordinates. Porosity ε of a medium is a dimensionless quantity measuring the total volume fraction occupied by void space and such definition takes for granted that interstices are in fact connected. A basic concept behind such average is the representative elementary volume (REV), whose characteristic length is much larger than the pore scale but significantly smaller than the macroscopic domain length. Governing equations are derived for quantities at REV center and resulting values are allegedly independent from REV size. A REV in a phosphogypsum-bearing porous material may contain solid grains (or lattice) as well as pore space filled with air and/or water (for wet conditions) so that its porosity ε may encompass both air-based ε_a and water-based ε_w counterparts ($\varepsilon = \varepsilon_a + \varepsilon_w$).

A basic radiological concept is the activity (of a radioactive source), which is the amount of radionuclides decaying during a given time. In radioactive decay, an unstable isotope attempts to reach stability by emitting radiation in the form of particles and/or electromagnetic waves. After decay, the former isotope is referred to as parent nuclide while the newer is known as daughter nuclide. The number of particles expected to decay (–dN) during a small time (dt) is proportional to the number of particles in the sample (N) and theses quantities are related to each other as:

$$-\frac{dN}{dt} = \lambda N \qquad (2)$$

where constant λ is referred to as decay constant, whose values is unique for each radioisotope. SI unit for radioactive decay is becquerel (Bq), which is defined as 1 Bq = 1 disintegration per second (dps); yet, curie (Ci) and disintegration per minute (dpm) have also been used.

Knowledge of ^{222}Rn activity concentration profile in phosphogypsum-bearing materials is essential to evaluate resultant exhalation rates. Accordingly, ^{222}Rn transport depends on related mobile activity in the REV, which is properly evaluated and expressed in terms of the so-called partition-corrected porosity ε_c and air-borne activity concentration c_a. For a dry medium without solid sorption, one may show that $\varepsilon_c = \varepsilon$ [87]. While internal sinks of ^{222}Rn activity refers to its decay, internal sources are associated to ^{226}Ra concentration, which can be inferred by assuming that such radioactive impurity is evenly distributed all over the phosphogypsum-bearing material. Corrected emanation rates into the pore system are assessed considering that some ^{222}Rn particles still undergo decay until they finally reach the interstices.

Valid for low-porosity materials, diffusion-dominant ^{222}Rn transport across porous medium layers is a relatively simple approach, in which convective transport is neglected while additional simplifying assumptions could include steady-state process and one-dimensional species transfer. The later implies that porous medium is stratified with respect to the coordinate axis parallel to the main and sole transport direction so that ^{222}Rn concentration becomes uniform at any normal plane. Further steps in the model framework include expansion of the solution domain up to two or three dimensions and time dependence. The later is appropriate for indoor ^{222}Rn accumulation while the former allows one to study edge effects.

Depending on the length scales, one may adopt distinct approaches for the mathematical role of ^{222}Rn exhaling material. A first rationale may assume that the phosphogypsum-bearing material is very slim such as housing panels or boards placed against walls or as part of the building envelope itself [89], [92], [93]. For constant ^{222}Rn diffusivity D_a in open air and disregarding air motion (i.e., neglecting convective mass transfer), ^{222}Rn activity distribution can be governed by the following time-varying diffusive-dominant transport equation [94]:

$$\frac{\partial c_a}{\partial t} = D_a \left(\frac{\partial^2 c_a}{\partial x^2} + \frac{\partial^2 c_a}{\partial y^2} + \frac{\partial^2 c_a}{\partial z^2} \right) - \lambda c_a \qquad (3)$$

where x, y and z are Cartesian coordinates, t is time, and sink term ($-\lambda c_a$) is due to ^{222}Rn self-decay. There is no source term (due to emanation) in Eq. (3) as air presumably lacks ^{226}Ra. One may assume that such later radionuclide is uniformly distributed in the phosphogypsum-bearing material thus yielding a fixed and homogeneous ^{222}Rn exhalation rate j_{Rn} into neighboring air. From the mathematical viewpoint, ^{222}Rn exhalation hence becomes a boundary condition rather than a source term in the governing species equation.

Conversely, a distinct approach should be adopted if the solution domain comprises a blunt (i.e., finite thickness) ^{222}Rn exhaling solid as, for instance, a phosphogypsum-bearing building block (brick) in a still-air detection test chamber [93]. In this case, the porous sample partially fills up the solution domain so that ^{222}Rn transport may occur under two distinct "conditions", namely, in open air and within the REV. Corresponding species diffusivities D_a and $\tilde{D}$ can be allegedly constant and conveniently related to each other as:

$$\tilde{D} = \delta D_a \quad \Leftrightarrow \quad \delta = \tilde{D}/D_a \tag{4}$$

where $\tilde{D}$ is bulk species diffusivity, which should be used whenever species flux j refers to the geometric cross-sectional area A. If Fick's law for ^{222}Rn diffusion is set up in terms of interstitial cross-sectional area, an "interstitial diffusivity"[5] should be used instead [21], [85].

Bearing in mind Eq. (4) and recalling that ^{222}Rn sources (^{226}Ra particles) exist only inside the solid matrix while ^{222}Rn sink (i.e., self-decay) occurs everywhere[6], Eq. (3) is extended to the following diffusive-dominant ^{222}Rn transport equation:

$$\varepsilon^n \frac{\partial c_a}{\partial t} = \delta^n D_a \left(\frac{\partial^2 c_a}{\partial x^2} + \frac{\partial^2 c_a}{\partial y^2} + \frac{\partial^2 c_a}{\partial z^2} \right) + n\tilde{G} - \varepsilon_c^n \lambda\, c_a \tag{5}$$

where $\tilde{G}$ is mobile ^{222}Rn activity generation rate per REV unit and the dimensionless parameter n denotes whether ^{222}Rn transport occurs outside ($n = 0$) or inside ($n = 1$) the porous matrix [95]. Uniform distribution of ^{226}Ra particles (yielding homogeneous generation rate $\tilde{G}$) and constant partition-corrected porosity ε_c are implicitly assumed in Eq. (5), which can then be regarded an extension to the governing equation proposed in [87].

Radon-222 can be additionally transported by convection, either forced or natural [96]. The former refers to the action of fans, pumps, blowers, or wind while the later is induced by fields (e.g., gravity) acting on density gradients due to thermal and/or solutal variations. Applications point to ^{222}Rn exhaling building envelopes subjected to indoor air currents (thermally induced or not) or to air flow over phosphogypsum-bearing embankments or stacks (piles) [97], [98]. Apart from ^{222}Rn activity governing equation suitably extended to include convective terms, the model framework must equally incorporate bulk fluid continuity and momentum equations to be solved for flow field velocity components. If thermal effects should be accounted for, energy equation is invoked to be usually solved for temperature field. If local thermodynamic equilibrium prevails inside the REV, temperatures are the same for all phases (e.g. solid, air, and water and so that $T_s = T_a = T_w = T$), which is reasonable if internal energy sources are negligible.

---

[5] Such species transport coefficient has also been referred to as "effective diffusivity" although such term may cause confusion with the usual meaning attributed to "effective" in the porous media literature, which refers the REV averaging procedure previously discussed. "Interstitial diffusivity" should be preferred as it is unambiguous.

[6] Yet, ^{222}Rn self-decay must be properly adjusted with respect to the interstitial volume content.

Aforesaid governing equations are typically coupled to one another and the corresponding solution domain may encompass both porous medium and open air. With respect to momentum equations, outside the porous material fluid flow can be governed by Navier-Stokes equations, which for Newtonian fluids and incompressible flow (constant density $\rho_a = \rho_\infty$) can be expressed in terms of fluid (air) dynamic viscosity μ_a as:

$$\rho_a\left(\frac{\partial \vec{v}}{\partial t} + \vec{v}\cdot\nabla\vec{v}\right) = \mu_a\nabla^2\vec{v} - \vec{\nabla}p + \rho_a\vec{g} \qquad (6)$$

where p is pressure and $\vec{g}$ is gravity acceleration. The so-called seepage (or filtration) velocity $\vec{v}$ has components assessed as $v = \dot{V}/A$, where $\dot{V}$ is the volumetric flow rate through a REV with geometric cross-sectional area A, while components of the so-called interstitial velocity are given as $v_{int} = \dot{V}/A_f$, based on the cross-sectional area $A_f = \varepsilon A$ related to the pore system. Referred to as Dupuit-Forchheimer relation [78], [79], these two velocities are related to each other as:

$$v = \varepsilon v_{int} \qquad \Leftrightarrow \qquad v_{int} = v/\varepsilon \qquad (7)$$

Regarding natural convection, Boussinesq approximation can be introduced so that thermo-physical properties become constant except for bulk air density ρ_a in buoyant forces [96]. Hence, incompressible flow presumably holds while a linear dependence of ρ_a upon local temperature T is supposed solely for the buoyancy term in momentum equations:

$$\rho_a(T) = \rho_\infty[1 - \beta_T(T - T_\infty)] \quad , \quad \beta_T = \frac{1}{\rho_\infty}\left(\frac{\partial \rho_a}{\partial T}\right)_\infty \qquad (8)$$

where β_T is the coefficient of thermal volumetric expansion and both density ρ_∞ and temperature T_∞ are reference values such as related to atmospheric air condition sufficiently far from the stack (pile or embankment) or related to indoor air sufficiently far from the building wall. Thinking of thermosolutal natural convection [99], one could claim for radon-induced bulk fluid (air) density variations but corresponding contributions are definitely smaller if compared to entire air mass content so that solutal variations in ρ_a can be safely ignored.

For fluid flow within the porous medium, a classical steady-state approach is Darcy's law relating fluid velocity $\vec{v}$ and pressure gradient $\vec{\nabla}p$, namely:

$$\vec{v} = -\frac{1}{\mu_a}\vec{\vec{K}}\cdot\vec{\nabla}p \qquad (9)$$

where $\vec{K}$ is the permeability tensor [78], [79]. For simplicity, phosphogypsum-bearing materials can be treated as isotropic so that permeability becomes a scalar K and Eq. (9) reduces to:

$$\vec{v} = -\frac{K}{\mu_a}\vec{\nabla}p \quad \Leftrightarrow \quad \vec{\nabla}p = -\frac{\mu_a}{K}\vec{v} \qquad (10)$$

It can be further assumed that K is constant throughout. In addition, porosity $\varepsilon = \varepsilon_c$ (dry porous medium without solid sorption) can be supposedly constant all over as well.

Extensions to Darcy's law have been attempted so as to bear some resemblance to Navier-Stokes equations [79]. For the so-called Darcy-Brinkman-Forchheimer approach and in line with [78], [95], [99], [100], [101], [102], one may put forward the following governing equations for bulk fluid mass (continuity), momentum, energy (with local thermal equilibrium but neglecting internal heat sources, viscous dissipation and radiative effects) and species (^{222}Rn activity):

$$\vec{\nabla}\cdot\vec{v} = 0 \qquad (11)$$

$$\frac{1}{\varepsilon^n}\frac{\partial \vec{v}}{\partial t} + \frac{1}{(\varepsilon^2)^n}\vec{v}\cdot\vec{\nabla}\vec{v} = \Gamma^n \upsilon_a \cdot \nabla^2\vec{v} - \frac{1}{\rho_\infty}\cdot\vec{\nabla}p - n\left(\frac{\upsilon_a}{K} + \frac{c_f|\vec{v}|}{\sqrt{K}}\right)\vec{v} + \vec{g}\beta_T(T_\infty - T) \qquad (12)$$

$$\left[(1-\varepsilon^n)\rho_s c_s + \varepsilon^n \rho_\infty c_{p,a}\right]\frac{\partial T}{\partial t} + \rho_\infty c_{p,a}\vec{v}\cdot\vec{\nabla}T = \left[(1-\varepsilon^n)k_s + \varepsilon^n k_a\right]\nabla^2 T \qquad (13)$$

$$\varepsilon^n \frac{\partial c_a}{\partial t} + \vec{v}\cdot\vec{\nabla}c_a = \delta^n D_a \nabla^2 c_a + n\widetilde{G} - \varepsilon^n \lambda\, c_a \qquad (14)$$

where $\upsilon_a = \mu_a/\rho_\infty$ is air kinematic viscosity, c_f is a dimensionless form-drag coefficient [79] also referred to as Ergun's coefficient [78], c_s is specific heat of solid matrix, $c_{p,a}$ is constant-pressure specific heat of air while k_s and k_a are thermal conductivities of solid matrix and air, respectively. Depending on porous medium characteristics [78], [79] [101], thermal conductivities of solid (k_s) and air (k_a) can be lumped in a REV-averaged (effective) thermal conductivity obeying a specific expression other than the linear relation in Eq. (13). Analogous to Eq. (4) for ^{222}Rn diffusivity, REV-to-fluid property ratios can also be introduced for kinematic viscosity, thermal conductivity, and heat capacity respectively as [100]:

$$\widetilde{\upsilon} = \Gamma \upsilon_a \Leftrightarrow \Gamma = \widetilde{\upsilon}/\upsilon_a \quad , \quad \widetilde{k} = \Lambda k_a \Leftrightarrow \Lambda = \widetilde{k}/k_a \quad ,$$
$$(\widetilde{\rho c}_p) = \sigma(\rho_\infty c_{p,a}) \Leftrightarrow \sigma = (\widetilde{\rho c}_p)/(\rho_\infty c_{p,a}) \qquad (15)$$

Modeling ^{222}rn Exhalation from Phosphogypsum Materials: Dimensionless Variables

Relying on Buckingham's Π-theorem and similarity rationale [94], transport phenomena models are expressed via dimensionless differential governing equations so that simultaneous influencing parameters can be lumped into fewer controlling parameters. Such practice may help to reduce the number of required experiments, tests, scale-up steps, or optimization procedures. If forced convection prevails regarding open air flow condition, one may choose the free stream velocity u_∞ as a reference value. Applying the previously presented model framework for ^{222}Rn generation and transfer for a two-dimensional domain (without loss of generality), dimensionless variables for time τ, Cartesian coordinates X and Y, velocity components U and V, pressure P, temperature θ, and ^{222}Rn activity concentration C can be defined as:

$$\tau = \frac{t}{\Delta t} \;,\; X = \frac{x}{L} \;,\; Y = \frac{y}{L} \;,\; U = \frac{u}{u_\infty} \;,\; V = \frac{v}{u_\infty} \;,\; P = \frac{p}{\Delta p} \;,\; \theta = \frac{T - T_\infty}{\Delta T} \;,$$

$$C = \frac{c_a - c_\infty}{\Delta c} \tag{16}$$

where L is a characteristic length of the domain whereas c_∞ is a reference level for ^{222}Rn activity concentration (e.g. in air away from the porous medium) allegedly to fulfill the condition $c_\infty > 0$. Depending on the problem physics, reference scales Δc and ΔT for ^{222}Rn activity concentration and temperature can be suitably defined while proper choices for time and pressure scales are:

$$\Delta t = L/u_\infty \text{ and } \Delta p = \rho_\infty u_\infty^2 \tag{17}$$

By introducing Eqs. (16) and (17) into Eqs. (11) to (14) and assuming $\vec{g} = -g\hat{j}$, one may write the following set of coupled dimensionless governing equations:

$$\frac{\partial U}{\partial X} + \frac{\partial V}{\partial Y} = 0 \tag{18}$$

$$\frac{1}{\varepsilon^n}\frac{\partial U}{\partial \tau} + \frac{1}{(\varepsilon^2)^n}\left(U\frac{\partial U}{\partial X} + V\frac{\partial U}{\partial Y}\right) = \frac{\Gamma^n}{Re}\left(\frac{\partial^2 U}{\partial X^2} + \frac{\partial^2 U}{\partial Y^2}\right) - \frac{\partial P}{\partial X} - n\left(\frac{1}{Re\,Da} + \frac{c_f|\vec{V}|}{\sqrt{Da}}\right)U \tag{19}$$

$$\frac{1}{\varepsilon^n}\frac{\partial V}{\partial \tau} + \frac{1}{(\varepsilon^2)^n}\left(U\frac{\partial V}{\partial X} + V\frac{\partial V}{\partial Y}\right) = \frac{\Gamma^n}{Re}\left(\frac{\partial^2 V}{\partial X^2} + \frac{\partial^2 V}{\partial Y^2}\right) - \frac{\partial P}{\partial Y} - n\left(\frac{1}{Re\,Da} + \frac{c_f|\vec{V}|}{\sqrt{Da}}\right)V + \frac{Gr}{Re^2}\theta \tag{20}$$

$$\sigma'' \frac{\partial \theta}{\partial \tau} + U \frac{\partial \theta}{\partial X} + V \frac{\partial \theta}{\partial Y} = \frac{\Lambda''}{\text{Re Pr}} \left(\frac{\partial^2 \theta}{\partial X^2} + \frac{\partial^2 \theta}{\partial Y^2} \right) \quad (19)$$

$$\varepsilon'' \frac{\partial C}{\partial \tau} + U \frac{\partial C}{\partial X} + V \frac{\partial C}{\partial Y} = \frac{\delta''}{\text{Re Sc}} \left(\frac{\partial^2 C}{\partial X^2} + \frac{\partial^2 C}{\partial Y^2} \right) + \frac{1}{\text{Re Sc}} \left[nS - \varepsilon'' R(C - C_\infty) \right] \quad (20)$$

where $C_\infty = -c_\infty/\Delta c$ is the C value given by the last of Eqs. (16) for null air-borne ^{222}Rn activity concentration ($c_a = 0$). Dimensionless parameters regarding convective heat and mass transfer in porous media arise as expected such as Reynolds (Re), Darcy (Da), Grashof (Gr), Prandtl (Pr) and Schmidt (Sc) numbers, expressed in terms of air properties respectively as:

$$\text{Re} = \frac{u_\infty L}{\upsilon_a}, \quad \text{Da} = \frac{K}{L^2}, \quad \text{Gr} = \frac{g \beta_T \Delta T L^3}{\upsilon_a^2}, \quad \text{Pr} = \frac{\upsilon_a}{\alpha_a}, \quad \text{Sc} = \frac{\upsilon_a}{D_a} \quad (21)$$

where $\alpha_a = k_a/(\rho_\infty c_{P,a})$ is air thermal diffusivity.

Besides those parameters, Eq. (20) introduces two unusual dimensionless numbers referred to as decay-to-diffusion (R) and emanation-to-diffusion (S) ratios[7], respectively related to ^{222}Rn decay and emanation processes [92], [93], [97], [98], [103], defined as:

$$R = \frac{\lambda L^2}{D_a} \quad \text{and} \quad S = \frac{\tilde{G} L^2}{D_a \Delta c} \quad (22)$$

In the model framework for time-dependent pressure-driven ^{222}Rn migration inside soil proposed in [21], two dimensionless groups were introduced in the species (^{222}Rn activity) concentration equation: a convection-to-decay ratio (N) and a mass-transfer Péclet number[8] (Pe$_m$, measuring the relative importance of convection with respect to diffusion). Recalling the previous definition for the decay-to-diffusion ratio R, Eq. (22), and bearing in mind definitions proposed[9] for N and Pe$_m$, it is interesting to verify that indeed:

$$\frac{\text{Pe}_m}{N} = \frac{convection/diffusion}{convection/decay} = \frac{decay}{diffusion} \Rightarrow \frac{\text{Pe}_m}{N} \cong R \quad (23)$$

A surrogate dimensionless group M can be additionally introduced[10], namely:

---

[7] R and S respectively assess the relative importance of ^{222}Rn decay and emanation in relation to its mass diffusion.
[8] There is also a corresponding and similar definition for heat-transfer Péclet number.
[9] Yet, the proposed definition for Péclet number should be apparently altered to Pe$_m$ = $K \Delta P_o (\varepsilon \mu D)^{-1}$, which seems to be the correct result after casting the proposed governing species transport equation into dimensionless form.
[10] In fact, original definitions for R, S and M included the porosity (either geometric or partition-corrected) [103].

$$M = \frac{S}{R} = \frac{\tilde{G}}{\lambda \Delta c} \qquad (24)$$

which might be interpreted as an emanation-to-decay ratio [93], [97], [98], [103]. Despite still sensitive to scale Δc for ^{222}Rn activity concentration, M number is independent from both open-air diffusivity D_a and the characteristic length L. In terms of R and M, Eq. (20) can be recast as:

$$\varepsilon^n \frac{\partial C}{\partial \tau} + U \frac{\partial C}{\partial X} + V \frac{\partial C}{\partial Y} = \frac{\delta^n}{Re\,Sc}\left(\frac{\partial^2 C}{\partial X^2} + \frac{\partial^2 C}{\partial Y^2}\right) + \frac{R}{Re\,Sc}\left[nM - \varepsilon^n(C - C_\infty)\right] \qquad (25)$$

In the absence of strong air currents (i.e., negligible forced convection), it might be rather cumbersome to identify a reference velocity u_∞, which can be awkwardly small. Dimensionless variables formerly proposed in Eqs. (16) basically remain the same but suitably "re-scaled" by $u_\infty = \upsilon_a/L$ wherever required. In Eqs.

to (20) or (25), such velocity scale implies that Reynolds number, as defined in line with the first of Eqs. (21), numerically[11] reduces to unity (Re = 1).

PHOSPHOGYPSUM PROPERTIES

Phosphogypsum was characterized from chemical and radiological viewpoint in [104]. One aspect about the later refers to ^{222}Rn exhalation, which is important parameter to assess radiation doses onto occupants within dwellings. Indoor ^{222}Rn activity concentration can be experimentally measured using activated charcoal assembled inside relatively simple samplers as the cylindrical one (diameter = 90 mm, height = 45 mm) shown in Figure 20(a). Charcoal for measurements must be dried at 75°C for at least 7 days and samplers must be exposed to indoor air for up to 30 days. Retained (adsorbed) ^{222}Rn is counted through gamma spectrometry with NaI detector based on the 609.3 keV peak, which refers to ^{214}Bi radionuclide (a decay product of ^{222}Rn).

Alternatively, solid state nuclear track detectors (SSNTD) have been largely employed due to relatively low cost (related to both detector itself and measurement process), with particular attention to CR-39 polycarbonate (allyl diglycol-cabonate, $C_{12}H_{18}O_7$) detectors presenting higher efficiency. They basically consist of a plastic diffusion chamber that is permeable solely to ^{222}Rn (i.e., decay products of such radionuclide are not able to penetrate through it). As depicted in Figure 20(b), SSNTD is placed inside the diffusion chamber in order to register alpha particle emissions occurred during ^{222}Rn decay. A calibration factor correlates the quantity of detected tracks and ^{222}Rn concentration in indoor air.

---

[11] It is important to stress this is only a numerical implication thanks to the choice of velocity scale.

(a)   (b)

Figure 20. Measurement devices for indoor ^{222}Rn activity concentration: (a) activated charcoal collector and (b) diffusion chamber with SSNTD.

As far as physical and mechanical properties of interest are concerned, preliminary tests were accomplished in [105] regarding phosphogypsum samples by-produced from three distinct phosphate fertilizer industries[12]. Properties were determined bearing in mind Brazilian standards (referred to as ABNT) for bulk density (NBR 12127), consistency and setting time (NBR 12128, employing a Vicat equipment), modulus of rupture (NBR 12129, utilizing EMIC hydraulic press model PCE 100D - 1 MN load), and free water / crystallization water content (NBR 12130).

Brazilian standards (NBR 13207) recommend the bulk density of 700 kg·m^{-3} for ordinary gypsum. As preliminary results for phosphogypsum pointed to approximately 570 kg·m^{-3}, results have been improved by properly separating small grains from samples.

Table 7. Crystallization water content results for phosphogypsum samples

Drying period	Sample #1	Sample #2	Sample #3
4 hours	14,43%	12,77%	n.a.**
5 hours	10,00%	9,54%	n.a.**
6 hours	9,36%	6,77%	n.a.**
7 hours	4,91%*	6,22%*	1,34%*
24 hours	5,90%*	6,19%*	1,62%*

* Values fulfilling standards recommended in NBR 13207; ** Not available or not measured.

The reference value for free water content is 1.3% as specified by NBR 13207. Samples were initially dried at 125°C for 4 hours but results were unsatisfactory for both ordinary gypsum (4.2%) and phosphogypsum (6.2%). After drying samples at the same temperature (125°C) for a longer period (5 hours), better results were obtained for gypsum (1.37%) and phosphogypsum (sample #1 = 0.60%, sample #2 = 0.66%, sample #1 = 1.12%). Results for crystallization water content are presented in Table 7 for different drying periods. One verifies that phosphogypsum samples dried up to 7 hours fulfill recommended standards (NBR 13207).

Table 8 shows test results for consistency whereas Table 9 shows results for setting time for both ordinary gypsum and phosphogypsum samples. Regarding the later test, setting starts

---

[12] They are here referred to as numbers, namely, sample #1, sample #2, and sample #3.

when tip remains 1 mm from base while test ends when tip no longer penetrates into the paste but it just leaves a slender imprint. Compared to ordinary gypsum and phosphogypsum sample #3, either sample #1 or sample #2 required an elevated water consumption, which jeopardized their performance in corresponding MOR tests (as shown ahead). If compared to ordinary gypsum, setting time occurred quite rapidly for phosphogypsum (in general basis), which may affect its handling as it loses its consistency faster.

Table 8. Consistency results for ordinary gypsum and phosphogypsum samples

Amount of gypsum (g)	Cone tip penetration (mm) Test #1	Test #2	Amount of phosphogypsum (g)	Cone tip penetration (mm) Sample #1	Sample #2	Sample #3
250.00	34	34	187.50	20	21	n.a.**
272.73	36	34	172.50	24	26	n.a.**
300.00	32*	34	150.00	34	33	n.a.**
319.15	29*	30*	157.90	31.5*	32*	n.a.**
333.33	24	26	200.00	n.a.**	n.a.**	38
			215.00	n.a.**	n.a.**	29*
			225.00	n.a.**	n.a.**	25

* Values fulfilling standards recommended in NBR 13207; ** Not available or not measured.

Table 9. Setting time results (min:sec) for ordinary gypsum and phosphogypsum samples

Gypsum	Test #1	Test #2	Phosphogypsum	Sample #1	Sample #2
Start	14:30*	16:00*	Start	06:00	08:30
End	31:00	31:00	End	09:00	16:26

* Values fulfilling standards recommended in NBR 13207.

Test bodies for MOR tests were prepared in molds with three cubic compartments so that three samples with 50 mm characteristic length were produced simultaneously for each material (ordinary gypsum and phosphogypsum samples). Accordingly, cross-sectional area then results 2500 mm^2 whereas water / phosphogypsum ratio was the same for consistency tests. Results for each test body are presented in Table 10. As already commented, higher water consumption of phosphogypsum samples #1 and #2 jeopardized their MOR performance. Ordinary gypsum was then added to phosphogypsum as an attempt to improve such property so that new MOR tests for 20% addition (mass basis) resulted in 8.7 MPa and 8.8 MPa for samples #1 and #2, respectively.

Table 10. MOR results for ordinary gypsum and phosphogypsum samples

Material	Test body	Load (N)	MOR (MPa)
Ordinary gypsum	1-G	31800	12.72*
	2-G	38900	15.56*
	3-G	30900	12.36*
Phosphogypsum sample #1	1-A	9300	3.72
	2-A	5400	2.16
	3-A	8800	3.52
Phosphogypsum sample #2	1-B	15000	6.00
	2-B	10100	4.04
	3-B	12300	4.92
Phosphogypsum sample #3	1-C	23800	9.52*
	2-C	24700	9.88*
	3-C	19800	7.92*

* Values fulfilling standards recommended in NBR 13207.

CONCLUSION

Aiming at low-cost dwellings in developing countries, non-conventional building materials have been extensively investigated. According, the present chapter presented and discussed some agroindustrial residues or wastes that are likely to provide a suitable and sustainable solution.

Both waste sisal and banana CTMPs were suitable for cement composite manufacturing via laboratory method similar to counterpart processes broadly used in commercial production. In addition, residual *Eucalyptus grandis* Kraft pulp presented similar behavior during fabrication steps with the advantage of being already available in pulp form and at relatively low costs. The incorporation of the aforesaid waste fibers at 8% (mass basis) into BFS-based matrix produced composites with fracture strength about 18 MPa slightly lower than the correspondent OPC-based materials. Both 12% (mass basis) incorporation of sisal and *Eucalyptus grandis* into BFS composites rendered tough composites (approximately 1.2 kJ·m^{-2} of toughness), which is a reasonable performance if compared to previous investigations carried out on sisal chemical pulp as reinforcement for BFS composites.

Microscopy images depicted the importance of proper linkage between composite phases, providing the coexistence of fiber fracture and pullout. Such major outcome could then explain the strength sustained as well as better toughness results achieved by sisal CTMP composites in comparison to corresponding performance of banana CTMP. Physical properties indicated poor packing of high-content fiber composites with consequent low density and high water absorption values, despite within acceptable standard limits. Both proposed waste fibers utilization and mechanical pulping methods together with low-energy cements as blast-furnace slag are likely to represent an attractive option for asbestos-free fiber-cements.

Results from phosphogypsum characterization and performance have been quite promising and encouraging. Among possible large-scale exploitation of such industrial waste one may point to building blocks or housing panels, provided that radiological issues related to ^{222}Rn exhalation and accumulation in indoor air have been properly overcome. Comprehensive understanding of related transport phenomena is likely to rely on experimental research supported by numerical simulation in order to yield reliable information concerning radiological impact and protection design of prospective safe scenarios.

REFERENCES

[1] Brazilian Government - Ministry of Cities - National Housing Secretary. *Housing Deficit in Brazil*; João Pinheiro Foundation: Belo Horizonte, Brazil, 2006; pp. 20-21.
[2] Agopyan, V.; Cincotto, M. A.; Derolle, A. In: *Proceedings of the 11th CIB Triennial Congress - CIB-89*; CIB: Paris, France, 1989; theme II, vol. I, pp. 353-61.
[3] Agopyan, V.; John, V. M. In: *Fibre reinforced cements and concretes: recent developments*; Swamy, R. N.; Barr, B.; Eds.; Elsevier: London, UK, 1989; pp. 296-305.
[4] Agopyan, V.; Savastano Jr., H.; John, V. M.; Cincotto, M. A. *Cem. Concr. Compos.* 2005, *27*, 527-536.
[5] Agopyan, V. In: *Natural fibre reinforced cement and concrete*; Swamy, R. N.; Ed.; Concrete Technology and Design 5; Blackie: Glasgow, UK, 1988; pp. 208-242.
[6] Coutts, R. S. P. *CSIRO For. Prod. Newsl.* 1986, *2*, 1-4.
[7] Heinricks, H.; Berkenkamp, R.; Lempfer, K.; Ferchland, H. J. In: *Proceedings of the 7th International Inorganic-Bonded Wood and Fiber Composite Materials Conference*; Moslemi, A. A.; Ed.; Siempelkamp Handling Systems Report; University of Idaho: Moscow, 2000; 12 p.
[8] Savastano Jr., H.; Agopyan, V.; Nolasco, A. M.; Pimentel, L. L. *Constr. Build. Mater.* 1999, *13*, 433-438.
[9] Coutts, R. S. P. In: *Proceedings of the 4th International Symposium Fibre Reinforced Cement and Concrete*; Swamy, R. N.; Ed.; E&FN Spon: London, UK, 1992; pp. 31-47.
[10] Gram, H. E. *Durability of natural fibres in concrete*; Swedish Cement and Concrete Research Institute: Stockholm, Sweden, 1983.
[11] Rehsi, S. S. In: Natural Fibre Reinforced Cement and Concrete (Concrete Technology and Design 5); Swamy, R. N.; Ed.; Blackie: Glasgow, UK, 1988; pp. 243-255.
[12] Fordos, Z. In: *Natural Fibre Reinforced Cement and Concrete (Concrete Technology and Design 5)*; Swamy, R. N.; Ed.; Blackie: Glasgow, UK, 1988; pp. 173-207.
[13] Guimarães, S. S. In: *Proceedings of the International Conference Development of Low-Cost and Energy Saving Construction Materials and Applications*; Envo: Rio de Janeiro, Brazil, 1984.
[14] Savastano Jr., H.; Warden, P. G.; Coutts, R. S. P. *Cem. Concr. Compos.* 2005, *27*, 583-592.
[15] Savastano Jr., H.; Turner, A.; Mercer, C.; Soboyejo, W.O. Mechanical behavior of cement-based materials reinforced with sisal fibers. J Mater Sci (2006) 41:6938–6948.

[16] Giannasi, F.; Thebaud-Mony, A. *Int J. Occup. Environ. Health.* 1997, *3*, 150-157.
[17] Brazilian Government - Ministry of Mining and Energy - National Department of Mineral Production (2008). Mineral Summary 2008 - Gypsite (in Portuguese). *http://www.dnpm.gov.br/assets/galeriaDocumento/SumarioMineral2008/gipsita.pdf.*
[18] Rutherford, P. M.; Dudas, M. J.; Samek, R.A. *Sci. Total Environ.* 1994, *149*, 1-38.
[19] UNSCEAR – United Nations Scientific Committee on the Effects of Atomic Radiation. *Sources and Effects of Ionizing Radiation*; United Nations: New York, UN, 2000.
[20] Nero, A. V. In: *Radon and Its Decay Products in Indoor Air*; Nazaroff, W. W.; Nero, A. V.; Eds.; John Wiley & Sons: New York, USA, 1988; pp. 1-53.
[21] Nazaroff, W. W.; Moed, B. A.; Sextro, R. G. In: *Radon and Its Decay Products in Indoor Air*; Nazaroff, W. W.; Nero, A. V.; Eds.; John Wiley & Sons: New York, USA, 1988, pp. 57-112.
[22] Stranden, E. In: *Radon and Its Decay Products in Indoor Air*; Nazaroff, W. W.; Nero, A. V.; Eds.; John Wiley & Sons: New York, USA, 1988, pp. 113-130.
[23] Mazzilli, B.; Palmiro, V.; Saueia, C.; Nisti, M. B. *J. Environ. Radioact.* 2000, *49*, 113-122.
[24] U.S. Government (1998). Code of Federal Regulations, Title 40, v. 7, pt. 61.202 (40CFR61.202). *http://www.epa.gov/epahome/topics.html.*
[25] Mazzilli, B.; Saueia, C. *Radiat. Prot. Dosim.* 1999, *86*, 63-67.
[26] ICRP – International Commission on Radiological Protection. *Protection Against Radon-222 at Home and at Work – ICRP Publication 65*; Annals of the ICRP; Pergamon Press: Oxford, UK, 1993; vol. 23/2.
[27] Bentur, A.; Akers, S. A. S. *Int. J. Cem. Compos. Lightweight Concr.* 1989, *11*, 99-109.
[28] Toledo Filho, R. D.; Scrivener, K.; England, G. L.; Ghavami, K. *Cem. Concr. Compos.* 2000, *22*, 127-43.
[29] John, V. M.; Agopyan, V.; Prado, T. A. In: *Proceedings of the 3rd Ibero-American Symposium on Roofing for Housing*; CYTED/USP: São Paulo, Brazil, 1998; pp. 51-59.
[30] Agopyan, V.; John, V. M. *Build. Res. Inf.* 1992, *20*, 233-235.
[31] Savastano Jr., H.; John, V. M.; Caldas, A. In: *Proceedings of the International Conference on Composites in Construction - CCC2001*; Figueiras, J.; Juvandes, L.; Faria, R.; Eds.; Swets & Zeitlinger: Lisse, The Netherlands, 2001; pp. 299-302.
[32] Tonoli, G. H. D.; Santos, S. F.; Joaquim, A. P.; Savastano Jr., H. In: *Proceeding of the 10th International Inorganic-Bonded Fiber Composites Conference*; University of Idaho: São Paulo, Brazil, 2006; 11 p.
[33] Oliveira, C.T.A.; John, V.M.; Agopyan, V. Pore water composition of activated granulated blast furnace slag cements pastes. In 2nd International Conference on Alkaline Cements and Concretes, ed. Kiev State Technical University of Construction and Architecture, May 1999, Kiev, 9p. (Accepted paper)
[34] John, V. M. *Slag cement activated with calcium silicates* (in Portuguese). Ph.D. thesis; University of São Paulo: São Paulo, Brazil, 1995.
[35] John, V. M.; Agopyan, V.; Derolle, In: *Proceedings of the 2nd International Symposium on Vegetable Plants and their Fibres as Building Materials*; Sobral, H. S.; Ed.; Chapman and Hall: London, UK, 1990; pp. 87-97.

[36] Appita - Australian Pulp and Paper Industry Technical Association. *Kappa Number of Pulp*. P201 m-86 (endorsed as part of AS 1301 by the Standards Association of Australia), 1986; 4 p.
[37] Higgins, H.G. *Paper physics in Australia*. CSIRO - Division of Forestry and Forest Products: Melbourne, Australia, 1996.
[38] Savastano Jr., H.; Warden, P. G.; Coutts, R. S. P. *Cem. Concr. Compos.* 2003, *25*, 311-319.
[39] Agopyan, V. *Fiber reinforced materials for civil construction in developing countries: use of vegetable fibers* (in Portuguese). University of São Paulo: São Paulo, Brazil, 1991.
[40] Richardson, I. G.; Wilding, C. R.; Dickson, M. J. *Adv. Cem. Res.* 1989, *2*, 147-157.
[41] Douglas, E.; Brandstetr, J. *Cem. Concr. Res.* 1990, *20*, 746-756.
[42] Fernandes, J. D.; Unkalkar, V. G.; Meshramkar, P. M.; Jaspal, N. S.; Didwania, H. P. In: *Nonwood Plant Fiber Pulping*; TAPPI Press: Atlanta, USA, 1981, v. 11, pp. 73-89.
[43] Misra, D. K. In: *Pulp and paper: chemistry and chemical technology*; 3rd. ed.; Casey, J. P.; Ed.; John Wiley & Sons: New York, USA, 1983, v. 1, pp. 504-530.
[44] Coutts, R. S. P.; Kightly, P. *J. Mater. Sci.* 1984, *19*, 3355-3359.
[45] Coutts, R. S. P.; Ridikas, V. *Appita*. 1982, *35*, 395-400.
[46] Soroushian, P.; Marinkute, S.; Won, J. P. *ACI Mater. J.* 1995, *92*, 172-180.
[47] Eusebio, D. A.; Cabangon, R. J.; Warden, P. G.; Coutts, R. S. P. In: *Proceedings of the 4th Pacific Rim Bio-Based Composites Symposium*, Bogor Agricultural University: Bogor, Indonesia, 1998; pp. 428-436.
[48] Soroushian, P.; Shah, Z.; Won, J. P. *ACI Mater. J.* 1995, *92*, 82-92.
[49] Abdelmouleh, M.; Boufi, S.; Salah, A.; Belgacem, M. N.; Gandini, A. *Langmuir*. 2002, *18*, 3203-3208.
[50] Blankenhorn, P. R.; Blankenhorn, B. D.; Silsbee, M. R.; Dicola, M. *Cem. Concr. Res.* 2001, *31*, 1049–1055.
[51] Pehanich, J. L.; Blankenhorn, P. R.; Silsbee, M. R. *Cem. Concr. Res.* 2004, *34*, 59-65.
[52] Xu, Y.; Chung, D. D. L. *Cem. Concr. Res.* 1999, *29*, 451-453.
[53] Britt, K. W.; *Handbook of Pulp and Paper Technology*; 2nd ed.; Van Nostrand Reinhold: New York, USA, 1970.
[54] Clark, J. d'A. *Pulp Technology and Treatment for Paper*; Miller Freeman: San Francisco, USA, 1987.
[55] Coutts, R. S. P. *Composites*. 1984, *15*, 139-143.
[56] Coutts, R. S. P. *J. Mater. Sci. Lett.* 1987, *6*, 140-142.
[57] Mohr, B. J.; Nanko, H.; Kurtis, K. E. *Cem. Concr. Compos.* 2005, *27*, 435-448.
[58] Bentur, A.; Mindess, S. *Fibre Reinforced Cementitious Composites*; 2nd ed.; Spon Press: London, UK, 2007.
[59] Akkaya, Y.; Picka, J.; Shah, S. P. *J. Mater. Civ. Eng.* 2000, *12*, 272–279.
[60] Chung, D. D. L. 2005. "Dispersion of Short Fibers in Cement". Jounal of Mat. in Civil Engineering 17(4) 379-383.
[61] Beraldo, A. L. In: *Nonconventional Materials for Rural Construction* (in Portuguese); Toledo Filho, R. D.; Nascimento, J. W. B.; Ghavami, K.; Eds.; Federal University of Paraíba / Brazilian Society of Agricultural Engineering: Campina Grande, Brazil, 1997, pp. 1-48.
[62] Lopes, W. G. R.; Valenciano, M. D. C. M.; Martins, S. C. F.; Beraldo, A. L.; Azzini, A. In: *Proceedings of the International Conference on Sustainable Construction into the*

Next Millennium: Environmentally Friendly and Innovative Cement Based Materials (in Portuguese); Barbosa, N. P.; Swamy, R. N.; Lynsdale, C.; Eds.; Federal University of Paraíba / University of Sheffield: João Pessoa, Brazil, 2000, pp. 379-393.

[63] Roma Jr., L. C. *Fibre-cement roofing tiles and cooling system: influence on performance of crossbreed and Holstein veal* (in Portuguese). M. Sc. dissertation; Faculty of Animal Science and Food Engineering, University of São Paulo: Pirassununga, Brazil, 2004.

[64] Savastano Jr., H.; Warden, P. G.; Coutts, R. S. P. *Cem. Concr. Compos.* 2000, *22*, 379-384.

[65] Silva, A. C. *Durability of cellulose fibres reinforced composites* (in Portuguese). M. Sc. dissertation; Polytechnic School, University of São Paulo: São Paulo, Brazil, 2002.

[66] Gram, H. E.; Gut, P. *FCR/MCR Toolkit Element 23 - Quality Control Guidelines*; SKAT, St. Gall, ILO: Geneva, Switzerland, 1991.

[67] Pereira, L. F. L. C.; Cincotto, M. A. *Chlorite determination in concrete of ordinary Portland cement: influence of type of cement* (in Portuguese). Technical Bulletin - BT / PCC / 294; Civil Engineering Department, Polytechnic School, University of São Paulo: São Paulo, Brazil, 2001.

[68] Taylor, H. F. W. *Cement Chemistry*; 2 ed.; Thomas Telford: London, UK, 1997.

[69] Toledo Filho, R. D.; Ghavami, K.; England, G. L.; Scrivener, K. *Cem. Concr. Compos.* 2003, *25*, 185-196.

[70] De Silva, P.; Bucea, L.; Moorehead, D. R.; Sirivivatnanon, V. *Cem. Concr. Compos.* 2006, *28*, 613-620.

[71] Tonoli, G. H. D.; Santos, S. F.; Rabi, J. A.; Santos, W. N.; Akiyoshi, M. M.; Savastano Jr., H. In: *Proceedings of the Brazilian Conference on Non-Conventional Materials and Technologies in Ecological and Sustainable Construction - BRAZIL-NOCMAT*; Ghavami, K.; Toledo Filho, R. D.; Carvalho, R. F.; Eds.; ABMTENC: Salvador, Brazil, 2006.

[72] Alves, S. M.; Pietrobon, C. L. R.; Pietrobon, C. E. In: *Proceedings of the 5th Brazilian Conference on Comfort of Ambient Spaces - ENCAC / 2nd Latin-American Conference on Comfort of Ambient Spaces - ELACAC* (in Portuguese); Pereira, O. R. P.; Eds.; ANTAC: Fortaleza, Brazil, 1999.

[73] Özişik, M. N. *Heat Transfer - A Basic Approach*. McGraw-Hill: New York, USA, 1985.

[74] Brazilian Committee of Civil Construction (CB-02, CE-02:135.07). *Thermal performance in buildings - Calculation methods of thermal transmittance, thermal capacity, thermal delay and solar heat factor of elements and components of buildings* (in Portuguese). ABNT: Rio de Janeiro, Brazil, 1998.

[75] Kawabata, C. Y. *Thermal performance from different types of roof in individual calf housing* (in Portuguese). M. Sc. dissertation; Faculty of Animal Science and Food Engineering, University of São Paulo: Pirassununga, Brazil, 2003.

[76] Devito, R. A. *Physical and mechanical studies of roofing tiles made of blast furnace slag cement reinforced with residual cellulose fibres* (in Portuguese). M. Sc. dissertation; Engineering School of São Carlos, University of São Paulo: São Carlos, Brazil, 2003.

[77] Ferreira, F. L.; Prado, R. T. A. In: *Proceedings of the 20th Conference on Passive and Low Energy Architecture*; Pontifical Catholic University of Chile - PLEA International: Santiago, Chile, 2003.

[78] Kaviany, M. *Principles of Heat Transfer in Porous Media*; Springer: New York, USA, 1995.

[79] Nield, D. A.; Bejan, A. *Convection in Porous Media*; Springer: New York, USA, 1998.
[80] Stranden, E.; Berteig, L. *Health Phys.* 1980, *39*, 275-284.
[81] Stranden, E.; Kolstad, K.; Lind, B. *Radiat. Prot. Dosim.* 1984, *7*, 55-58.
[82] Edwards, J. C.; Bates, R. C. *Health Phys.* 1980, *39*, 263-274.
[83] Loureiro, C. O. *Simulation of the steady-state transport of radon from soil into houses with basements under constant negative pressure*. Ph.D. thesis; Environmental Health Sciences, University of Michigan: Ann Arbor, USA, 1987.
[84] Nazaroff, W. W. *Rev. Geophys.* 1992, *30*, 137-160.
[85] Yu, C.; Loureiro, C.; Cheng, J. J.; Jones, L. G.; Wang, Y. Y.; Chia, Y. P.; Faillace, E. *Data collection handbook to support modeling impacts of radioactive materials in soil*. Environmental Assessment and Information Sciences Division, Argonne National Laboratory: Argonne, USA, 1993.
[86] Riley, W. J.; Robinson, A. L.; Gadgil, A. J.; Nazaroff, W. W. *Atmos. Environ.* 1999, *33*, 2157-2168.
[87] Andersen, C. E. *Radon transport modelling: user's guide to RnMod3d*. Riso-R-1201(EN); Riso National Laboratory: Roskilde, Denmark, 2000.
[88] Nazaroff, W. W., Teichman, K. *Environ. Sci. Technol.* 1990, *24*, 774-782.
[89] Jang, M.; Kang, C. S.; Moon, J. H. *J. Environ. Radioact.* 2005, *80*, 153-160.
[90] Nikezic, D.; Urosevic, V. *Nucl. Instrum. Meth. A.* 1998, *406*, 486-498.
[91] Piliposian, G. T.; Appleby, P. G. *Continuum Mech. Therm.* 2003, *15*, 503-518.
[92] Rabi, J. A.; Silva, N.C. *J. Environ. Radioact.* 2006, *86*, 164-175.
[93] Rabi, J. A.; Silva, N.C. In: *Proceedings of the 11th Brazilian Congress of Thermal Sciences and Engineering – ENCIT 2006*; ABCM – Brazilian Society of Mechanical Sciences and Engineering; Ed.; CD-ROM: Curitiba, Brazil, 2006, CIT06-0428.
[94] Bird, R. B.; Stewart, W. E.; Lightfoot, E. N. *Transport Phenomena*. John Wiley & Sons: New York, USA, 1960.
[95] Mohamad, A.A. *Int. J. Therm. Sci.* 2003, *42*, 385-395.
[96] Kays, W. M.; Crawford, M. E. *Convective Heat and Mass Transfer*. McGraw-Hill: New York, USA, 1993.
[97] Rabi, J. A.; Mohamad, A. A. *J. Porous Media.* 2005, *8*, 175-191.
[98] Rabi, J. A.; Mohamad, A. A. *Appl. Math. Model.* 2006, *30*, 1546-1560.
[99] Bennacer, R.; Beji, H.; Mohamad, A. A. *Int. J. Therm. Sci.* 2003, *42*, 141-151.
[100] Merrikh, A. A.; Mohamad, A. A. *Int. J. Heat Mass Transfer.* 2002, *45*, 4305-4313.
[101] Ingham, D. B. In: *Current Issues on Heat and Mass Transfer in Porous Media*; Ingham, D. B.; Ed.; NATO Advanced Study Institute on Porous Media; Ovidius University Press: Constanţa, Romania, 2003, pp. 1-10.
[102] Askri, F.; Ben Salah, M.; Jemni, A.; Ben Nasrallah, S. In: *Applications of Porous Media*; Reis, A. H.; Miguel, A. F.; Eds.; ICAMP 2004; CGE: Évora, Portugal, 2004, pp. 315-318.
[103] Rabi, J. A.; Mohamad, A. A. In: *Applications of Porous Media*; Reis, A. H.; Miguel, A. F.; Eds.; ICAMP 2004; CGE: Évora, Portugal, 2004, pp. 471-478.
[104] Silva, N. C. *Natural radionuclides and toxic elements within phosphogypsum stacks in Brazil: characterization and lixiviation* (in Portuguese). Ph.D. thesis; University of São Paulo: Piracicaba, Brazil, 2001.
[105] Rabi, J. A.; Silva, N. C.; Soares, S. M.; Prado, A. S.; Santos, A. R. In: *Proceedings of the Inter-American Conference on Non-Conventional Materials and Technologies in*

Ecological and Sustainable Construction - IAC-NOCMAT; Ghavami, K.; Savastano Jr., H.; Joaquim, A. P.; Eds.; ABMTENC: Rio de Janeiro, Brazil, 2005.

In: Agricultural Wastes
Eds: Geoffrey S. Ashworth and Pablo Azevedo

ISBN 978-1-60741-305-9
© 2009 Nova Science Publishers, Inc.

Chapter 11

FROM SOLID BIOWASTES TO LIQUID BIOFUELS

Leandro S. Oliveira[1] and Adriana S. Franca

DEMEC (Department of Mechanical Engineering)/UFMG
Av. Antônio Carlos, 6627, 31270-901 Belo Horizonte, MG, Brasil

ABSTRACT

In a time when a foreseeable complete transmutation from a petroleum-based economy to a bio-based global economy finds itself in its early infancy, agricultural wastes, in the majority currently seen as low-valued materials, are already beginning their own transformation from high-volume waste disposal environmental problems to constituting natural resources for the production of a variety of eco-friendly and sustainable products, with second generation liquid biofuels being the leading ones. Agricultural wastes contain high levels of cellulose, hemicellulose, starch, proteins, and, some of them, also lipids, and as such constitute inexpensive candidates for the biotechnological production of liquid biofuels (e.g., bioethanol, biodiesel, dimethyl ether and dimethyl furan) without competing directly with the ever-growing need for world food supply. Since agricultural wastes are generated in large scales (in the range of billions of kilograms per year), thus being largely available and rather inexpensive, these materials have been considered potential sources for the production of biofuels for quite some time and have been thoroughly studied as such. In the last decades, a significant amount of information has been published on the potentiality of agricultural wastes to be suitably processed into biofuels, with bioethanol as the main research subject. Thus, it is the aim of this chapter to critically analyze the current situation and future needs for technological developments in the area of producing liquid biofuels from solid biowastes. The state-of-the-art in producing bioethanol, bio-oil and biodiesel from agricultural wastes will be discussed together with the new trends in the area. The emerging biowaste-based liquid biofuels (e.g., biogasoline, dimethyl ether and dimethyl furan) currently being studied will also be discussed.

---

[1] Corresponding author. E-mail: leandro@demec.ufmg.br. Tel:+55-31-34093512. Fax:+55-31-3443-3783.

1. INTRODUCTION

The strongly negative scenario that has been continually painted by both media reports and debates and by the skeptical scientific community regarding the foreseeable future of petroleum-derived energy has positively contributed to the envisaging of a prospective prolific future for biomass energy. Also, owing to the facts that the agricultural and food industries generate large volumes of wastes worldwide annually and that there is a growing demand for proper waste disposal management due to environmental impact concerns, tremendous pressure has been placed on the research, governmental and industrial communities to adequately study and formulate proposals for the recovery, recycling and upgrading of such biological wastes. Furthermore, an inappropriate shift from food producing to biofuel producing using agricultural food resources (e.g., sugar and vegetable oil) has been a major concern in several countries where the economies are largely (and sometimes solely) based on agriculture. Thus, the idea of producing biofuels from biosolid wastes (waste biomass) comes naturally as a solution for problems related to a much-feared new energy crisis, to the current high-volume agricultural and food waste disposal management problems, and to the avoidance of a future food supply shortage. The generic term 'biosolid wastes' will be herein adopted to describe all biomass wastes that are solid in their natural states and also require physical and/or biochemical treatment to be converted into a liquid biofuel. This chapter is restricted to the analysis of liquid fuels produced from waste biomass due to its relevance to the transportation sector (the majority of transportation systems operate with internal combustion engines that run on volatile liquid fuels), and to the fact that liquid transportation fuels account for approximately 30% of the carbon emissions in industrialized countries (Gomez et al., 2008). A switch to alternative non-liquid fuels is foreseen for the transportation sector only in the long-term future and, in the short to medium term, a more sustainable means of producing liquid transportation fuels ought to be sought.

The alternatives currently being sought for liquid biofuel production from waste biomass are as diverse as are the types of biomass suitable for such conversion (Figure 1). Regarding the types of waste biomass, a more generic categorization would be a division into lignocellulosic, starchy and oily wastes, representative of residues from forestry and agricultural operations and practices.

2. WASTE BIOMASS

Defining biomass is quite a difficult task, because all of the definitions available are somewhat tied to the application to which the biomass is destined, and, more recently, due to the increasing economic relevance of biomass, the definition has evolved to include the concept of sustainable production (Goldemberg and Coelho, 2004; de Vos, 2006). Thus, the definition to be presented in this chapter will be tied to the application at hand, which is the production of liquid biofuels. Biomass is herein defined as the biodegradable fraction of products, waste and residues from organic non-fossil material of biological origin that is readily available in a renewable or recurring sustainable basis and also can be used as a liquid energy source by means of thermal/chemical/biological conversion.

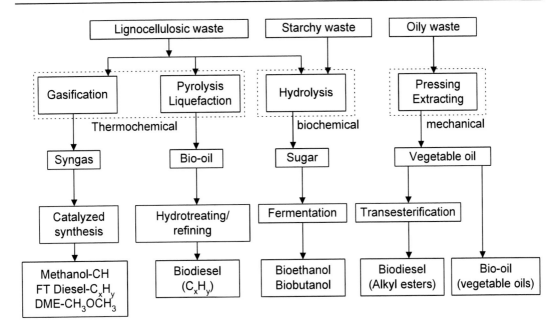

Figure 1. Pathways commonly studied for the conversion of biomass to liquid biofuels (adapted from Champagne, 2008; FT = Fischer-Tropsch).

Biomass is usually comprised of lignocellulosic materials and, as such, is comprised of high contents of cellulose, hemicellulose, lignin and proteins, thus constituting renewable natural resources for a plethora of inexpensive eco-friendly and sustainable materials. For the most part, these biomass materials are usually readily available as wood residues (left over from forestry operations), municipal wastes, agricultural and food wastes, and dedicated energy crops. Wood residues are the largest source of biomass energy in the world, followed by municipal wastes, agricultural residues and dedicated energy crops (Lin and Tanaka, 2006). In biomass, cellulose is usually the largest fraction, comprising about 40 to 50% of the total biomass weight, whereas hemicellulose is usually in the range of 20 to 40% by weight (Saxena et al., 2009) and lignin in the range of 5 to 30% by weight depending on the nature of the plant, whether it is herbaceous or woody. Woody plant materials are composed of tightly-bound fibers due to a higher content of lignin compared to herbaceous plants and grasses, which are comprised of more loosely-bound fibers, indicating lower lignin contents (McKendry, 2002).

Cellulose is the most abundant natural polymer in nature, with an average molecular weight of 100,000, and comprised of linear chains of (1,4)-D-glucopyranose units linked in a (1-4) beta configuration. Hemicellulose is a heterogeneous branched oligomer, with a molecular weight <30,000, and comprised of both C_6- and C_5-monosaccharides, primarily xylose and glucose. Hemicellulose is present in plant materials tightly (but not co-valently) bound to the surface of cellulose microfibrils. Lignin, the third most abundant polymer in the plant kingdom, is a highly cross-linked amorphous polymer built up of phenylpropane units, with three aromatic alcohols, termed monolignols, as precursors, namely p-coumaryl, coniferyl and sinapyl alcohols. The respective aromatic constituents of these alcohols in the polymer are p-hydroxyphenyl, guaiacyl and syringyl moieties (Buranov and Mazza, 2008). As a major cell wall component, it provides rigidity, internal transport of water and nutrients and resistance against microbial and chemical assaults on the structural sugars, namely cellulose and hemicellulose. These characteristics are one of the contributors to the set of factors precluding the complete technological exploitation of plant materials for the production of biofuels and other biomaterials, thus being a component of what is generically termed 'biomass recalcitrance' (Himmel et al., 2007).

Based solely on its application as bioenergy feedstock, waste biomass can be categorized into high- and low-moisture content biomass (McKendry, 2002). This categorization is of relevance because, aside from the inherent chemical constituency, the selection of the form of energy conversion process for a specific type of biomass will be dictated by its moisture content. High-moisture content biomass is more appropriate for a wet conversion process, such as fermentation, i.e., processes involving biochemically mediated reactions, whereas low-moisture content biomass is more economically suited to conversion processes such as combustion, pyrolysis or gasification (McKendry, 2002). Agricultural and food solid wastes, the biomass types to be discussed in this essay, are generated in high quantities worldwide (Laufenberg et al., 2003; Lal, 2008; Rodríguez-Couto, 2008; Lora and Andrade, 2009), and, depending on the source and pretreatment they might have been subjected to, they could either fall into the high- or the low-moisture content category, hence, being appropriate for either wet conversion processes or thermochemical conversion processes.

Agricultural wastes (or residues), in their majority, are currently considered low value materials that are highly prone to microbial spoilage, hence, limiting their exploitation. Further exploitation can also be precluded by legal restrictions and the costs of collection, drying, storage and transportation. Thus, for the most part, currently, these materials are either used as animal feed (not always adequate due to difficulties in properly balancing the animal nutritional requirements), combustion feedstock or disposed to landfill causing major environmental issues (e.g., emission of large quantities of volatile organic compounds in the case of combustion, and contamination of groundwater in the case of landfill) (Oliveira and Franca, 2008). Estimating the correct amount of residue produced by a specific crop is not an easy task, because the availability of a residue is dependent on the specific variety of the crop, on the seasonal variations and on the geographical location of the crop, amongst other less relevant factors. Among the diversity of agricultural crops, cereals crops are the ones currently producing the largest amounts of biomass residues, followed by sugar crops, legumes and oil crops (Lal, 2008). Straw is the dominant residue (Prasad et al., 2007), usually presenting low contents of lignin (5–20% by weight) and, thus, being considered a low-density residue.

The availability of agricultural residues as energy feedstock is a function of the amount produced of the associated crop, of the residue-to-crop ratio, of the collection efficiency and of the amount used in other competing applications, such as fodder for livestock, feedstock for fertilizer, materials for construction and direct burning in boilers and furnaces (Purohit et al., 2006). It has been suggested that, after accounting for seasonal variation and, also, for the use of agricultural residues for soil conservation and livestock feed, an average of only 15% of the total residue production would be available for industrial energy generation (Bowyer and Stockmann, 2001). The amount of residue produced by a specific crop is usually measured by the Residue Coefficient (RC), which is calculated as the residue-to-crop ratio, weight by weight, considering a percent recovery fraction ranging from 15 to 70% depending on the agricultural practices adopted (harvesting, collection, transportation and storage efficiencies) and on the intensity of other competing applications on location (Purohit et al., 2006; Lora and Andrade, 2009). Some examples of CR values for a diversity of perennial crops are presented in Table 1.

Table 1. Residue Coefficients (CR) for perennial crops around the world

Crop-Residue	Residue Coefficients (CR= weight of available residue/weight of related crop produced)			
	Cuiping et al. (2004)	Purohit et al. (2006)	Haq and Easterly (2006)	Lora and Andrade (2009)
Rice-straw	0.623	1.530	0.740	—
Rice-straw + husk	—	—	—	1.700
Sorghum-straw	1.000	—	—	—
Sorghum-stover	—	—	0.740	—
Wheat-straw	1.366	1.470	-	1.300
Barley, rye, oats-straw	—	—	1.670	—
Soybean-stems and leaves	1.500	—	—	1.400
Sunflower-stalks	2.000	—	—	—
Rapeseed-stalks	2.000	—	—	—
Mustard-stalks	—	1.850	—	—
Cotton-stalks	—	3.000	—	—
Cotton-stalks and leaves	3.000	—	—	—
Cotton-gin trash	—	—	0.900	—
Cotton-field trash	—	—	0.600	—
Sugarcane-leaves	0.100	—	—	—
Sugarcane-bagasse	—	0.250	0.250	—
Corn-stalk and cob	2.000	1.860		1.000
Corn-stover	—		1.100	—
Arhar-stalk	—	1.320	—	—
Hemp-stems and leaves	2.500	—	—	—
Groundnut-shell	—	0.330	—	—
Jute-stick	—	1.850	—	—
Manioc-stems and leaves	—	—	—	0.800
Coffee-husks	—	—	—	0.210

The exact amounts of residues produced for a specific crop are very difficult to predict due to differences in agricultural practices adopted in different geographical locations, seasonal and cultivar variability, and the lack of sufficient data collected to be statistically representative of a specific crop. Collection of statistical data on production of agricultural residues became an established practice only recently, when both the prospects of adding value to this type of biomass residue by producing biofuels and other biomaterials and the environmental issues arising from the large-scale disposal management problems became of relevance.

3. LIQUID BIOFUELS

When considering the alternatives currently available, or that are under intensive investigation for the near-future supply of the global energy market, the ones that represent truly sustainable options are the solar, wind, hydro, biomass and geothermal energies. However, for direct use in the transportation sector, which represents about a third of the total energy consumed worldwide, none of these primary energy sources are suited and conversion processes are needed to transform them into energy carriers such as electricity, hydrogen and biofuels (Petrus and Noordermeer, 2006). Although commercialization of electrical vehicles has been pushed forward already in some developed countries, a major technical barrier, associated to the efficient storage of large amounts of electricity, constituted the main (but not the only one) factor that prevented its rapid spread in the transportation market. Hydrogen, although representing the best alternative as a sustainable fuel due to the fact that it generates water as the waste product, presents two major limitations in terms of the associated technology currently available: the conversion routes from primary energy sources to hydrogen are quite inefficient, with some of them having a worse carbon footprint than burning fossil fuels (Gomez et al., 2008); and significant developments are required regarding enhancing energy densities in hydrogen storage and improving safety issues regarding hydrogen transportation and storage. For both the aforementioned fuel options, there is also the precluding aspect of the overall inertia of the vehicles manufacturers in developing and embracing innovative technology, mainly when it means manufacturing products that would last longer, which would be the case of electrical vehicles that usually have less wearable parts than an internal combustion engine vehicle. Thus, considering the aforementioned aspects and the fact that current technologies for promoting motion in the transportation sector are based on internal combustion engines that run on liquid fuels, biomass-based liquid biofuels become, in the short term, the most suitable alternative to be sought for transportation purposes.

The definition of biofuel should encompass all forms of fuels generated from renewable biological sources. Hence, biohydrogen, charcoal, biomethane, biomethanol, bioethanol, biobutanol (often called "biogasoline"), biodiesel, bio-oil, dimethyl ether (DME) and dimethyl furan (DMF) are all possible forms of biofuels that can be produced from waste biomass. When talking about biomass-derived biofuels, one should differentiate between what is termed 'first' and 'second generation' biofuel. First generation biofuels are those which are produced from conventional technological conversion of sugar, starch or vegetable oil as feedstock yielded from sugarcane, grains and oilseeds, respectively. Sugar and starch are usually converted into bioalcohols (mostly bioethanol) by fermentation processes, with the starch materials requiring a pretreatment step to release the fermentable sugars from the starch. Vegetable oils are usually cold-pressed from oilseeds and reacted with short chain alcohols (e.g., methanol and ethanol) to produce biodiesel. The terminology 'first generation' is directly related to the fact that the referred feedstock (sugarcane, oilseeds and grains) are products of cultivated crops that could instead be directed to the animal or human food chain. Current problematic issues regarding the production of first generation biofuels are the nonsensical (at least from a scientific point-of-view) food vs. fuel debate (Muller et al., 2007; Erickson et al., 2008), and the skeptical professionals (scientists and economists) claims that, not only first generation biofuels are not economically viable, but they also have a negative impact in climate mitigation and carbon cycles, and, also, the serious limiting capacity of biofuel production from the usual feedstock (Zabaniotou et al., 2008).

Second generation biofuels are those which are produced from cheap and abundant plant biomass (Gomez et al., 2008), where, ideally, the whole plant material, consisting mainly of cellulose, hemicellulose and lignin, is used in the conversion process, thus, producing little or no residue at all. Agricultural and food wastes are therefore perfect candidates for feedstock in the production of second generation biofuels and are envisioned as an attractive solution to the aforementioned problems associated with the production of first generation biofuels. The major advantages of second generation biofuels are that greater amounts of biofuel can be produced per unit mass of raw material and, consequently, per cultivated hectare; the combustion of such fuels has the potential to be carbon neutral; and, also, the process can be completely dissociated from food and fiber production, thus, not affecting the increasingly demand for food and fiber supplies. However, a number of technological hurdles, mostly related to lignin and cellulose recalcitrance to chemicals and enzymes, must still be overcome in the processing of such biomass materials into second generation biofuels in large scale production (Himmel et al., 2007; Gomez et al., 2008). These technological barriers are more prominent when considering the conversion of such materials into bioalcohols, e.g., bioethanol, which is presently the most relevant energy carrier for use in the transportation sector, used directly as fuel or in vegetable oils transesterification reactions for the production of biodiesel. In summary, the chemical and structural features of biomass (imparted by its major components, namely cellulose, hemicellulose and lignin) constitute a barrier to liquid penetration and/or enzyme activity, thus precluding the release of soluble sugars which are the natural substrates for microorganisms used for alcoholic fermentation.

Regarding transportation fuels, currently, approximately 98% of the fuels used in combustion engines are derived from petroleum, carrying with them the consequent negative aspects of net carbon emissions (accounting for approximately 30% of the carbon emissions in industrialized countries), of fuels security (associated with the usual political volatility of the major petroleum-producing countries), and a prospective exhaustion of oil reserves in a foreseeable future. As such, the respective automotive engines are designed to run on liquid fuels that meet a set of strict specifications associated to the physicochemical characteristics of those fuels. Thus, in the short term, to meet the transportation sector demands and specifications, the alternative choices available for fuels are either to design new processes that will convert biomass to liquid biofuels that meet the set of specifications established for the ones currently in use, or to redesign the engines to run on other types of liquid fuels (or on blends with fuels other than petroleum-derived ones), preferably renewable ones.

To meet the aforementioned specifications, the choices are either to use biodiesel in engines that run on fossil Diesel (not requiring any modifications in the engines currently in use) or to convert the lignocellulosic materials into hydrocarbon-like fuels such as gasoline and diesel (e.g., pyrolysis-derived bio-oils, Fischer-Tropsch Diesel, DME). The conversion of carbohydrates (basically cellulose and hemicellulose) to hydrocarbons should be focused on the removal of the heteroatoms, such as oxygen, from the carbohydrate polymers, since the energy content per mass unit of a fuel increases with decreasing oxygen content (Petrus and Noordermeer, 2006). However, a certain amount of oxygen is desirable to assure a clean burning, i.e., to guarantee the maximum conversion (stoichiometric) of the fuel into carbon dioxide and water during its combustion.

The redesigning of the engines to run on liquid biofuels, such as bioethanol, is an accomplished target. Since the mid 1970s, Brazil has successfully invested in such endeavor and since the early 1980s has been using bioethanol, produced primarily from sugarcane, as a fuel for hydrated ethanol-dedicated spark-ignition engines (Geller, 1985). The dedicated spark-ignition engines using neat unblended bioethanol have undergone minor modifications relative to gasoline engines. Currently, the majority of the vehicles being manufactured in Brazil use the 'flexible' spark-ignition engine technology ('flexible vehicles') that can run solely on gasoline, solely on bioethanol, or on an unrestricted range of blends of gasoline and bioethanol. Presently, approximately 60% of the worldwide production of fuel ethanol is sugar based (primarily produced from sugarcane in Brazil) and 40% is starch based (primarily produced from corn in the United States). The technologies for producing bioethanol from both sugarcane and corn, although far from ideal, with each presenting their own bottlenecks, can be considered mature and under constant evolution. Reviews on the state-of-the-art of the processing technology and discussions (sometimes controversial) on the energetic performance, economic and environmental aspects of bioethanol fuel from sugarcane and corn are aplenty in the literature (Hatzis et al., 1996; Kheshgi et al., 2000; Pimentel, 2003; Pimentel and Patzek, 2005; Agarwal, 2007; Neves et al., 2007; Demirbaş, 2007a; Balat et al., 2008; Dale, 2008; Luque et al., 2008). This type of review is outside the scope of this essay and the reader is encouraged to assess the cited references for a more comprehensive view of the subject.

3.1. Biodiesel from Agri-Food Residues

Biodiesel is a name applied to fuels manufactured by the transesterification of renewable oils and fats and by the esterification of free fatty acids. Biodiesel can be employed as a fuel for unmodified diesel engines (Graboski and McCormick, 1998). This type of fuel has been commercially produced in Europe since the early 1990s (Barnwal and Sharma, 2005). Biodiesel can be blended in any proportion with fossil diesel or it can be used in its pure form without causing harmful effects to an unmodified compression ignition (diesel) engine, while simultaneously reducing harmful exhaust emissions, such as sulfur, unburned hydrocarbons, carbon monoxide and particulate. In addition, biodiesel is completely miscible with petroleum diesel fuel and can be employed as a blend. Several reviews on the production, analysis, and use of biodiesel in internal combustion engines are available in the literature and the reader is encouraged to read them for a more comprehensive view of the subject (Graboski and McCormick, 1998; Ma and Hanna, 1999; Fukuda et al., 2001; Pinto et al., 2005; Meher et al., 2006; Sharma et al., 2008; Murugesan et al., 2009; Balat and Balat, 2008; Demirbaş, 2009).

Even though research on biodiesel has gained a tremendous momentum since the early 1990s, due to the growing concern about fossil fuel increasing prices, petroleum reserves depletion and associated environmental issues, biodiesel production is still considered not economically feasible in comparison to petroleum diesel, with the contributors to the cost of biodiesel being the raw material, the type of process used for purification, the reactants and the storage costs. Feedstock (e.g., soybean, rapeseed, sunflower seeds) production costs are responsible for approximately 70 to 80% of total fuel production costs (Sharma et al., 2008). Such drawbacks could be minimized by the use of a "less valuable" feedstock, such as oily agricultural and food residues.

Considering only solid residues (the focus of this essay), the first attempt to propose the use of an agricultural residue to produce biodiesel published in the literature was done by Giannelo et al. (2002), where tobacco (*Nicotiana tabacum* L.) seeds, a residue from the tobacco leaves production, was analyzed in terms of their oil content and respective composition and also in terms of the oil physical and chemical properties as related to fuel quality. The crop-to-residue ratio for tobacco is about 1.2, i.e., a CR value of 0.833 kg of seeds available as residue for every kg of tobacco leaves produced (Fujimori and Matsuoka, 2007). Since the oil of tobacco seeds is a non-edible oil, it is not used as a commercial product in the food industry sector, and most of the seeds are left unused in the fields (Usta, 2005). As a proposal for an alternative use for these seeds, Usta (2005) prepared and tested a biodiesel fuel from the seeds untreated oil, which was determined to comprise approximately 38% of the total seed weight. Methanol was used as the esterifying agent in a 6:1 alcohol-to-oil ratio and NaOH was used as alkaline catalyst. The reaction was carried out in a rotary evaporator at 55°C for a period of 90 min, with the resulting mixture being continuously stirred for another 90 min after the heating was turned off. About 86% of the oil was converted to biodiesel, which was considered an acceptable yield for untreated oil. The performance of the produced biodiesel was tested in a turbocharged indirect injection diesel engine and the results demonstrated the tobacco seed oil biodiesel to be suited for blends of 25 to 30% with fossil diesel. Later, Veljković et al. (2006) investigated the production of fatty acid methyl ester (FAME) from crude tobacco seed oil presenting high amounts of free fatty acids (FFA) (acid value higher than 35). The biodiesel production was carried out in a two-step process, an acid-catalyzed esterification step followed by a base-catalyzed transesterification step. The maximum yield of FAME was about 91% in a 30-minute reaction in the second step. The tobacco seed oil biodiesel obtained presented fuel properties within the specifications of both American and European standards, with the exception of a somewhat higher acid value than that specified by the European Standard.

Cotton seeds are used for production of commercial oil in some countries, but in the majority of the cotton producing countries, the seeds constitute solid residues from cotton crops due to the presence of gossypol, a toxic agent. Meneghetti et al. (2006) performed a comparative study of the ethanolysis of the oils of both castor beans and cotton seeds using classical alkaline and acid catalysts (NaOH, CH_3ONa and HCl). Regarding the different types of catalysts used for the transesterification of cottonseed oil, the ester yields were far superior for the alkaline catalysts than for the acidic one, with sodium methoxide being the most prominent (~90% yield). Georgogianni et al. (2008) also investigated the production of biodiesel from cottonseed oil by both alkaline conventional and in situ transesterification reactions. Both methanol and ethanol were investigated as esterifying agents using mechanical stirring and ultrasonication. In situ transesterification, ultrasonication and mechanical stirring were equally efficient in producing methyl esters. In all cases, the use of methanol provided shorter reaction times and higher yields of esters than the use of ethanol, and, for both alcohols, alkali-catalyzed in situ transesterification promoted an increase in yield of esters in a short time span. In situ transesterification of cottonseed oil was also investigated by Qian et al. (2008), with the objectives of simultaneously producing biodiesel and a gossypol-free cottonseed meal that can be used as animal protein feed resources. Both objectives were successfully achieved with an extraction of 99% of the total oil in the cottonseeds, a 98% conversion in methyl esters in three hours of alkaline-catalyzed in situ transesterification reaction and a content of gossypol in the cottonseed meal far below the FAO standard. However, a high alcohol-to-oil molar ratio (135:1) was used to achieve those goals.

Demirbaş (2008) performed studies on preparation of biodiesel by non-catalytic transesterification of cottonseed oil with methanol and ethanol in supercritical conditions. The ester yields were higher when using methanol at high temperatures (~99% yield at 523K) than when using ethanol (~84% yield at 523K), with the alcohol-to-oil molar ratio of 41:1 in both cases. Zhang and Jiang (2008) investigated the acid-catalyzed esterification of Zanthoxylum bungeanum seed oil for biodiesel production. Z. bungeanum is a fruit widely used in China for its medicinal and flavor characteristics and its seed oil, representing 27 to 31% in weight, is non-edible. Although the authors estimated an annual production potential of one million tons of seeds in China, an unfortunate fact related to the seed oil is that it has a high content of free fatty acids (~25%). Hence, a two-step process was employed for the production of fatty acid methyl esters, with an initial acid-catalyzed esterification step to convert the FFA into methyl esters being followed by an alkali-catalyzed transesterification step. The total yield of fatty acid methyl ester was ~98%.

The potential of defective coffee beans oil as a feedstock for biodiesel production was evaluated by Oliveira et al. (2008). Defective coffee beans impart a negative quality to the aroma and flavor of coffee beverage and were considered as solid residues of coffee production (representing an average of 20% of the total production of coffee in Brazil). Conventional alkaline-catalyzed (sodium methoxide as a catalyst) transesterification of the unrefined defective coffee oil (24% unsaponifiable matter) was carried out in a mechanically-stirred reactor using both methanol and ethanol as esterifying agents (alcohol-to-oil molar ratio of 6:1 in both cases). The yields (~65%) for the reactions with the oil of non-defective coffee beans were lower than those for the oil of defective beans (~70%), indicating the need for correction of the amount of catalyst to be used due to the content of free fatty acids of the oil (~3% w/w) not previously esterified or neutralized. For the defective coffee oil, the FFAs (~10% w/w) were previously neutralized, thus, favoring the higher yield in the alkaline transesterification reaction. In a follow-up to that study, Nunes and Oliveira (2008) removed the unsaponifiable fraction of the defective coffee oil prior to the alkali-catalyzed transesterification reaction with ethanol, raising the fatty acids ethyl ester yields to 96%. Panoutsou et al. (2008) performed an economic analysis of the biodiesel options in Greece and pointed out that, based solely on the available numbers for tomato seeds production as solid residues from the tomato processing industry, the biodiesel yield (kg/ha) would be higher than for any of the currently produced oilseed crops in Greece. Also, the estimated land requirements would be very low when compared to the requirements of the oilseed crops for the same average biodiesel yield.

3.2. Bioalcohols from Agri-Food Residues

Agri-food wastes, in its majority, are comprised of lignocellulosic materials and as such are potential candidates for the production of bioalcohols (biomethanol, bioethanol and biobutanol) as liquid biofuels. Since most lignocellulosic materials fall under the category of high-moisture content biomass (McKendry, 2002), a wet conversion process is preferred. The basic process of conversion of lignocellulosic materials into bioethanol and biobutanol is based on the degradation of lignin and hemicellulose to release the cellulose (pre-treatment), followed by the enzymatic hydrolysis of cellulose into fermentable sugars and the subsequent fermentation of these sugars by a suitable microorganism. The fermentation process is usually carried out in aqueous media and, thus, a processing step is required to separate the bioalcohol produced by the microorganisms from the water used. Biomethanol is usually produced by the distillation of liquid products from the pyrolysis of wood and other lignocellulosic materials.

3.2.1. Bioethanol

Bioethanol is the leading alternative liquid biofuel used in the transportation sector with Brazil and United States being the world leaders both in terms of consumption (be it neatly or as a blend with gasoline) and of production (Neves et al., 2007; Balat et al., 2008). Bioethanol (CH_3CH_2OH) is an oxygenated liquid fuel (35% oxygen) that presents a potential to reduce particulate emissions in compression-ignition engines. As a fuel, it also presents a higher octane number, broader flammability limits, higher flame speeds and higher heats of vaporization than gasoline, thus, allowing for higher compression ratio, shorter burn times and leaner burn engines. These properties lead to theoretical efficiency advantages over gasoline in an internal combustion engine (Balat et al., 2008). However, when used as a neat fuel, bioethanol presents a lower energy density than gasoline and cold-start problems exist due to its lower vapor pressure (Neves et al., 2007). Other disadvantages include its corrosiveness, low flame luminosity, miscibility with water, and toxicity to ecosystems (Balat et al., 2008).

Bioethanol, for fuel applications, is currently being worldwide produced by fermentation of either molasses (e.g., sugarcane) or starchy materials (e.g., corn). The sugar in sugarcane is readily available and, thus, its conversion into bioethanol is straightforward by employing fermentation microorganisms such as baker's yeast (*Saccharomyces cerevisae*). In the case of starchy materials, starch has to be broken down to its sugary components first (liquefaction/saccharification) and then the monosaccharides are used by the fermenting microorganisms to convert them into bioethanol. The technology for liquefaction and saccharification of starch can be considered mature, and, hence, also the technology for bioethanol production from starchy materials such as corn. On the other hand, the process is not so easy for lignocellulosic biomass, such as agricultural residues. Crystalline cellulose has to be firstly freed from the covalently linked matrix of hemicellulose and lignin in which it is embedded, and then hydrolyzed to release its fermentable sugars. Dilute acid (usually H_2SO_4) can hydrolyze hemicellulose to yield pentose and hexose sugars. However, with the exception of a few microorganisms, such as some Clostridia strains (e.g., *C. acetobutylicum*) commonly used for butanol production in acetone-butanol-ethanol (ABE) integrated fermentation processes (Marchal et al., 1992; Zverlov et al., 2006), no naturally occurring microorganisms yet cultured were shown to efficiently convert both pentoses and hexoses into a single product such as bioethanol (Bohlmann, 2006). Consequently, production of ethanol from lignocellulosic biomass has not yet been attained at commercial scale.

Several reviews on the subject of production of bioethanol from lignocellulosic materials (e.g., agricultural residues) are available in the literature (Lynd, 1996; Wyman, 1999; Sun and Cheng, 2002; Hahn-Hägerdal et al., 2006; Lin and Tanaka, 2006; Demirbaş, 2007a; Neves et al., 2007; Prasad et al., 2007; Jørgensen et al., 2007; Antizar-Ladislao and Turrion-Gomez, 2008; Balat et al., 2008; Hayes, 2008; Sánchez and Cardona, 2008; Yang and Wyman, 2008; Saxena et al., 2009), thus, allowing for a comprehensive view of the trends, opportunities and challenges faced in the area. The reader is encouraged to assess the cited references for a more in-depth view of the subject.

Regardless of the biochemical process selected for the lignocellulosic biomass-to-bioethanol conversion (Wyman, 1999; Saxena et al., 2009), a few crucial issues must still be addressed, when compared to the well-established technologies employed for sugar- or starch-based ethanol production (Hahn-Hägerdal et al., 2006): (i) a cost-effective and efficient de-polymerization process of cellulose and hemicellulose to fermentable sugars must yet be developed, i.e., a process that economically overcomes the problem of biomass recalcitrance to saccharification; (ii) industrially viable strains of microorganisms must be engineered to allow for the efficient fermentation of mixed-sugar hydrolyzates containing both hexoses and pentoses sugars, and that would also be resilient to the presence of inhibitory compounds, such as those produced by the delignification process and by the fermenting microorganisms themselves; (iii) advanced process integration must be accomplished to minimize the energy demand for the conversion process; and (iv) the concept of biorefinery must be integrated to the process to reduce costs and increase revenues by efficiently using the co-products generated in the pretreatment and post-treatment steps (lignin, lignin degradation products such as furfural, and others).

Regarding the issue of eliminating or reducing biomass recalcitrance to saccharification, there are two main approaches under thorough investigation: (i) the transgenic approach of lignin modification, by genetically reducing the lignin content of biomass which is directly proportional to the recalcitrance to both acid pretreatment and enzymatic digestion (Chapple et al., 2007; Chen and Dixon, 2007); and (ii) the development of innovative technologies for the pre-treatment of lignocellulosic biomass (Jørgensen et al., 2007; Balat et al., 2008). The re-engineering of microbes has been widely suggested as a means of efficiently utilizing the wide range of mixed-sugar hydrolyzates produced in the pre-treatment processing steps for the production of ethanol (Vertès et al., 2007; Stephanopoulos, 2007). Advanced process integration has been approached in several ways and has been recently reviewed by Cardona and Sánchez (2007). Process integration can be done in any of the process steps and is classified accordingly. Examples of process integration for ethanol production from lignocellulosic biomass are: co-fermentation of lignocellulosic hydrolyzates, where the complete assimilation of all the sugars released during pre-treatment and hydrolysis steps is targeted by utilizing mixed cultures of microorganisms capable of assimilating both pentoses and hexoses; simultaneous saccharification and fermentation in which the enzymatic degradation of cellulose is combined with the fermentation of the hydrolyzed sugars in one single vessel; and simultaneous fermentation and ethanol removal from the fermenting broth, in which the removal/recovery of ethanol is done to avoid inhibition of the fermentation microorganisms by the high concentrations of ethanol in the fermentation medium. The removal of ethanol can be done by vacuum extraction, gas stripping, pervaporation, liquid extraction and other techniques (Cardona and Sánchez, 2007). Integrated biorefinery has not yet surpassed the stage of being a rather appealing concept and, thus, still has a long way to go before it can be fully implemented in any scale (Taylor, 2008).

3.2.2. Biobutanol

Biobutanol is currently being posed as the best alternative liquid biofuel for the transportation sector (Ramey, 2007; Dürre, 2007; Ezeji et al., 2007), with alleged claims of better engine performance than ethanol and gasoline (Ramey, 2007). However, no rigorous scientific work has been presented to date on the performance of biobutanol as a transportation fuel and researchers have presented their point of view with caution since many questions about butanol as a fuel remain unanswered (Dürre, 2007; Ezeji et al., 2007). On the other hand, purely empirical tests, like that of Ramey (2007), where a 100% butanol-fueled 1992 model automobile was driven across the United States (about 10,000 miles) in 2005, have demonstrated a promising future for this alcohol as a fuel.

The current processes used for the production of butanol are mostly targeted to a market not related to fuel production and are based on chemical synthesis from fossil-oil-derived ethylene, propylene, and triethyl-aluminum or carbon monoxide and hydrogen (Zverlov et al., 2006). However, the classical process for producing butanol is the acetone-butanol-ethanol (ABE) fermentation process, carried out by bacteria belonging to the genus *Clostridium* (Lenz and Moreira, 1980; Dürre, 2007). Recently, with the intensive search for an alternative renewable fuel, the interest in butanol as a fuel was renewed and several attempts to improve the ABE fermentation process were reported in the literature, most of them related to trying to eliminate the problem of fermentation inhibition of the bacteria by the products formed (e.g., butanol) and to the recovery of the solvents produced. The optimization studies have dealt with recovery of butanol or inhibitors removal by gas stripping (Ennis et al., 1986; Mollah and Stuckey, 1993; Maddox et al., 1995; Qureshi et al., 1992, 2007; Qureshi and Blaschek, 2001), by liquid-liquid extraction (Qureshi and Maddox, 1995), by pervaporation using silicalite-filled GFT PDMS composite membrane (Jonquières and Fane, 1997), by perstraction membrane with oleyl alcohol as the perstraction solvent (Qureshi and Maddox, 2005), by electrodialysis (Qureshi et al., 2008a). A comparative study of the several product removal techniques integrated to the fermentation process was performed by Qureshi et al. (1992) with the conclusion that gas stripping and pervaporation were the most promising techniques in terms of both technical and economical performances (Ezeji et al., 2007). The use of gas stripping for the integrated removal of fermentation products was later revisited (Qureshi and Blaschek, 2001; Ezeji et al., 2007) and confirmed that this technique resulted in reduced butanol inhibition, thereby improving total solvent productivity and yield. Successful test results on the production of acetone-butanol by fermentation in pilot and pre-industrial scale plants were compiled and discussed by Nimcevic and Gapes (2000).

Like bioethanol, biobutanol can be produced from a diversity of feedstock, with waste biomass being a potential candidate. In fact, a few studies on the use of agri-food residues have been published in relation to the production of butanol (Lenz and Moreira, 1980; Marchal et al., 1992; Grobben et al., 1993; Zverlov et al., 2006; Qureshi et al., 2008abc). Lenz and Moreira (1980) studied the economic feasibility of using liquid whey as an alternative feed for the ABE fermentation process. The conclusions were that a superior economic position was occupied by the feedstock used when compared to other options such as molasses. Several observations were made regarding the major drawbacks of such process, with the very low levels of butanol observed in the final broth being the leading one when it comes to making attempts for commercial production. Other observed difficulties were the need for strict anaerobic conditions, delicate culture maintenance and propagation, and a tendency for bacteriophage infection and *Lactobacilli* contamination. Marchal et al. (1992) developed a large-scale process for the conversion of lignocellulosic biomass to acetone-butanol involving the steam-explosion pretreatment of corncobs, the enzymatic hydrolysis of the pretreated material and the acetone-butanol fermentation of the hydrolyzate. In the two-step enzymatic hydrolysis and fermentation process, a *Clostridium acetobutylicum* strain capable of utilizing xylose and presenting limited sensitivity to the fermentation inhibitors was used. The process was carried out in batch reactors in facilities for biomass conversion in France. Potato wastes were used by Grobben et al. (1993) as a substrate (14% w/v) for *Clostridium acetobutylicum* DSM1731 strain in an ABE fermentation process integrated with membrane extraction. With the use of a polypropylene perstraction system and an oleyl alcohol/decane mixture as the extractant, the product yield was about two-fold higher than that of the conventional system. Fouling of the membrane system occurred after 50 h of operation.

Zverlov et al. (2006) reported the development of the ABE industrial process in the former USSR, where the original feedstock (pure starch) used in the fermentation process was partially replaced by molasses and partial hydrolyzates of lignocellulosic residues, containing mainly xylose and arabinose ("pentose hydrolyzate"). The lignocellulosic residues used were corn cobs, sunflower shells and hemp waste. The commonly reported problems of *Clostridium acetobutylicum* degeneration was solved by development of a continuous fermentation process adapted to the number of generations possible in one cycle so that degeneration did not occur. Also, according to Zverlov et al. (2006), the common problem of bacteriophage infections reported by the Western fermentation plants were not experienced by the Russian ABE plants due to the use of independently isolated *C. acetobutylicum* strains that were not susceptible to bacteriophage infections and to the employment of a rigorous sterilization scheme in the process. The Clostridia strains used were reported to produce solvents at higher temperatures (37°C). Problems with infection by lactic acid bacteria were reported frequently. Butanol was successfully produced from wheat straw hydrolyzate using *Clostridium beijerinckii* P260 by Qureshi et al. (2007). Wheat straw was pretreated with dilute sulfuric acid and subsequently hydrolyzed by a mix of enzymes. Fermentation inhibition due to salts or inhibitory products was not observed. Fermentations of sulfuric acid and enzyme treated corn fiber hydrolysates were studied by Qureshi et al. (2008c) for the production of butanol with *Clostridium beijerinckii* BA101. In the case of dilute acid hydrolysates, inhibition of the cell growth and butanol production were observed. Butanol production was improved (from 1.7 to 9.3 g/L) by removal of the inhibitors with XAD-4 resin. A lower butanol production (8.6 g/L) was obtained for the enzyme hydrolysate of corn fiber. The *Clostridium* strain used was demonstrated capable of utilizing xylose as efficiently as glucose in the fermentation process. A recent review on the progress in technology for the production of butanol from agricultural residues was published by Qureshi and Ezeji (2008).

3.2.3. Biomethanol

Biomethanol is produced from synthesis gas utilizing conventional gasification of biomass at high temperatures (800–1000°C) and, subsequently, catalytic synthesis of the produced mixture of CO_2 and H_2 with a molar ratio of 1:2, under elevated pressures (4–10 MPa) (Luque et al., 2008). Biomethanol is considerably easier to recover than ethanol for it does not form an azeotropic mixture with water. However, as a fuel, it presents several shortcomings, including low vapor pressure, low energy density, high toxicity, high flammability and incompatibility with current technologies used in compression-ignition engines. The literature on the production and characterization of biomethanol as a fuel is scarce.

Güllü and Demirbaş (2001) have performed pyrolysis studies of biomass residues such as hazelnut shells and compared the methanol yield with the results for pyrolysis of both softwood and hardwood. The hazelnut shells yielded about 8 times more methanol than softwood and about 4.5 times more methanol than hardwood. The production of acetic acid and acetone were also far superior for the hazelnut shells than for both softwood and hardwood.

3.3. Dimethyl Ether

Among the diversity of fuels that can be produced from biomass, dimethyl ether (DME) is considered one of the most prominent fuels in the replacement of petroleum-derived liquid fuels (Semelberger et al., 2006). DME is the ether with the shortest carbon chain (CH_3OCH_3), with an oxygen content of 35%, and presenting a soot-free burning. Its physical properties are similar to those of liquefied petroleum gas (LPG). DME is an odorless and also colorless gas at ambient temperature, which is a major disadvantage in terms of handling and storage. It becomes a thin fluid when slightly pressurized (above 0.5 MPa), with a viscosity lower than that of fossil diesel by a factor of about 20, which is currently the most challenging aspect of a diesel engine running on DME (Semelberger et al., 2006). It requires a higher injected volume in a compression-ignition engine to supply the same amount of energy as the fossil diesel fuel, due to its lower density and combustion enthalpy (Arcoumanis et al., 2008). The advantages of DME over conventional diesel are low-combustion noise, decreased emissions of NO_x, hydrocarbons and carbon monoxide, and the projected decreasing of global warming potential in a 500-year time horizon (Arcoumanis et al., 2008).

Currently, DME is primarily produced from natural gas in a two-step process where syngas (typically generated from the steam reforming of methane) is first converted to methanol which, in turn, is subsequently dehydrated to dimethyl ether (Semelberger et al., 2006; Ahlgren et al., 2008). Its primary use is as a propellant in spray cans. DME can be produced from a variety of other sources, such as coal and biomass (Gruden, 2003). However, only a few applications related to the production of DME from waste biomass have been reported in the literature (Wang et al., 2007; Ahlgren et al., 2008). Wang et al. (2007) have used a Cu-Zn-Al methanol catalyst combined with HZSM-5 to produce DME from a biomass-derived syngas containing nitrogen. The syngas was produced by air-steam gasification of pine sawdust in a bubbling fluidized bed biomass gasifier. Ahlgren et al. (2008) have evaluated the feasibility of on-farm production of either DME or Fischer-Tropsch diesel (FTD) from wheat straw and have concluded that FTD diesel was a more likely alternative.

3.4. Bio-Oil

The term bio-oil is usually applied to the oil produced from the pyrolysis of biomass. Given the diversity of biomass that can be used in this kind of process, the chemical composition of the produced oil, and, hence, its properties as a fuel are inherent to both the type of biomass and the type of pyrolysis process being used. The pyrolysis processes currently under research and development are based on two distinct concepts: slow pyrolysis; and fast or flash pyrolysis (Şensöz et al., 2000). The differences between the two concepts are in terms of the chemistry of the final product (bio-oil), and its overall yield and quality. High conversions are reported to be obtained with fast pyrolysis processes (Bridgwater and Peacoke, 2000), whereas with slow pyrolysis conversion of the biomass yields more useful and value-added energy products (Karaosmanoğlu et al., 1999). Plenty of published literature is available on the study of production of bio-oils utilizing agri-food residues as feedstock, and several comprehensive reviews are also available for an in-depth view of the subject (Bridgwater et al., 1999; Bridgwater and Peacoke, 2000; Demirbaş, 2000; Demirbaş and Arin, 2002; Kirubakaran et al., 2009). Just a handful of the works on waste biomass pyrolysis will be described here for the sake of illustration only. However, it is noteworthy to mention that reports on the use of bio-oil in compression-ignition engines are scarce.

Flash pyrolysis products were obtained by Demirbaş (2002) for hazenut shells and tea factory waste. An increase in yields of both acetic acid and methanol was observed with alkali (sodium carbonate) treatment of the biomass and, also, with an increase in pyrolysis final temperature from 675 to 875 K. The production of acetic acid and methanol in the pyrolysis of tea factory waste followed the same behavior as in the pyrolysis of hazelnut shells, but with lower yield values for both products. Vacuum pyrolysis tests at bench and pilot plant scales were performed by Garcìa-Pèrez et al. (2002) using sugarcane bagasse as feedstock. The pilot-scale reactor yielded less oil than the lab-scale reactor (30.1% w/w against 34.4% w/w, respectively), and also more charcoal (25.7% w/w against 19.4% w/w, respectively). The resulting oils presented interesting liquid fuel properties, such as low ash content (0.05% w/w), high calorific value (22.4 MJ/kg) and relatively low viscosity (4.1 cSt at 90°C).

The pyrolysis of cashew nut shells in a fixed bed vacuum pyrolysis reactor was studied by Das and Ganesh (2003). The oil-to-liquid ratio in the pyrolysis products was found to remain almost constant in the range between 400°C and 550°C and calorific value of the bio-oil produced was as high as 40 MJ kg^{-1}. In a subsequent study, Das et al. (2004) presented a detailed characterization of the bio-oil produced in the vacuum pyrolysis reactor with the oil having a unique characteristic of being comprised of long (C_6 to C_{15}) linear carbon chains providing an excellent solubility in fossil diesel oils. The oil had a high C/H ratio which contributed to its high heating value of 40 MJ kg^{-1}. Also, due to its inherent antioxidant properties (mainly due to the presence of strongly polar phenol groups), it presented good stability when stored at room temperature (little change in viscosity over time). Hazenut shells, olive husks and beech and spruce wood samples were studied by Demirbaş (2007b) as feedstock for the production of bio-oil in a fast pyrolysis apparatus. The ratios of acetic acid, methanol and acetone in the aqueous phase were higher than those in the non-aqueous phase. The yield of neutral oils increased from 18 to 33% with increasing temperature, while the methoxyl content decreased.

Qiang et al. (2008) have performed a fast pyrolysis of rice husks with an intermediate auto-thermal pyrolysis set to produce bio-oil. The chemical analysis of the bio-oil revealed that it contained high amounts of nitrogen and inorganic elements, which were feedstock dependent. The bio-oil presented a somewhat non-Newtonian fluid behavior which was attributed to the existence of extractives. It was determined to be thermally unstable, presenting significant increases in viscosity and undergoing phase separation (due to its water content) at elevated temperatures. It was further concluded that, when compared to internationally proposed fuel specifications for bio-oils, the bio-oil produced by the authors presented undesirable properties such as high contents of water, solids and ash, and, thus, an upgrading would be necessary for it to become commercially accepted. The fast pyrolysis of washed and unwashed empty fruit bunches, a waste of the palm oil industry, was investigated by Abdullah and Gerhauser (2008). A fluidized-bed bench scale fast pyrolysis reactor was used in the study. The maximum yield of liquids produced from washed empty fruit bunches was increased by more than 22% points when compared to the yield for the unwashed feedstock. It was determined that the ash content of the feedstock had a significant influence on the yield of organics, with lower yields of pyrolysis liquid as the concentration of ash became higher.

3.5. Other Liquid Biofuels

Other prominent candidates for liquid biofuels that can be produced using waste biomass as feedstock are Fischer-Tropsch diesel (Ahlgren et al., 2008), Dimethylfuran (Román-Leshkov et al., 2007) and lignin-derived fuels (Gellerstedt et al., 2008). Fischer-Tropsch diesel (FTD) is a diesel-like liquid fuel produced from syngas (obtained by the gasification of coal or biomass) by means of Fischer-Tropsch reactions. This type of fuel is well-suited for direct use in diesel engines as it presents a high cetane number and does not contain sulfur. Ahlgren et al. (2008) evaluated the possibility for organic farms to be self-sufficient in renewable fuel in a long term perspective by producing either DME or FTD from wheat straw. Their main conclusion was that FTD was a more likely alternative since its annual cost was lower and its impact on global warming potential presented a small difference to that of DME. Also, FTD is a liquid fuel in its natural state and DME requires a pressurized infrastructure system and engine modifications.

DMF (2,5-dimethylfuran) is liquid fuel obtained from the selective removal of five oxygen atoms from fructose. This selective removal of oxygen atoms is accomplished by first removing three atoms by dehydration to produce 5-hydroxymethylfurfural and subsequently removing two oxygen atoms by hydrogenolysis to produce DMF (Román-Leshkov et al., 2007). Compared to bioethanol, DMF presents 40% higher energy density, 20 K higher boiling point, and it is not soluble in water. No work on the use of waste biomass for the production of DMF was reported in the literature up to this point. Gellerstedt et al. (2008) reported on the possibility of producing liquid oil as a fuel by a one-step pyrolysis of lignin at temperatures below 400°C and in the presence of formic acid and an alcohol. This oil contained alkylated phenols together with alkane and alkene structures with low O/C ratios (30–35%). The heating value was also found to be similar to that of petroleum-based fuels. Lignins from a commercial spruce sodium lignosulfonate and from steam-exploded birch wood were used as feedstock in the pyrolysis process.

4. CONCLUSIONS

The agricultural and food sectors are overflowing with issues to be properly addressed, with the majority of them related to an increasing demand for production, with its consequent problems of an increasing consumption of energy, an increasing production of residues, and an increasing disregard for soil abatement. The transportation liquid fuel sector is currently flooded with many problems with just a few, if none, plausible ready-to-implement solutions. The major problem in the transportation liquid fuel sector is its strong dependence on petroleum-derived fuels imparted by the currently available technology for automotive engines. Thus, agri-food wastes, mostly represented by lignocellulosic residues, come into this scenario as a prospective renewable source of liquid biofuels and, hence, a plausible short- to mid-term solution to the main issue faced by the transportation sector.

A variety of biofuels can be produced utilizing lignocellulosic residues as feedstock. Among the several options, bioethanol is the lead fuel currently being studied in terms of technology development for both production (pre-treatment, microbe and enzyme engineering, and process integration) and engine use and performance. In this regard, several issues must still be addressed in order for commercial units to be built and become fully operational. Optimistic predictions estimate a 10- to 15-year horizon for the lignocellulosic bioethanol technology to be mature and fully operational (Taylor, 2008). Integration processes, such as fermentation-product recovery processes, have been successfully developed in attempts to optimize the existing technologies. The current major issue in bioethanol production that will make lignocellulosic bioethanol production facilities commercially viable relies on the development of microorganism strains capable of utilization of both pentoses and hexoses as substrates (produced in the biomass pre-treatment operations). Also, these 'new' microorganisms should be capable of overcoming the problem of fermentation inhibition promoted by the products of pre-treatment operations and by the presence of concentrated bioethanol in the fermentation media. The accomplishment of these goals requires an array of research areas to be brought together, such as chemistry, biology, engineering and economics.

Research on the production and use of biodiesel as a transportation fuel has gained momentum since it was first implemented in Europe in the early 1990s, and it is now the second alternative biofuel in terms of production and use. The production technology has evolved from the use of homogeneous alkali catalyzed transesterification reactions to the employment of ultrasound and microwave irradiation together with heterogeneous solid catalysts. The feasibility of using a few oily agricultural residues (e.g., tobacco seeds, defective coffee beans, etc.) as feedstock has been demonstrated, with great potential expected for others, such as tomato seeds. In situ transesterification reactions have been successfully carried out on the laboratory scale. However, the major shortcomings of this prospective technology rely on the necessity to use high alcohol-to-oil molar ratios (e.g., 135:1) to accomplish reasonable yields (>85%) of fatty acid alkyl esters. The successful implementation of this technology will bring significant savings regarding the production costs. Since the production of biodiesel is strongly dependent on the availability of short chain alcohols, major developments in bioethanol production technology will consequently promote biodiesel production and use. The possibility of producing both bioethanol and biodiesel from distinct residues from a single crop is very appealing; one such example would be the use of defective beans for extraction of vegetable oil (Oliveira et al., 2008) and use of coffee husks for bioethanol fermentation (Franca et al., 2008). The solid residues of both the oil pressing (press-cake) and fermentation of coffee husks could then be destined to the production of low-cost adsorbents in an integrated biorefinery approach (Nunes et al., 2009).

Biobutanol, as an alcohol fuel, presents distinct advantages over bioethanol in the sense that it can be used as is in unmodified spark-ignition engines (hence, it was coined the term 'biogasoline') and as a fuel it presents higher energy density than bioethanol. Major advancements in the production technology are related to the integrated fermentation-product recovery processes that also partly deal with the problem of fermentation inhibition caused by the presence of high concentrations of butanol in the fermentation medium. However, in terms of its production, several issues must still be addressed, such as the susceptibility of the *Clostridia* strains currently used to phage, *lactobacilli* attacks, and the susceptibility of some strains to fermentation inhibition by the presence of undesired compounds produced in the pre-treatment steps. Another major issue is the lack of rigorous engine tests with this fuel. As a fuel for the transportation sector, butanol still has a long (but rather promising) research road ahead.

Biomethanol has still to be proven a worthy liquid fuel for the transportation sector, with its high toxicity and very low vapor pressure being the most critical issues to be properly tackled. Advancements in its production from solid waste biomass are related to the advancements of biomass gasification and pyrolysis technologies. Scientific knowledge on the pyrolysis of biomass has come to a point where, based on the chemical composition of the biomass and the selected process parameters, the amounts of gases, liquid and char can be predicted, thus allowing for more rigorous control of the process in terms of the desired products to be obtained. Another prospective fuel produced in the pyrolysis of waste biomass is bio-oil. Bio-oil was demonstrated to be an interesting liquid fuel for power generation, for it presents high energy densities and adequate viscosities for use in burners when methanol is present in small amounts. However, when it comes to transportation liquid fuels, there is a lack of rigorous scientific work to demonstrate its feasibility as such.

Other types of liquid fuels that can be produced from solid waste biomass are dimethyl ether, Fischer-Tropsch diesel, dimethyl furan and lignin-derived fuels. The processes for producing dimethyl ether, Fischer-Tropsch diesel and dimethyl furan all rely on the production of syngas first and then, for each of the fuels, different paths to the end product are taken. As a consequence, the interest in Fischer-Tropsch reaction has been rekindled, and major advancements in this area are expected for the next 10 to 15 years. However, since these fuels can be produced using natural gas and coal as feedstock, which in turn are produced in very large quantities and at lower costs than petroleum-derived products, the time horizon for the technology of producing such fuels from waste biomass to become feasible and mature can be greatly stretched. Lignin-derived fuels are still in early infancy and their future is quite uncertain at this point.

In summary, agricultural and food solid residues are the best candidates for feedstock in the production of alternative liquid fuels in a short- to mid-term time horizon. A great deal of research funding has already been directed to this area in the past recent years and a great deal more is expected in the least for the next 20 years. The main drive for this is the predicted upcoming major crisis in the petroleum-derived fuel sector. Research in this area still has a long road ahead and ephemeral fashions and professional vanities should be put aside if consolidated innovative technologies are to be brought to life in a short time span.

ACKNOWLEDGMENTS

The authors gratefully acknowledge financial support from the following Brazilian Government Agencies: CNPq and FAPEMIG.

REFERENCES

Abdullah, N. and Gerhauser, H. (2008) Bio-oil derived from empty fruit bunches, *Fuel*, 87, 2606-2613.

Agarwal, A.K. (2007) Biofuels (alcohols and biodiesel) applications as fuels for internal combustion engines, *Progress in Energy and Combustion Science*, 33, 233-271.

Ahlgren, S., Baky, A., Bernesson, S., Nordberg, Å., Norén, O. and Hansson, P.-A. (2008) Future fuel supply systems for organic production based on Fischer-Tropsch diesel and dimethyl ether from on-farm-grown biomass, *Biosystems Engineering*, 99, 145-155.

Antizar-Ladislao, B. and Turrion-Gomez, J.L. (2008) Second-generation biofuels and local bioenergy systems, *Biofuels, Bioproducts and Biorefining,* 2, 455-469.

Arcoumanis, C., Bae, C., Crookes, R. and Kinoshita, E. (2008) The potential of di-methyl ether (DME) as an alternative fuel for compression-ignition engines: A review, *Fuel,* 87, 1014-1030.

Balat, M. and Balat, H. (2008) A critical review of bio-diesel as a vehicular fuel, *Energy Conversion and Management,* 49, 2727–2741.

Balat, M., Balat, H. and Öz, C. (2008) Progress in bioethanol processing, *Progress in Energy and Combustion Science,* 34, 551-573.

Barnwal, B.K. and Sharma, M.P. (2005) Prospects of biodiesel production from vegetable oils in India, *Renewable and Sustainable Energy Reviews,* 9, 363-378.

Bohlmann, G.M. (2006) Process economic considerations for production of ethanol from biomass feedstocks, *Industrial Biotechnology,* 2, 14-20.

Bowyer, J.L. and Stockmann, V.E. (2001) Agricultural residues, *Forest Products Journal,* 51, 10-21.

Bridgwater, A.V., Meier, D. and Radlein, D. (1999) An overview of fast pyrolysis of biomass, *Organic Geochemistry,* 30, 1479-1493.

Bridgwater, A.V. and Peacocke, G.V.C. (2000) Fast pyrolysis processes for biomass, *Renewable and Sustainable Energy Reviews,* 4, 1-73.

Buranov, A.U. and Mazza, G. (2008) Lignin in straw of herbaceous crops, *Industrial Crops and Products,* 28, 237–259.

Cardona, C.A. and Sánchez, O.J. (2007) Fuel ethanol production: Process design trends and integration opportunities, *Bioresource Technology,* 98, 2415–2457.

Champagne, P. (2008) Biomass. In T. Letcher (Ed.) *Bio-mass in Future Energy: Sustainable and Clean Energy Alternatives for our Planet* (pp. 151-170) London: Elsevier Applied Science.

Chapple, C., Ladisch, M. and Meilan, R. (2007) Loosening lignin's grip on biofuel production, *Nature Biotechnology,* 25, 746-748.

Chen, F. and Dixon, R.A. (2007) Lignin modification improves fermentable sugar yields for biofuel production, *Nature Biotechnology,* 25, 759-761.

Cuiping, L., Yanyongjie, Chuangzhi, W. and Haitao, H. (2004) Study on the distribution and quantity of biomass residues resource in China, *Biomass and Bioenergy,* 27, 111 – 117.

Dale, B. (2008) Biofuels: Thinking clearly about the issues, *Journal of Agricultural and Food Chemistry,* 56, 3885–3891.

Das, P. and Ganesh, A. (2003) Bio-oil from pyrolysis of cashew nut shell—a near fuel, *Biomass and Bioenergy,* 25, 113 – 117.

Das, P., Sreelatha, T. and Ganesh, A. (2004) Bio oil from pyrolysis of cashew nut shell-characterisation and related properties, *Biomass and Bioenergy,* 27, 265-275.

Demirbaş A. (2000) Recent advances in biomass conversion technologies. *Energy Edu. Sci. Technol.,* 6, 19-40.

Demirbaş, A. (2002) Partly chemical analysis of liquid fraction of flash pyrolysis products from biomass in the presence of sodium carbonate, *Energy Conversion and Management,* 43, 1801–1809.

Demirbaş A. and Arin, G. (2002) An overview of biomass pyrolysis, *Energy Sources,* 5, 471-482, 2002.

Demirbaş, A. (2007a) Progress and recent trends in biofuels, *Progress in Energy and Combustion Science,* 33, 1–18.

Demirbaş, A. (2007b) The influence of temperature on the yields of compounds existing in bio-oils obtained from biomass samples via pyrolysis, *Fuel Processing Technology,* 88, 591–597.

Demirbaş, A. (2008) Studies on cottonseed oil biodiesel prepared in non-catalytic SCF conditions, *Bioresource Technology*, 99, 1125-1130.

Demirbaş, A. (2009) Progress and recent trends in biodiesel fuels, *Energy Conversion and Management*, 50, 14-34.

de Vos, R. (2006) Defining Biomass. Which types of biomass will count as renewable energy sources?, *Refocus*, September/October, 58-59.

Dürre, P. (2007) Biobutanol: An attractive biofuel, *Biotechnology Journal*, 2, 1525–1534.

Ennis, B.M., Marshall, C.T., Maddox, I.S. and Paterson, A.H.J. (1986) Continuous product recovery by in-situ gas stripping/condensation during solvent production from whey permeate using *Clostridium acetobutylicum*, *Biotechnology Letters*, 8, 725–30.

Erickson, B., Lindenboom, K., Ramsey, N. and West, T. (2008) Industry perspectives on the food-versus-fuel debate, *Industrial Biotechnology*, 4, 224-229.

Ezeji, T.C., Qureshi, N. and Blaschek, H.P. (2007) Bioproduction of butanol from biomass: from genes to bioreactors, *Current Opinion in Biotechnology*, 18, 220–227.

Franca, A.S., Gouvea, B.M., Torres, C., Oliveira, L.S. and Oliveira, E.S. (2008) Feasibility of ethanol production from coffee husks, *Journal of Biotechnology*, 136S, S269–S275.

Fujimori, S. and Matsuoka, Y. (2007) Development of estimating method of global carbon, nitrogen, and phosphorus flows caused by human activity, *Ecological Economics*, 62, 399–418.

Fukuda, H., Kondo, A. and Noda, H. (2001) Biodiesel fuel production by transesterification of oils, *Journal of Bioscience and Bioengineering*, 92, 405-416.

Garcìa-Pèrez, M., Chaala, A. and Roy, C. (2002) Vacuum pyrolysis of sugarcane bagasse, *Journal of Analytical and Applied Pyrolysis*, 65, 111–136.

Geller, H.S. (1985) Ethanol fuel from sugar cane in Brazil, *Annual Review of Energy*, 10, 135-164.

Gellerstedt, G., Li, J., Eide, I., Kleinert, M. and Barth, T. (2008) Chemical structures present in biofuel obtained from lignin, *Energy and Fuels*, 22, 4240-4244.

Georgogianni, K.G., Kontominas, M.G., Pomonis, P. J., Avlonitis, D. and Gergis, V. (2008) Alkaline conventional and in situ transesterification of cottonseed oil for the production of biodiesel, *Energy and Fuels*, 22, 2110–2115.

Giannelo, P.N., Zannikos, F., Stournas, S., Lois, E. and Anastopoulos, G. (2002) Tobacco seed oil as an alternative diesel fuel: physical and chemical properties, *Industrial Crops and Products*, 16, 1–9.

Goldemberg, J. and Coelho, S.T. (2004) Renewable energy—traditional biomass vs. modern biomass, Energy Policy, 32, 711–714.

Gomez, L.D., Steele-King, C.G. and McQueen-Mason, S.J. (2008) Sustainable liquid biofuels from biomass: the writing's on the walls, *New Phytologist*, 178, 473-485.

Graboski, M.S. and McCormick, R.L. (1998) Combustion of fat and vegetable oil derived fuels in diesel engines, *Progress in Energy and Combustion Science*, 24, 125-164.

Grobben, N.G., Eggink, G., Petrus Cuperus, F. and Huizing, H.J. (1993) Production of acetone, butanol and ethanol (ABE) from potato wastes: fermentation with integrated membrane extraction, *Applied Microbiology and Biotechnology*, 39, 494-498.

Gruden, D. (2003) Fuels, In *The Handbook of Environmental Chemistry*, Vol. 3, Part T, (pp. 255-288) Berlin: Springer-verlag.

Güllü, D. and Demirbaş, A. (2001) Biomass to methanol via pyrolysis process, *Energy Conversion and Management*, 42, 1349-1356.

Hahn-Hägerdal, B., Galbe, M., Gorwa-Grauslund, M.F., Lidén, G and Zacchi, G. (2006) Bio-ethanol – the fuel of tomorrow from the residues of today, *Trends in Biotechnology*, 24, 549-556.

Haq, Z. and Easterly, J.L. (2006) Agricultural residue availability in the United States, *Applied Biochemistry and Biotechnology*, 129-132, 3-21.

Hatzis, C., Riley, C. and Philippidis, G.P. (1996) Detailed material balance and ethanol yield calculations for the biomass-to-ethanol conversion process, *Applied Biochemistry and Biotechnology*, 57-58, 443-459.

Hayes, D.J. (2008) An examination of biorefining processes, catalysts and challenges, *Catalysis Today*, In press, doi:10.1016/j.cattod.2008.04.017.

Himmel, M.E., Ding, S.Y., Johnson, D.K., Adney, W.S., Nimlos, M.R., Brady, J.W. and Foust, T.D. (2007) Biomass recalcitrance: Engineering plants and enzymes for biofuels production, *Science*, 315, 804-807.

Jonquières, A. and Fane, A. (1997) Filled and unfilled composite GFT PDMS membranes for the recovery of butanols from dilute aqueous solutions: influence of alcohol polarity, *Journal of Membrane Science*, 125, 245-255.

Jørgensen, H., Kristensen, J.B. and Felby, C. (2007) Enzymatic conversion of lignocellulose into fermentable sugars: Challenges and opportunities, *Biofuels, Bioproducts and Biorefining* 1, 119-134.

Karaosmanoğlu, F., Tetik, E. and Göllu, E. (1999) Biofuel production using slow pyrolysis of the straw and stalk of the rapeseed plant, *Fuel Processing Technology*, 59, 1–12.

Kheshgi, H.S., Prince, R.C. and Marland, G. (2000) The potential of biomass fuels in the context of global climate change: Focus on transportation fuels, *Annual Review of Energy and the Environment*, 25, 199-244.

Kirubakaran, V., Sivaramakrishnan, V., Nalini, R., Sekar, T., Premalatha, M. and Subramanian, P. (2009) A review on gasification of biomass, *Renewable and Sustainable Energy Reviews*, 13, 179–186.

Lal, R. (2008) Crop residues as soil amendments and feedstock for bioethanol production, *Waste Management*, 28, 747–758.

Laufenberg, G., Kunz, B. and Nystroem, M. (2003) Transformation of vegetable waste into value added products: (A) the upgrading concept; (B) practical implementations, *Bioresource Technology*, 87, 167–198.

Lenz, T.G. and Moreira, A.R. (1980) Economic evaluation of the acetone-butanol fermentation, *Ind. Eng. Chem. Prod. Res. Dev.*, 19, 478-483.

Lin, Y. and Tanaka, S. (2006) Ethanol fermentation from biomass resources: current state and prospects, *Applied Microbiology and Biotechnology*, 69, 627-642.

Lora, E.S. and Andrade, R.V. (2009) Biomass as energy source in Brazil, *Renewable and Sustainable Energy Reviews*, 13, 777-788.

Luque, R., Herrero-Davila, L., Campelo, J.M., Clark, J.H., Hidalgo, J.M., Luna, D., Marinas, J.M. and Romero, A.A. (2008) Biofuels: a technological perspective, *Energy and Environmental Science*, 1, 542-564.

Lynd, L.R. (1996) Overview and evaluation of fuel ethanol from cellulosic biomass: Technology, Economics, the Environment, and Policy, *Annual Review of Energy and the Environment*, 21, 403–465.

Ma, F. and Hanna, M.A. (1999) Biodiesel production: a review, *Bioresource Technology*, 70, 1-15.

Maddox, I.S., Qureshi, N. and Roberts-Thomson, K. (1995) Production of acetone-butanol-ethanol from concentrated substrates using *Clostridium acetobutylicum* in an integrated fermentation-product removal process, *Process Biochemistry*, 30, 209-215.

Marchal, R., Ropars, M., Pourquié, J., Fayolle, F. and Vandecasteele, J.P. (1992) Large-scale enzymatic hydrolysis of agricultural lignocellulosic biomass. Part 2: Conversion into acetone-butanol, *Bioresource Technology*, 42, 205-217.

McKendry, P. (2002) Energy production from biomass (part 1): Overview of biomass, *Bioresource Technology*, 83, 37-46.

Meher, L.C., Sagar, D.V. and Naik, S.N. (2006) Technical aspects of biodiesel production by transesterification - a review. *Renewable and Sustainable Energy Reviews*, 10, 248-268.

Meneghetti, S.M.P., Meneghetti, M.R., Wolf, C.R., Silva, E.C., Lima, G.E.S., Coimbra, M.A., Soletti, J.I. and Carvalho, S.H.V. (2006) Ethanolysis of castor and cottonseed oil: A systematic study using classical catalysts, *JAOCS*, 83, 819–822.

Mollah, A.H. and Stuckey, D.C. (1993) Feasibility of *in situ* gas stripping for continuous acetone-butanol fermentation by *Clostridium acetobutylicum*, *Enzyme and Microbial Technology*, 15, 200-207.

Muller, M., Yelden, T. and Schoonover, H. (2007) *Food versus fuel in the United States, can both win in the era of ethanol?*, Institute for Agriculture and Trade Policy, Minneapolis, MN, September.

Murugesan, A., Umarani, C., Chinnusamy, T.R., Krishnan, M., Subramanian, R. and Neduzchezhain, N. (2009) Production and analysis of bio-diesel from non-edible oils—A review, *Renewable and Sustainable Energy Reviews*, 13, 825-834.

Neves, M.A., Kimura, T., Shimizu, N. and Nakajima, M. (2007) State of the art and future trends in bioethanol production, Dynamic Biochemistry, *Process Biotechnology and Molecular Biology*, 1, 1-14.

Nimcevic, D. and Gapes, J.R. (2000) The acetone-butanol fermentation in pilot plant and pre-industrial scale, *Journal of Molecular Microbiology and Biotechnology*, 2, 15-20.

Nunes, D.L. and Oliveira, L.S. (2008) Potencial de óleo de café para produção de biodiesel (Potential of coffee oil for the production of biodiesel). In: 5o Congresso Brasileiro de Plantas Oleaginosas, Óleos, Gorduras e Biodiesel, 2008, Lavras. (in Portuguese)

Nunes, A.A., Franca, A.S. and Oliveira, L.S. (2009) Activated carbons from waste biomass: An alternative use for biodiesel production solid residues, *Bioresource Technology*, 100, 1786-1792.

Oliveira, L.S. and Franca, A.S. (2008) Low-Cost Adsorbents from Agri-Food Wastes, In L. V. Greco, M. N. Bruno (Eds.), *Food Science and Technology: New Research* (pp. 171-209) New York: Nova Publishers.

Oliveira, L.S., Franca, A.S., Camargos, R.R.S. and Ferraz, V.P. (2008) Coffee oil as a potential feedstock for biodiesel production, *Bioresource Technology*, 99, 3244-3250.

Panoutsou, C., Namatov, I., Lychnaras, V. and Nikolaou, A. (2008) Biodiesel options in Greece, *Biomass and Bioenergy*, 32, 473-481.

Petrus, L. and Noordermeer, M.A. (2006) Biomass to biofuels, a chemical perspective, *Green Chemistry*, 8, 861–867.

Pimentel, D. (2003) Ethanol fuels: Energy balance, economics, and environmental impacts are negative, *Natural Resources Research*, 12, 127-134.

Pimentel, D. and Patzek, T.W. (2005) Ethanol production using corn, switchgrass, and wood; biodiesel production using soybean and sunflower, *Natural Resources Research*, 14, 65-76.

Pinto, A.C., Guarieiro, L.L.N., Rezende, M.J.C., Ribeiro, N.M., Torres, E.A. and Lopes, W.A. (2005) Biodiesel: an overview, *Journal of the Brazilian Chemical Society*, 16, 1313–30.

Prasad, S., Singh, A. and Joshi, H.C. (2007) Ethanol as an alternative fuel from agricultural, industrial and urban residues, *Resources, Conservation and Recycling*, 50, 1–39.

Purohit, P., Tripathi, A.K., and Kandpal, T.C. (2006) Energetics of coal substitution by briquettes of agricultural residues, *Energy*, 31, 1321-1331.

Qian, J., Wang, F., Liu, S. and Yun, Z. (2008) In situ alkaline transesterification of cottonseed oil for production of biodiesel and nontoxic cottonseed meal, *Bioresource Technology*, 99, 9009–9012.

Qiang, L., Xu-lai, Y. and Xi-feng, Z. (2008) Analysis on chemical and physical properties of bio-oil pyrolyzed from rice husk, *Journal of Analytical and Applied Pyrolysis*, 82, 191-198.

Qureshi, N., Maddox, I.S. and Friedl, A. (1992) Application of continuous substrate feeding to the ABE fermentation: Relief of product inhibition using extraction, perstraction, stripping, and pervaporation, *Biotechnology Progress*, 8, 382–90.

Qureshi, N. and Maddox, I.S. (1995) Continuous production of acetone-butanol-ethanol using immobilized cells of Clostridium acetobutylicum and integration with product removal by liquid-liquid extraction, *Journal of Fermentation and Bioengineering*, 80, 185-189.

Qureshi, N. and Blaschek, H.P. (2001) Recovery of butanol from fermentation broth by gas stripping, *Renewable Energy*, 22, 557–564.

Qureshi, N. and Maddox, I.S. (2005) Reduction in butanol inhibition by perstraction: Utilization of concentrated lactose/whey permeate by *Clostridium acetobutylicum* to enhance butanol fermentation economics, *Trans. IChemE, Part C*, 83(C1), 43–52.

Qureshi, N., Saha, B.C. and Cotta, M.A. (2007) Butanol production from wheat straw hydrolysate using *Clostridium beijerinckii*, *Bioprocess and Biosystems Engineering*, 30, 419-427.

Qureshi, N. and Ezeji, T.C. (2008a) Butanol, 'a superior biofuel' production from agricultural residues (renewable biomass): Recent progress in technology, *Biofuels, Bioproducts and Biorefining*, 2, 319-330.

Qureshi, N., Ezeji, T.C., Ebener, J., Dien, B.S., Cotta., M.A. and Blaschek, H.P. (2008b) Butanol production by *Clostridium beijerinckii*. Part I: Use of acid and enzyme hydrolyzed corn fiber, *Bioresource Technology*, 99, 5915–5922.

Qureshi, N., Saha, B.C., Hector, R.E. and Cotta, M.A. (2008c) Removal of fermentation inhibitors from alkaline peroxide pretreated and enzymatically hydrolyzed wheat straw: Production of butanol from hydrolysate using *Clostridium beijerinckii* in batch reactors, *Biomass and Bioenergy*, 32, 1253-1358.

Ramey, D.E. (2007) Butanol: The other alternative fuel, Proceedings of the 19th Annual Conference of the National Agricultural Biotechnology Council (NABC19), South Dakota State University, Brookings, SD, May 22-24, 2007, p. 137-147.

Rodriguez-Couto, S. (2008) Exploitation of biological wastes for the production of value-added products under solid-state fermentation conditions, *Biotechnology Journal*, 3, 859-870.

Román-Leshkov, Y., Barrett, C.J., Liu, Z.Y., Dumesic, J.A. (2007) Production of dimethylfuran for liquid fuels from biomass-derived carbohydrates, *Nature*, 447, 982-985.

Sánchez, O.J. and Cardona, C.A. (2008) Trends in biotechnological production of fuel ethanol from different feedstocks, *Bioresource Technology*, 99, 5270–5295.

Saxena, R.C., Adhikari, D.K. and Goyal, H.B. (2009) Biomass-based energy fuel through biochemical routes: A review, *Renewable and Sustainable Energy Reviews*, 13, 167-178.

Semelberger, T.A., Borup, R.L. and Greene, H.L. (2006) Dimethyl ether (DME) as an alternative fuel, *Journal of Power Sources*, 156, 497-511.

Şensöz, S., Angin, D. and Yorgun, S. (2000) Influence of particle size on the pyrolysis of rapeseed (Brassica napus L.): fuel properties of bio-oil, *Biomass and Bioenergy*, 19, 271-279.

Sharma, Y.C., Singh, B. and Upadhyay, S.N. (2008) Advancements in development and characterization of biodiesel: A review, *Fuel*, 87, 2355-2373.

Stephanopoulos, G. (2007) Challenges in engineering microbes for biofuels production, *Science*, 315, 801-804.

Sun, Y. and Cheng, J. (2002) Hydrolysis of lignocellulosic materials for ethanol production: a review, *Bioresource Technology*, 83, 1–11.

Taylor, G. (2008) Biofuels and the biorefinery concept, *Energy Policy*, 36, 4406–4409.

Usta, N. (2005) Use of tobacco seed oil methyl ester in a turbocharged indirect injection diesel engine, *Biomass and Bioenergy*, 28, 77-86.

Veljković,V.B., Lakićević, S.H., Stamenković, O.S., Todorović, Z.B. and Lazić, M.L. (2006) Biodiesel production from tobacco (Nicotiana tabacum L.) seed oil with a high content of free fatty acids, *Fuel*, 85, 2671–2675.

Vertès, A.A., Inui, M. and Yukawa, H. (2007) Alternative technologies for biotechnological fuel ethanol manufacturing, *Journal of Chemical Technology and Biotechnology*, 82, 693-697.

Wang, T., Chang, J., Fu, Y., Zhang, Q. and Li, Y. (2007) An integrated biomass-derived syngas/dimethyl ether process, *Korean Journal of Chemical Engineering*, 24, 181-185.

Wyman, C.E. (1999) Biomass ethanol: Technical progress, opportunities, and commercial challenges, *Annual Reviews in Energy and the Environment*, 24, 189-226.

Yang, B. and Wyman, C.E. (2008) Pretreatment: the key to unlocking low-cost cellulosic ethanol, *Biofuels, Bioproducts and Biorefining*, 2, 26–40.

Zabaniotou, A., Ioannidou, O. and Skoulou, V. (2008) Rapeseed residues utilization for energy and 2nd generation biofuels, *Fuel*, 87, 1492-1502.

Zhang, J. and Jiang, L. (2008) Acid-catalyzed esterification of Zanthoxylum bungeanum seed oil with high free fatty acids for biodiesel production, *Bioresource Technology*, 99, 8995–8998.

Zverlov, V.V., Berezina, O., Velikodvorskaya, G.A. and Schwarz, W.H. (2006) Bacterial acetone and butanol production by industrial fermentation in the Soviet Union: use of hydrolyzed agricultural waste for biorefinery, *Applied Microbiology and Biotechnology*, 71, 587–597.

INDEX

A

abatement, 281
access, 26
accessibility, 135
accounting, 246, 247, 268, 271
acetic acid, 76, 278, 280
acetone, 275, 277, 278, 280, 285, 286, 287, 288, 289
achievement, 227
acid, 32, 34, 36, 37, 38, 50, 58, 59, 61, 63, 70, 73, 74, 84, 114, 130, 131, 132, 134, 137, 138, 139, 144, 146, 147, 148, 150, 151, 166, 168, 169, 171, 178, 187, 189, 202, 204, 273, 274, 275, 276, 278, 280, 281, 282, 288
acidity, 90, 173, 206, 213
activated carbon, 3, 112, 125, 126, 156, 172, 175, 177, 178, 182, 184, 185, 186
activation, xii, 126, 137, 172, 175, 177, 185, 219
adaptation, 45, 54, 65, 70, 73, 163
ADC, 161
additives, ix, 61, 142, 167, 168, 182, 185, 199
adhesion, 233, 234
adsorption, x, xi, 109, 110, 111, 113, 115, 116, 117, 118, 119, 120, 121, 123, 125, 126, 156, 172, 175, 177, 178, 182, 183, 185, 239, 240, 246
adsorption isotherms, 111, 115
advocacy, 27
aerobic bacteria, 38, 61, 166
AFM, 234, 235
Africa, 133, 139, 216
afternoon, 244
age, 31, 33, 67, 162, 225, 227, 228, 239
ageing, 225, 231, 232, 234, 235, 236, 237, 238, 239, 240, 241, 242
agent, 170, 172, 178, 203, 273
aggregation, 86
agricultural market, 14
agricultural sector, 2

agriculture, vii, viii, ix, 1, 8, 16, 19, 20, 23, 24, 25, 26, 27, 31, 60, 68, 71, 77, 79, 88, 89, 90, 91, 125, 134, 183, 194, 210, 213, 216, 217, 218, 224, 266
Air Compliance Agreement, 23, 24, 28
air pollutants, 20
air quality, 23, 24
alanine, 139
alcohol, 135, 273, 274, 277, 278, 281, 282, 286
alcohols, 9, 47, 267, 270, 273, 282, 283
alfalfa, 86
algae, 130, 132, 133, 134
alkaline hydrolysis, 137
alkaloids, 180
alkane, 281
alternative, xi, xii, 8, 10, 12, 14, 40, 87, 132, 138, 149, 151, 155, 157, 170, 171, 172, 173, 174, 175, 176, 179, 180, 184, 186, 187, 204, 211, 219, 225, 230, 247, 266, 270, 271, 273, 275, 276, 277, 279, 281, 282, 283, 284, 285, 287, 288
alternative energy, 14
alternatives, 173, 180, 266, 270
alters, 185
aluminum, 277
ambient air, x, 109, 120, 125
amendments, 9, 12, 14, 16, 193, 286
amines, 189
amino acids, 47, 130, 131, 132, 133, 134, 138, 139, 142, 143, 149
ammonia, viii, x, 2, 3, 19, 20, 22, 26, 37, 38, 57, 59, 86, 109, 110, 111, 112, 113, 114, 116, 117, 118, 119, 120, 121, 122, 123, 124, 125
ammonium, 167
amplitude, 244
amylase, 64, 74
anaerobic bacteria, 73
anaerobic digesters, 36, 73, 171
anaerobic sludge, 34, 36, 70, 72, 73, 75
androgen, 4
animal agriculture, viii, 20, 24, 25, 26, 27

292   Index

animal feeding operations, viii, 19, 20
animals, viii, 2, 3, 4, 5, 6, 7, 8, 11, 19, 20, 26, 56, 57, 86, 135, 139, 143, 144, 146, 162, 164, 165, 193, 209, 246
anthocyanin, 173
antibiotic, 8
antibiotic resistance, 8
antioxidant, 280
aqueous solutions, 172, 177, 178, 186, 286
arginine, 132
Aristotle, 195
aromatic hydrocarbons, 8
arrest, 49
arsenic, 85
asbestos, 221, 223, 236, 243, 244, 245, 258
ash, ix, 10, 77, 78, 79, 80, 81, 82, 83, 84, 85, 86, 87, 88, 89, 90, 91, 110, 111, 112, 115, 116, 125, 126, 166, 168, 199, 211, 215, 280
Asia, 133, 139, 149
aspartate, 150
assessment, xi, 75, 129, 185, 223
assets, 259
assimilation, 138, 276
assumptions, 247, 248
atomic force, 234
atoms, 281
ATP, 53
attacks, 282
attention, viii, 19, 20, 21, 23, 27
Attorney General, viii, 20, 23
Australia, 13, 260
authority, 26, 27
availability, 14, 16, 17, 32, 34, 84, 130, 135, 150, 157, 161, 165, 196, 204, 214, 217, 222, 268, 282, 286
averaging, 249
avoidance, 266
awareness, x, 179

B

bacillus, 70
Bacillus subtilis, 140
bacteria, 7, 8, 11, 17, 33, 35, 38, 39, 49, 50, 55, 59, 61, 62, 64, 66, 68, 70, 72, 73, 74, 75, 76, 123, 130, 132, 133, 135, 136, 141, 143, 144, 146, 149, 210, 277, 278
bacteriocins, 38
bacteriophage, 277, 278
bacterium, 63, 76, 151
banks, 220
barley, 75, 87
barriers, 271

basic needs, 130
beef, 6
beer, 76
behavior, 221, 229, 230, 232, 240, 243, 244, 247, 257, 258, 280
Beijing, 15, 16
bending, 222, 230
beneficial effect, 85
benign, 198, 199
benzene, 8
beverages, x, 129
binding, x, 177
bioavailability, 14, 56
biocatalysts, 56
biochemistry, 39, 144
bioconversion, xi, 71, 143, 191, 198, 204, 217
biodegradability, 194
biodegradable materials, 206
biodegradation, 68, 139, 187, 194
biodiesel, xi, xiii, 156, 170, 174, 175, 180, 186, 265, 270, 271, 272, 273, 274, 282, 283, 284, 285, 287, 288, 289
biodiversity, 210
biofuel, 147, 156, 170, 266, 270, 271, 275, 276, 282, 284, 285, 288
biological activity, 34
biological processes, 24
biomass, viii, 9, 10, 11, 13, 17, 29, 30, 36, 38, 39, 40, 45, 47, 48, 56, 57, 58, 70, 72, 73, 75, 118, 119, 125, 126, 132, 133, 134, 138, 149, 150, 152, 161, 170, 171, 177, 186, 187, 188, 189, 198, 201, 203, 204, 215, 266, 267, 268, 269, 270, 271, 274, 275, 276, 277, 278, 279, 280, 281, 282, 283, 284, 285, 286, 287, 288, 289
biomass materials, 126, 188, 267, 271
biomaterials, 268, 269
bioremediation, 34, 66, 67
biosynthesis, 136, 144, 148, 150, 169
biosynthetic pathways, 136
biotechnology, 40, 60, 71, 146, 147, 149, 156, 213, 215, 217
bleaching, 222, 226, 227, 233
blends, xi, 155, 158, 174, 225, 271, 272, 273
blocks, 224
blood, 137
bloodstream, 5
body fluid, 206
boilers, 10, 175, 268
bonding, 225, 226
bonds, 239
bowel, 123
Brazil, xi, 147, 150, 155, 156, 157, 158, 174, 179, 182, 219, 220, 221, 222, 223, 224, 225, 236, 242,

258, 259, 260, 261, 262, 263, 272, 274, 275, 285, 286
breakdown, 34, 49, 54, 56, 195
breeding, vii, 1
buffalo, 193
building blocks, 258
bureaucracy, 220
burn, 10, 275
burning, 3, 31, 168, 174, 175, 176, 268, 270, 271, 279
Bush Administration, 27
butyl ether, 8, 218
by-products, ix, xii, 2, 37, 39, 57, 59, 68, 74, 126, 142, 143, 157, 178, 193, 219, 220, 222, 225

C

cabbage, 64, 79, 81
cadmium, 172
caffeine, 160, 161, 165, 166, 169, 171, 179, 182, 183, 185, 188
calcium, 177, 202, 223, 225, 240, 242, 259
calcium carbonate, 240, 242
calibration, 254
Canada, 11, 41, 206
cancer, 223
candidates, xiii, 265, 271, 274, 281, 283
carbohydrate, 45, 57, 75, 163, 181, 271
carbohydrates, 38, 50, 57, 85, 130, 132, 134, 160, 163, 164, 170, 180, 186, 206, 235, 271, 288
carbon, ix, 4, 8, 11, 12, 13, 16, 25, 33, 34, 35, 47, 48, 57, 59, 63, 65, 70, 77, 89, 111, 112, 115, 125, 126, 133, 134, 137, 141, 147, 167, 168, 175, 177, 178, 179, 198, 209, 210, 216, 233, 237, 239, 266, 270, 271, 272, 277, 279, 280, 285
carbon dioxide, 11, 25, 57, 147, 177, 209, 237, 239, 271
carbon materials, 126
carbon monoxide, 272, 277, 279
carbonization, 110, 111, 115, 118, 119, 126, 172, 175, 179
carbonyl groups, 119
carboxylic acids, 32
carcinogenicity, 144
carob, 139
carrier, 271
case study, 17
cast, 198, 217
casting, 203, 218, 253
catalyst, 36, 174, 177, 273, 274, 279
catalytic activity, 49, 177, 189
catchments, 6
categorization, 266, 268
catfish, 161, 182

cation, 65, 202
cattle, 4, 5, 7, 9, 11, 17, 58, 64, 65, 68, 71, 86, 161, 162, 179, 193, 203, 213, 217
CBS, 141, 149
cell, ix, x, 30, 40, 47, 50, 69, 73, 129, 130, 132, 135, 138, 140, 141, 146, 147, 148, 149, 150, 151, 152, 153, 183, 233, 234, 267, 278
cell death, 50
cellulose, xii, xiii, 9, 32, 34, 37, 38, 52, 65, 70, 135, 143, 161, 166, 206, 219, 221, 222, 226, 227, 231, 239, 243, 261, 265, 267, 271, 274, 275, 276
cellulose fibre, 261
Census, 194, 215
ceramic, 126, 243, 244
cheese, x, 129, 132, 139, 141, 148, 207
chemical properties, 15, 83, 86, 126, 180, 184, 194, 218, 273, 285
chemisorption, 178
chicken, 161, 179, 182
children, 131, 132
Chile, 261
China, 12, 16, 17, 35, 130, 133, 274, 284
chitin, 71
chlorine, 233
chromium, 172, 202
chromosome, 136
circulation, 36, 125
cladding, 41
classes, 146
classification, 75, 115
Clean Air Act, 20, 22, 25, 26
cleaning, 141, 158, 222
climate change, 25, 209
cloning, 136, 138, 149
clusters, 136
CO2, 12, 133, 168, 171, 202, 209, 216, 225, 237, 240, 278
coal, ix, 15, 77, 84, 87, 90, 126, 168, 172, 279, 281, 283, 287
cocoa, 139, 172
cocoon, 202, 203
coffee, xi, 37, 139, 151, 155, 156, 157, 158, 159, 160, 161, 162, 163, 164, 165, 166, 167, 168, 169, 170, 171, 172, 173, 174, 175, 176, 177, 178, 179, 180, 181, 182, 183, 184, 185, 186, 187, 188, 189, 217, 274, 282, 285, 287
Colombia, 12, 156
colonisation, 213
combined effect, 165
combustion, 9, 10, 12, 90, 110, 126, 156, 167, 168, 175, 176, 177, 179, 211, 266, 268, 270, 271, 272, 275, 279, 283
commodity, 156, 179

community, 12, 23, 26, 204, 215, 266
compatibility, 224
complement, 158
complexity, 184
compliance, 156
components, vii, xii, 1, 2, 26, 38, 39, 42, 52, 57, 114, 122, 136, 220, 221, 225, 244, 249, 250, 252, 261, 271, 275
composites, 221, 222, 224, 225, 227, 228, 229, 230, 231, 232, 233, 234, 235, 236, 242, 243, 257, 261
composition, ix, 6, 14, 49, 50, 52, 59, 61, 67, 70, 73, 74, 77, 89, 111, 112, 115, 130, 134, 140, 144, 153, 158, 159, 160, 162, 163, 175, 176, 185, 186, 188, 189, 198, 225, 233, 259, 273, 279, 283
compost, x, 33, 34, 51, 55, 58, 59, 62, 64, 67, 69, 71, 75, 109, 110, 113, 120, 121, 122, 125, 167, 179, 198, 199, 202, 205, 206, 209, 211, 213, 214
composting, ix, xi, 3, 29, 33, 34, 39, 44, 45, 51, 53, 55, 58, 59, 60, 61, 62, 63, 64, 65, 67, 68, 69, 70, 71, 72, 73, 139, 167, 178, 185, 191, 193, 194, 195, 197, 198, 202, 205, 207, 208, 209, 211, 212, 213, 214, 215, 216, 217, 218
compounds, 5, 7, 8, 25, 33, 34, 39, 47, 57, 123, 130, 144, 160, 165, 166, 169, 180, 200, 205, 227, 233, 276, 282, 284
concentrates, 88, 137, 141, 150, 151, 161
concentration, x, xii, 5, 6, 7, 20, 31, 35, 37, 38, 44, 45, 52, 75, 85, 88, 109, 112, 113, 116, 120, 123, 124, 125, 138, 146, 160, 164, 170, 171, 177, 178, 199, 203, 204, 220, 223, 224, 246, 247, 248, 252, 253, 254, 255, 280
concrete, 41, 158, 161, 163, 221, 258, 261
condensation, 285
conductivity, 79, 202, 204, 242, 243, 244, 251
confidence, ix, 30, 58
configuration, 267
confinement, viii, 19, 20
confusion, 24, 249
Congress, iv, viii, 20, 21, 24, 25, 27, 186, 258, 262
consensus, 42
consent, 6, 201
conservation, 38, 57, 59, 156, 211, 268
construction, xii, 211, 220, 221, 225, 260, 268
consumers, 37
consumption, x, xi, 32, 36, 42, 44, 65, 129, 134, 135, 136, 137, 143, 146, 155, 169, 188, 207, 221, 223, 240, 256, 275, 281
contact time, 112, 113, 117, 119, 120, 121, 122, 123, 178
contaminant, viii, 20, 24
contaminants, viii, 19, 21
contaminated soils, 34
contamination, ix, x, 2, 3, 6, 7, 10, 32, 85, 141, 199, 209, 211, 268, 277
continuity, 249, 251
control, viii, 29, 30, 36, 37, 39, 43, 51, 53, 54, 56, 59, 62, 68, 69, 70, 79, 82, 83, 86, 88, 110, 116, 120, 136, 150, 161, 162, 167, 173, 179, 185, 201, 211, 227, 283
conversion, ix, 9, 30, 57, 67, 90, 110, 132, 152, 161, 162, 169, 171, 174, 176, 192, 197, 201, 205, 207, 266, 267, 268, 270, 271, 273, 274, 275, 276, 277, 279, 284, 286
conversion rate, 161
cooking, 12, 16, 226
cooling, 39, 41, 261
copper, 172, 202, 206
corn, 9, 59, 74, 148, 161, 162, 163, 164, 167, 186, 189, 272, 275, 278, 287, 288
correlation, 62
corrosion, 168, 179
costs, xii, 21, 23, 26, 75, 142, 156, 158, 161, 162, 168, 170, 180, 204, 205, 211, 212, 219, 220, 222, 223, 225, 229, 257, 268, 272, 276, 283
cotton, 78, 79, 81, 83, 213, 273
Council of Europe, 32, 62
coupling, 231
Court of Appeals, 23
covering, x, 109, 113, 120, 125, 157, 223, 243
CPC, 199
crack, 230
credit, 31
critical analysis, 189
crop production, 3
crops, vii, 1, 2, 9, 16, 20, 25, 26, 79, 80, 81, 82, 83, 85, 86, 88, 89, 167, 184, 222, 267, 268, 269, 270, 273, 274, 284
CRS, 19, 23, 27, 28
crystalline, 49, 242
crystallization, xii, 220, 255
crystals, 137, 242
CSF, 234
cultivation, xi, 47, 48, 58, 68, 85, 129, 130, 132, 150, 171, 179, 183
culture, 39, 45, 47, 48, 63, 67, 136, 137, 138, 139, 141, 143, 147, 149, 150, 153, 169, 171, 179, 189, 212, 216, 277
culture conditions, 137, 141
curing, 158, 207, 237
currency, 156, 179
cyanide, 72
cycles, 231, 232, 233, 234, 235, 238, 239, 240, 241, 271
cyst, 73
Czech Republic, 16

D

dairies, 23
dairy, 24
dairy industry, 6, 11
danger, 21
data collection, 163
database, 125
death, 37, 48, 56, 67, 68, 207, 216
death rate, 48, 67
decay, xii, 48, 50, 135, 206, 209, 220, 223, 227, 246, 247, 248, 249, 253, 254
decision making, 204, 211
decisions, 24
decomposition, 32, 59, 166, 167, 198, 207
defects, 173
defendants, 26
defense, 136, 153
deficiency, 86
definition, 2, 21, 22, 24, 25, 42, 156, 247, 253, 266, 270
degradation, xii, 4, 32, 39, 42, 52, 53, 54, 56, 57, 61, 63, 67, 68, 72, 138, 143, 150, 166, 171, 173, 179, 192, 194, 197, 198, 206, 220, 224, 274, 276
degradation mechanism, xii, 220
degradation process, 68
dehydration, 223, 281
Denmark, 187, 262
density, ix, xii, 47, 70, 77, 78, 84, 158, 160, 168, 217, 220, 229, 231, 232, 233, 238, 249, 250, 255, 257, 268, 279
Department of Agriculture, 20, 27
Department of Commerce, 90
Department of Energy, 15, 90
Department of Justice, 22
Department of the Interior, 27
deposition, 162
derivatives, 137, 150, 164
destruction, 21, 33, 37, 41, 52, 55, 56
detection, 6, 173, 249
developed countries, 32, 130, 221, 270
developing countries, x, 2, 12, 130, 220, 222, 223, 257, 260
diarrhea, 143
diesel engines, 272, 281, 285
diesel fuel, 272, 279, 285
diet, x, 129, 130, 144, 146, 147, 161, 162, 163, 164, 188, 194
differentiation, 173
diffusion, 34, 178, 207, 240, 246, 248, 249, 253, 254, 255
diffusivities, 249

diffusivity, 41, 242, 243, 244, 246, 248, 249, 251, 253, 254
digestibility, 59, 70, 142, 143, 150, 161, 162, 163, 164, 165, 185, 186, 188
digestion, 11, 34, 35, 36, 39, 40, 41, 42, 43, 44, 45, 46, 48, 49, 51, 52, 53, 55, 57, 58, 59, 60, 61, 62, 63, 64, 65, 66, 67, 68, 69, 70, 71, 72, 73, 74, 75, 163, 171, 181, 183, 187, 188, 276
digestive enzymes, 165
direct cost, 211
discharges, 20, 21, 26
disclosure, 21
discomfort, 207, 224
dispersion, 236, 239
displacement, 228
disposition, 57
dissolved oxygen, 2, 53
distillation, 170, 275
distilled water, 231
distribution, 78, 150, 168, 181, 227, 236, 239, 246, 248, 249, 284
diuretic, 165
diversity, 46, 51, 53, 64, 67, 71, 72, 146, 180, 268, 277, 279
division, 266
DMF, 270, 281
DNA, 50, 62, 138
doors, 224
dosage, 178
dosing, 58
drainage, 7, 207, 226, 229
drinking water, 3, 178, 189
drugs, 8
dry matter, 9, 11, 14, 79, 143, 157, 161, 162, 163, 164, 167, 169, 176
drying, xi, 111, 149, 155, 156, 157, 158, 161, 168, 173, 176, 180, 184, 211, 217, 255, 268
dumping, 21, 25, 31, 90, 205, 221
durability, 220, 222, 224, 225, 233
duration, 33, 38, 44
dyes, 172, 175, 178, 200
dynamic viscosity, 250

E

earth, 46, 134, 195
earthworm, 192, 198, 202, 203, 204, 209, 210, 213, 214, 215, 216, 217, 218
earthworms, xii, 191, 192, 194, 195, 197, 198, 199, 201, 202, 203, 204, 206, 207, 208, 213, 214, 216, 217, 218
economic efficiency, vii, 19, 20
economics, 40, 146, 282, 287, 288

ecosystem, 218
edema, 144
Education, 16
EEG, 184
effluent, 2, 3, 6, 7, 36, 41, 48, 54, 62, 71, 204, 226
effluents, 6, 7, 16, 17, 35, 36, 37, 62, 63, 89, 143, 213, 215, 216
elaboration, 52, 171
elasticity, 221, 232
electricity, 9, 10, 11, 270
electromagnetic, 248
electromagnetic waves, 248
electron, 125, 228, 231, 242
emergency planning, 21
emission, 8, 12, 20, 34, 168, 211, 268
employment, 162, 165, 278, 282
endocrine, 4, 17, 65, 70
endonuclease, 137, 150
endothermic, 9, 172
endotoxins, 135, 136
energy, x, 9, 10, 11, 14, 16, 17, 30, 34, 35, 37, 40, 44, 49, 52, 57, 110, 125, 129, 130, 133, 135, 139, 143, 144, 158, 163, 168, 170, 171, 174, 175, 180, 188, 193, 199, 217, 223, 226, 227, 229, 230, 231, 236, 249, 251, 258, 266, 267, 268, 270, 271, 275, 276, 278, 279, 281, 282, 283, 285, 286, 288, 289
energy consumption, 135, 226, 236
energy density, 275, 278, 281, 282
energy supply, 11
England, 210, 259, 261
entrepreneurs, 208
environment, vii, viii, ix, 1, 2, 4, 13, 14, 16, 17, 19, 21, 25, 30, 31, 32, 46, 47, 48, 72, 119, 120, 122, 123, 126, 192, 198, 224, 225, 237, 238, 244
environmental effects, 10, 63, 65
environmental impact, 20, 90, 266, 287
environmental issues, 156, 179, 223, 268, 269, 272
Environmental Protection Agency (EPA), 2, 8, 21, 22, 23, 24, 25, 26, 27, 28, 53, 55, 56, 63, 65, 192, 209, 223
enzymes, 9, 34, 38, 41, 47, 49, 50, 51, 52, 54, 56, 61, 64, 75, 135, 143, 152, 168, 169, 197, 198, 271, 278, 286
EPCRA, viii, 19, 20, 21, 22, 23, 24, 25, 26, 27
equilibrium, 44, 178, 183, 251
erosion, 20, 167
ESI, 185
esophagus, 163
ester, 174, 273, 274, 289
estimating, 285
estriol, 4, 7
estrogen, 5, 6

ethanol, 8, 9, 16, 139, 147, 170, 174, 180, 182, 183, 189, 222, 231, 237, 270, 272, 273, 274, 275, 276, 277, 278, 284, 285, 286, 287, 288, 289
ethyl acetate, 169
ethylene, 277
eucalyptus, xii, 15, 62, 219, 222, 239
Europe, 10, 16, 56, 272, 282
European Union, 69
evaporation, 41, 167
evening, 114
evidence, 25, 26
evolution, 40, 43, 45, 62, 74, 89, 195, 272
excretion, 4, 5, 164
exercise, 192
exotoxins, 135
experimental design, 144
exploitation, xii, 49, 57, 64, 146, 156, 220, 224, 246, 258, 268
exporter, 156, 179
exposure, 56, 118, 217, 223, 237, 238, 246
extinction, 210
extraction, 9, 45, 135, 137, 180, 194, 222, 233, 273, 276, 277, 278, 282, 285, 288

F

fabrication, 227, 229, 237, 257
failure, 22, 26, 36, 199, 221, 247
family, 12, 194
farm land, 209
farmers, 6, 9, 86, 110, 192, 195, 208, 209
farms, xi, 2, 6, 7, 10, 20, 22, 23, 110, 156, 157, 174, 193, 209, 281
fat, 162, 163, 164, 165, 166, 285
fatty acids, 35, 48, 49, 67, 74, 114, 123, 124, 125, 174, 176, 186, 272, 273, 274, 289
fauna, 213
fear, 89, 134, 210
feces, 164
federal courts, viii, 19, 22, 23
federal law, 23
Federal Register, 26, 27
feed additives, 2, 8, 152
feedback, 139
feedback inhibition, 139
feet, 144
fermentation, ix, xi, 29, 35, 37, 38, 41, 59, 61, 63, 64, 65, 66, 70, 72, 75, 120, 130, 138, 139, 143, 146, 147, 148, 150, 151, 152, 156, 158, 160, 166, 167, 168, 169, 170, 173, 179, 180, 181, 182, 185, 186, 187, 188, 189, 268, 270, 271, 275, 276, 277, 278, 282, 285, 286, 287, 288, 289
fertility, 84, 86, 130, 192, 193, 195, 210, 211, 215

Index

fertilizers, 23, 89, 111, 167, 171, 176, 180, 207, 209, 211, 223, 224
fever, 135
fiber content, 160, 225, 229, 230, 243
fibers, xii, 163, 219, 220, 221, 222, 224, 225, 226, 227, 229, 230, 231, 233, 234, 235, 236, 239, 240, 243, 257, 258, 260, 267
fibrillation, 226, 234
field crops, 193
field trials, 14, 83, 86
filament, 221
filters, 36, 58, 75
filtration, 36, 226, 250
financial crisis, 220
financial support, 181, 283
Finland, 142
firms, 193
first generation, 270, 271
first responders, 26
fish, 2, 6, 16, 37, 38, 64, 74, 142, 146, 161, 163, 179, 210
fisheries, viii, 29
fishing, 74
fixation, 147
flame, 275
flammability, 275, 278
flammability limit, 275
flatulence, 143
flavor, 169, 274
flooding, 206
flora, 62, 72
flow field, 249
fluctuations, 9, 51, 72, 167, 246, 247
flue gas, 86, 168, 177, 179
fluid, 36, 49, 50, 247, 249, 250, 251, 279, 280
fluidized bed, 59, 168, 176, 279
food, viii, ix, x, xiii, 2, 29, 30, 31, 35, 39, 42, 45, 54, 57, 61, 63, 64, 70, 72, 75, 88, 110, 129, 130, 132, 133, 134, 135, 137, 139, 142, 143, 146, 150, 156, 161, 162, 163, 170, 171, 172, 185, 194, 200, 210, 265, 266, 267, 268, 270, 271, 272, 273, 274, 277, 279, 281, 283, 285
food industry, viii, 29, 30, 31, 39, 54, 57, 142, 273
food processing industry, 161
food production, x, 30, 129, 146, 185
formaldehyde, 135, 172, 177, 182
fossil, 14, 89, 170, 266, 270, 271, 272, 273, 277, 279, 280
fouling, 168, 179
fragility, 36
France, 142, 258, 277
fructose, 281
fruits, xi, 1, 155, 157, 173

FTIR, 177
fuel, xi, 8, 9, 10, 11, 12, 14, 15, 16, 17, 89, 91, 110, 116, 156, 167, 168, 170, 172, 174, 175, 176, 177, 179, 180, 183, 184, 187, 209, 270, 271, 272, 273, 275, 276, 277, 278, 279, 280, 281, 282, 283, 284, 285, 286, 287, 288, 289
funding, 220
fungi, 33, 34, 130, 132, 133, 134, 135, 143, 146, 148, 152, 171, 182
fungus, 70, 143, 150, 179
furan, xiii, 265, 270, 283

G

garbage, 67
gases, x, 9, 20, 25, 109, 110, 123, 168, 283
gasification, 9, 10, 16, 176, 177, 189, 268, 278, 279, 281, 283, 286
gasoline, 11, 170, 271, 272, 275, 276
gel, 49, 112, 116
gene, 62, 136, 138, 149, 153
generation, xi, 9, 10, 17, 30, 32, 35, 36, 38, 41, 45, 47, 60, 157, 158, 185, 191, 193, 199, 206, 207, 209, 211, 221, 227, 246, 249, 252, 268, 270, 271, 283, 284, 289
genes, 136, 138, 139, 285
genetics, 148
genome, 49
genre, 47
Germany, 112, 142, 151, 153
germination, 79, 83, 202
ginger, 76
global climate change, 286
global demand, viii, 29
global economy, xii, 265
glucose, 5, 47, 48, 67, 169, 170, 267, 278
glutamate, 138
glutamic acid, 148
glycoside, 56
goals, 274, 282
gout, 134, 137
gouty arthritis, 137
government, iv, viii, 20, 21, 22, 23, 25, 26, 27, 192
gracilis, 147
grading, 158
grains, ix, 142, 170, 222, 247, 255, 270
granules, 36
grass, 6, 8, 12, 81, 162
grasses, 134, 267
gravity, 249, 250
grazing, vii, 1, 2, 4, 6, 86, 209
Greece, 274, 287
greenhouse gases, 8

groundwater, 2, 6, 25, 199, 268
groups, x, 8, 23, 25, 26, 27, 32, 44, 46, 64, 109, 118, 119, 123, 125, 136, 163, 177, 181, 253, 280
growth, x, 3, 8, 14, 33, 36, 37, 38, 42, 43, 44, 45, 46, 47, 49, 50, 51, 58, 62, 63, 67, 69, 70, 72, 73, 79, 83, 86, 90, 91, 129, 130, 132, 134, 139, 141, 142, 143, 146, 161, 163, 166, 171, 172, 179, 180, 183, 195, 201, 203, 204, 209, 216, 278
growth rate, x, 36, 37, 45, 47, 49, 51, 67, 129, 132, 134, 146, 161, 201, 203, 204
growth temperature, 42, 45, 46, 47, 49, 50, 67
guidance, 25
guidelines, 207
Guinea, 189
gut, 197, 198, 208, 210

H

half-life, 223
hardness, 223
harmful effects, 272
harvesting, xi, 9, 134, 155, 157, 173, 174, 180, 181, 268
hazardous substances, viii, 19, 20, 21, 22, 23, 26, 27
hazards, x, 3, 221, 223
health, vii, ix, x, 1, 16, 20, 21, 23, 24, 25, 26, 123, 124, 125, 143, 148, 192, 194, 195, 223, 246
heat, 38, 39, 40, 41, 43, 44, 45, 49, 50, 55, 62, 64, 126, 137, 168, 177, 199, 206, 207, 243, 244, 245, 251, 253, 261
heat capacity, 251
heat loss, 39, 44, 45
heat release, 44, 168
heat transfer, 243, 244, 245
heating, ix, 10, 12, 15, 29, 37, 43, 110, 141, 168, 176, 184, 273, 280, 281
heating rate, 15
heavy metals, 82, 85, 87, 88, 89, 182, 186, 202, 204, 208
height, 254
hematology, 144
hemicellulose, xiii, 9, 32, 38, 135, 206, 222, 265, 267, 271, 274, 275, 276
hemp, 278
heterogeneity, 42, 49, 221, 227
hexane, 174
histamine, 173
Honda, 71, 141, 149
hormone, 6
host, 135
House, 21, 24, 25, 27, 90, 151, 215, 217
House Appropriations Committee, 24
households, 202

housing, xii, 110, 220, 248, 258, 261
human activity, 285
humidity, 113, 118, 120, 179, 237
hybrid, 76
hydrocarbons, 47, 133, 143, 271, 272, 279
hydrogen, x, 20, 22, 26, 35, 37, 59, 62, 63, 109, 111, 114, 115, 116, 117, 119, 123, 124, 125, 183, 184, 198, 270, 277
hydrolysis, 39, 135, 148, 170, 186, 275, 276, 277, 286
hydrothermal system, 59
hydroxide, 225, 240
hydroxyl, 118, 177
hygiene, 211

I

ICAM, 262
identification, 75, 174, 231
images, 228, 231, 232, 233, 235, 242, 257
immersion, 225, 237
immunization, 135
impact energy, 221
implementation, xii, 146, 170, 184, 192, 246, 282, 286
impregnation, 177, 178
in vitro, 187
in vivo, 179
incidence, 16, 73
inclusion, 144, 161, 162, 163, 164, 165, 230
income, 162, 222
incompatibility, 278
independence, 146
India, 11, 12, 16, 17, 35, 72, 77, 78, 79, 129, 130, 146, 149, 152, 156, 191, 192, 193, 194, 196, 199, 201, 203, 204, 211, 212, 213, 214, 215, 216, 217, 218, 284
indication, 144, 168, 171, 172, 180, 229
indicators, 55, 213
indigenous, 45, 203
Indonesia, 156, 260
industrial wastes, 31, 135, 143, 213, 216, 218
industrialisation, 31
industrialized countries, 266, 271
industry, vii, x, xii, 10, 11, 15, 19, 20, 23, 25, 30, 54, 57, 60, 68, 71, 72, 90, 110, 124, 125, 135, 142, 143, 158, 174, 175, 176, 179, 187, 192, 199, 200, 203, 207, 208, 213, 216, 217, 222, 274, 280
inertia, 244, 270
infancy, xii, 14, 265, 283
infants, 131
infection, 56, 277, 278
infectious disease, 15

infestations, 173, 184
infrastructure, 212, 281
inhibition, 59, 165, 178, 185, 276, 277, 278, 282, 288
initiation, 236
injuries, viii, 19, 23, 24, 26
injury, 21, 23
inoculation, 32, 166, 171, 179
inoculum, 63, 139, 198
instability, 37
institutions, 192
insulation, 39, 41, 221
integration, 276, 281, 284, 288
interaction, 35, 51, 73, 229, 232
interactions, 153
interest groups, 26
interest rates, 220
interface, 234, 235
international standards, 224
interpretation, viii, 19, 22
interval, 230
intestine, 72, 198
invertebrates, 217
investment, xi, 155, 174, 180
ion adsorption, 178
ionization, 185
ionizing radiation, 224
ions, 50, 172, 177, 178, 189
iron, xi, xii, 191, 208, 219, 225
irradiation, 282
isotope, 247
Italy, 142, 215, 217

J

Japan, 109, 110, 111, 112, 113, 114, 125
joints, 137

K

Kenya, 172
kerosene, 177, 189
kidney, 134, 137
kidney stones, 134
kidneys, 5, 136
kinetic parameters, 42
kinetic studies, 178
kinetics, 75, 152, 163, 175, 178

L

labo(u)r, 37, 141, 207, 212, 220
lactation, 5, 7

lactic acid, 38, 64, 74, 278
lactose, 141, 288
lakes, 6, 8
land, vii, 1, 2, 3, 4, 6, 10, 17, 21, 22, 31, 32, 39, 40, 53, 54, 56, 82, 85, 89, 90, 146, 167, 185, 192, 204, 209, 211, 274
land disposal, 192, 211
landfills, 31, 32, 59, 63, 70, 192
language, 24, 25
laws, viii, 19, 20, 21, 22, 24, 25, 26
leaching, 2, 6, 12, 16, 20, 21, 85, 87, 172, 182, 239
legislation, viii, 20, 21, 24, 25, 26, 27, 31
legislative proposals, 21
legume, 215
leucine, 169
LFA, 83
life span, 32, 144
ligand, 178
lignin, 135, 143, 149, 152, 161, 222, 226, 227, 229, 233, 235, 267, 268, 271, 274, 275, 276, 281, 283, 284, 285
limitation, 39, 144
linear dependence, 250
linkage, 257
lipases, 52
lipids, xiii, 34, 49, 50, 52, 57, 62, 69, 130, 265
liquid crystals, 49
liquid fuels, 266, 270, 271, 279, 283, 288
liquids, 280
litigation, 25
liver, 132, 136, 162
liver cancer, 136
livestock, vii, viii, x, 1, 2, 3, 6, 7, 8, 11, 19, 20, 21, 22, 24, 25, 29, 109, 110, 124, 125, 143, 144, 161, 179, 180, 187, 209, 215, 268
local authorities, 21
local government, 25, 27
location, vii, 1, 30, 58, 194, 268
logistics, xi, 191, 223
long distance, 168, 209
love, 195
low temperatures, 66, 138, 168
lower prices, 174
LPG, 279
lumen, 240
luminosity, 275
lutein, 125
lysis, 68, 135
lysozyme, 51

M

machinery, 181

macromolecules, 50, 51, 72
macronutrients, 85
macropores, 112
Malaysia, 129
management, viii, x, 17, 19, 20, 32, 34, 38, 39, 69, 126, 167, 173, 179, 180, 184, 192, 193, 194, 198, 203, 204, 211, 213, 215, 246, 266, 269
manganese, 8, 208
manufacturing, xii, 149, 199, 219, 234, 257, 270, 289
manure, viii, xii, 2, 3, 6, 7, 8, 9, 11, 12, 15, 16, 17, 20, 21, 24, 25, 26, 31, 32, 33, 38, 58, 59, 64, 65, 66, 67, 68, 71, 72, 73, 89, 133, 171, 188, 191, 192, 193, 194, 195, 197, 199, 201, 202, 203, 206, 208, 209, 211, 213, 214, 217
mapping, 136
market, vii, xi, 1, 68, 142, 155, 158, 174, 180, 193, 221, 222, 243, 270, 277
marketing, 135, 193
markets, 158, 174, 180
mass spectrometry, 185
material resources, 247
materials science, 126
matrix, 177, 220, 222, 224, 225, 226, 227, 228, 229, 230, 231, 233, 234, 235, 236, 239, 240, 246, 249, 251, 257, 275
maturation, 33, 34
maximum specific growth rate, 47
meals, 11
measurement, 111, 112, 113, 118, 120, 150, 209, 254
measures, 192, 208
meat, 1, 35, 60, 68, 72, 165
mechanical performances, 237
mechanical properties, xii, 220, 221, 227, 230, 239
media, 4, 26, 74, 132, 210, 224, 266, 275, 282
medium composition, 50
melting, 49, 50, 168
melting temperature, 168
membranes, 49, 50, 68, 286
mercury, 235
metabolism, 33, 35, 37, 39, 40, 48, 57, 61, 137, 149, 162, 163
metabolites, 3, 4, 5, 7, 36, 47, 48, 132, 135, 136, 138, 143
metal oxides, 88
metals, 26, 88, 178, 192, 203
methane, 25
methanol, 75, 148, 173, 174, 270, 273, 274, 278, 279, 280, 283, 285
methyl tertiary, 8
Mexico, 156
mice, 181, 182
microbial cells, 39, 44, 68, 137, 138, 143
microbial communities, 37

microbial community, 210
microorganism, 275, 282
microscope, 125, 228, 231
microscopy, 234
microstructure, 242, 244
migration, 253
milk, ix, 163, 164, 188, 207
mine soil, 87, 90
MIP, 235
missions, viii, 19, 22, 24
mixing, x, xii, 40, 66, 109, 113, 120, 125, 141, 195, 203, 219, 222, 227
mobility, 206
modeling, 262
models, 246, 247, 252
modulus, xii, 220, 230, 231, 232, 255
moisture, x, xi, 34, 38, 44, 73, 78, 111, 113, 118, 119, 120, 121, 125, 134, 155, 157, 158, 159, 161, 162, 171, 175, 176, 177, 179, 202, 206, 207, 216, 222, 227, 229, 243, 246, 268, 274
moisture content, 34, 73, 111, 113, 120, 121, 134, 157, 158, 159, 162, 171, 175, 176, 177, 179, 202, 216, 222, 227, 246, 268, 274
molar ratios, 282
molasses, 72, 143, 166, 275, 277, 278
mole, 47, 48
molecular oxygen, 34
molecular weight, 34, 38, 267
molecules, 34, 39, 118
momentum, 246, 249, 250, 251, 272, 282
money, 192, 207
Moon, 262
morning, 114, 244
morphology, 126, 242
mortality, 203, 204
motion, 248, 270
mRNA, 50
mutagenesis, 138
mutant, 138, 169
mutation, 139
mycelium, 143, 199
mycology, 149

N

NaCl, 75
NADH, 66
nation, 22
NATO, 262
natural food, 163, 173
natural gas, 139, 147, 279, 283
natural habitats, 46
natural resources, vii, viii, xiii, 1, 19, 21, 265, 267

Natural Resources Conservation Service, 27
nausea, 143
Netherlands, 74, 161, 214, 215, 259
neurotoxicity, 136
New York, iv
New Zealand, 1, 6, 7, 11, 16, 17
Nicaragua, 68, 166
Nigeria, 29
Nile, 163, 185
nitrates, 32
nitrification, 41, 57
nitrogen, ix, 2, 3, 12, 14, 20, 30, 34, 38, 57, 65, 73, 77, 89, 111, 115, 120, 132, 134, 137, 138, 162, 164, 165, 167, 177, 179, 188, 194, 197, 198, 202, 203, 209, 216, 217, 218, 279, 280, 285
nitrogen gas, 111, 120
nitrous oxide, 12, 209
N-N, 167
noise, 279
North America, 56
nucleic acid, 130, 132, 134, 136, 137, 144, 146, 150, 153
nucleus, 136
nutrients, 2, 3, 8, 10, 20, 25, 26, 34, 38, 79, 85, 89, 132, 142, 143, 160, 163, 164, 167, 193, 197, 198, 199, 209, 216, 217, 267
nutrition, viii, ix, 29, 30, 39, 46, 52, 54, 57, 58, 74, 86, 142, 188
nutritional assessment, 148

O

obligate, 44, 46
observations, 49, 50, 277
octane, 8, 275
octane number, 275
OECD, 2, 17
oil, xiii, 72, 73, 110, 125, 135, 138, 143, 146, 168, 169, 170, 174, 175, 177, 180, 186, 193, 194, 213, 214, 265, 268, 270, 271, 273, 274, 277, 279, 280, 281, 282, 283, 284, 285, 287, 288, 289
oils, 17, 174, 270, 271, 272, 273, 279, 280, 284, 285, 287
oilseed, 194, 274
Oklahoma, viii, 20, 22, 23
olive oil, 60, 68
operator, 21, 22, 44
optimization, 39, 138, 252, 277
organ, 55
organic chemicals, 2, 20
organic compounds, 5, 40, 180, 206, 268
organic matter, x, 12, 15, 25, 33, 43, 44, 65, 163, 164, 171, 172, 192, 199, 203, 209

organic solvents, 222
organism, 36, 49, 50, 133, 138, 141
osmotic pressure, 130
outsourcing, 204
oxidation, 3, 6, 7, 14, 40, 71, 87, 126, 173
oxides, ix, 77, 87
oxygen, 2, 9, 11, 12, 32, 33, 34, 38, 39, 40, 41, 43, 44, 47, 48, 62, 69, 75, 115, 116, 117, 175, 233, 271, 275, 279, 281
oxygen consumption, 41, 62
oxygen consumption rate, 41
ozone, 233

P

Pacific, 260
packaging, x, 3, 193, 240
pain, 144
palm oil, 71, 280
pancreas, 162
parameter, 69, 179, 236, 249, 254
parasite, 55
particles, ix, 20, 77, 168, 227, 229, 234, 236, 239, 248, 249
particulate matter, 25
partition, 248, 249, 253
pasteurization, ix, 29
pasture, 3, 6, 31, 56, 218
pathogens, 2, 3, 7, 25, 26, 31, 39, 55, 56, 63, 64, 66, 75, 192, 199, 206, 208
pathology, 144
pathways, 136
peat, 15
penalties, 22, 26
permeability, 3, 229, 246, 251
permit, 26
peroxide, 288
personal communication, 212
pesticide, ix, 173
pesticides, 25, 26
pests, ix
petrifaction, xii, 220
pH, 12, 14, 32, 36, 37, 38, 46, 47, 48, 53, 55, 57, 63, 65, 74, 79, 84, 85, 86, 87, 90, 137, 140, 162, 167, 172, 177, 178, 179, 185, 202, 203, 206, 207, 208
phage, 282
pharmaceuticals, xi, 191
phase decomposition, 75
phenol, 70, 178, 280
phenolphthalein, 237, 238
phenylalanine, 138, 139, 148
phosphate, 23

phosphor(o)us, viii, 2, 3, 17, 19, 20, 22, 23, 137, 198, 202, 204, 285
physical and mechanical properties, xii, 220, 255
physical health, 210
physical properties, 84, 86, 229, 230, 232, 238, 239, 250, 279, 288
physics, 125, 252, 260
physiology, 69
pigs, x, 4, 11, 109, 110, 114, 123, 124, 125, 139, 144, 161, 162, 164, 165, 179, 186, 187
planning, 21, 211
plants, xi, 6, 10, 12, 35, 56, 68, 79, 86, 87, 88, 90, 110, 129, 133, 171, 187, 191, 192, 198, 199, 200, 207, 210, 224, 267, 277, 278, 286
plasma, 165
plasmid, 136
plastics, 193
poison, ix
Poisson equation, 246
polarity, 286
policymakers, 20
political aspects, 130
pollutants, viii, xi, 2, 19, 20, 21, 22, 26, 90, 191, 192, 193
pollution, viii, 2, 7, 8, 26, 29, 30, 31, 60, 61, 66, 73, 88, 90, 110, 127, 146, 148, 160, 201, 221
polycarbonate, 254
polymer, 54, 58, 221, 267
polymerization, 276
polymerization process, 276
polymers, 32, 33, 34, 35, 53, 54, 72, 199, 271
polyphenols, 160, 166, 180
polypropylene, 278
poor, 36, 37, 85, 142, 220, 227, 231, 234, 257
population, x, 6, 11, 30, 31, 33, 37, 38, 40, 41, 42, 43, 45, 46, 51, 52, 53, 56, 58, 59, 60, 62, 69, 129, 130, 131, 147, 149, 194, 220
population growth, x, 129, 130, 149
pork, 22
porosity, 78, 84, 111, 116, 126, 229, 232, 239, 240, 246, 247, 248, 249, 251, 253
porous media, 223, 246, 249, 253
Portugal, 262
potassium, 85, 167, 172, 179, 198, 201
potato, 37, 46, 47, 48, 51, 52, 53, 54, 61, 68, 70, 73, 74, 79, 81, 82, 85, 285
poultry, viii, 1, 2, 4, 6, 7, 10, 11, 12, 14, 15, 19, 20, 21, 22, 23, 24, 25, 72, 144, 161, 187, 193, 201, 203, 214
power, ix, xi, 9, 10, 11, 17, 77, 86, 87, 89, 90, 110, 167, 185, 191, 283
power plants, xi, 10, 86, 191
precipitation, 90

pregnancy, 5, 7
pressure, 44, 111, 175, 226, 246, 247, 250, 251, 252, 253, 262, 266, 275, 278, 282
prevention, 147, 182
prices, 142, 165, 272
private firms, 193
probability, 230
producers, viii, xi, 19, 22, 23, 24, 25, 155, 174, 180, 222
production costs, 170, 272, 282
production technology, 282
productivity, viii, 12, 29, 30, 57, 152, 210, 216, 277
profitability, 171
program, 146, 192
proliferation, 45, 51, 158, 167
propagation, 230, 277
proportionality, 232
proposition, 211
propylene, 277
protein synthesis, 164
proteins, xiii, 34, 49, 50, 57, 130, 132, 138, 142, 143, 161, 178, 180, 216, 265, 267
proteolytic enzyme, 66
prototype, 177
public health, 20, 21, 23, 24, 25, 26
pulse, 67
pumps, 249
purification, 126, 138, 178, 272
pyrite, 87
pyrolysis, 9, 12, 13, 14, 119, 126, 168, 177, 179, 183, 268, 271, 275, 278, 279, 280, 281, 283, 284, 285, 286, 288

R

race, 26
radiation, 90, 91, 206, 217, 223, 244, 245, 246, 248, 254, 259
radioisotope, 248
radon, 91, 223, 250, 262
rain, 90, 173, 246
rainfall, 87, 207
range, ix, xi, xiii, 4, 5, 12, 14, 33, 37, 39, 42, 43, 44, 45, 46, 48, 49, 51, 78, 81, 82, 83, 87, 88, 119, 138, 139, 140, 143, 155, 159, 167, 172, 177, 178, 197, 204, 221, 225, 229, 235, 265, 267, 272, 276, 280
raw materials, xi, xii, 74, 129, 142, 143, 144, 156, 172, 219, 227, 237
reactants, 272
reaction medium, 41
reaction rate, 42, 138
reaction temperature, 40
reaction time, 273

reactivity, 15
reality, 73
reasoning, 23, 210
recalling, 249
recombinant DNA, 139
recombination, 138, 139
recovery, 3, 25, 26, 45, 62, 73, 138, 149, 167, 203, 217, 224, 266, 268, 276, 277, 282, 285, 286
recovery processes, 282
recycling, ix, xi, xii, 29, 30, 31, 39, 46, 53, 56, 57, 110, 135, 188, 191, 193, 204, 216, 217, 219, 266
refining, 234
regenerate, 199
regional, 20
regrowth, 213
regulation, viii, x, 19, 20, 62
regulations, 24, 60, 71, 211, 224
regulators, 49
reinforcement, xii, 219, 220, 221, 222, 227, 230, 236, 237, 257
relationship, 42, 44, 69, 71
relationships, 146
relevance, 62, 179, 266, 268, 269
remediation, 26
renewable energy, 8, 16, 285
repackaging, x
repair, 130
reprocessing, ix, 29, 30, 31, 38, 39, 54, 56, 58
reproduction, 136, 201, 203
research funding, 283
reserves, 271, 272
residues, viii, ix, xi, xii, 1, 2, 8, 15, 16, 29, 30, 33, 38, 59, 72, 73, 74, 135, 139, 142, 146, 148, 150, 152, 155, 156, 157, 159, 160, 161, 166, 167, 170, 171, 172, 174, 175, 179, 180, 182, 184, 185, 186, 187, 191, 193, 196, 197, 199, 201, 206, 213, 216, 217, 218, 219, 220, 221, 222, 223, 237, 257, 266, 267, 268, 269, 272, 273, 274, 275, 277, 278, 279, 281, 282, 283, 284, 285, 286, 287, 288, 289
resistance, 8, 14, 38, 126, 244, 267
resolution, 24
resources, viii, x, 2, 8, 16, 19, 20, 21, 30, 57, 58, 129, 130, 150, 266, 273, 286
respiration, 44, 202
retention, 12, 34, 36, 37, 38, 45, 51, 52, 53, 55, 61, 63, 78, 162, 167, 180, 234
reticulum, 162
ribonucleic acid, 138, 148
rice, x, 67, 81, 82, 83, 84, 85, 86, 109, 110, 111, 112, 125, 126, 134, 135, 140, 146, 151, 172, 215, 280, 288
rice husk, x, 109, 110, 125, 126, 172, 280, 288

risk, xi, 2, 8, 12, 23, 27, 32, 130, 143, 144, 145, 176, 189, 209, 223, 229
risk assessment, xi, 130, 145, 189
RNA, 50, 134, 137, 138
Romania, 262
room temperature, 111, 175, 203, 243, 280
runoff, 2, 7, 17, 20, 23
rural development, 17, 193

S

safety, ix, 25, 30, 32, 137, 247, 270
salinity, 79, 87
salmon, 147
salmonella, 7
salt, 2, 72, 87
salts, 3, 86, 87, 169, 206, 207, 278
sample, 7, 111, 112, 113, 116, 118, 120, 204, 248, 249, 255, 256, 257
sampling, 112
saturation, 120, 206
savings, 209, 282
sawdust, 65, 73, 279
scaling, 35, 180
scarcity, 57
scientific knowledge, 30
SCP, x, 129, 130, 132, 133, 134, 135, 136, 137, 138, 139, 141, 142, 143, 144, 145, 146, 148, 149, 150, 151
search, 217, 277
searching, 142
second generation, xiii, 265, 270, 271
security, 31, 57, 271
sediment, 12, 16
seed, vii, 1, 82, 140, 169, 174, 273, 274, 285, 289
selecting, 181
selenium, 85
self-regulation, 44
SEM micrographs, 231, 234
Senate, 24, 25, 27, 28
sensitivity, 37, 55, 277
separation, 132, 158, 174, 226, 280
sequencing, 76, 136
series, 65, 88, 237, 240
severity, 7
sewage, 7, 17, 31, 33, 40, 41, 44, 45, 46, 51, 52, 53, 54, 55, 59, 60, 61, 63, 64, 65, 68, 69, 70, 71, 72, 73, 75, 197, 198, 199, 203, 213, 214, 216, 217
sex, 163
SFS, 181
shade, 207
shape, 242, 243
sheep, 6, 11, 62, 161, 163, 179, 182, 188, 197, 213

shellfish, 210
shock, 138
shortage, 204, 266
silane, 231, 232
silica, 116, 125, 126, 161
similarity, 252
simulation, 42, 66, 222, 247, 258
Singapore, 146
SiO_2, 111, 112, 115, 119, 126, 225
sites, 21, 23
skilled personnel, 204
skin, xi, 136, 144, 155, 157, 158, 159, 160, 163
slag, xii, 216, 219, 225, 238, 244, 258, 259, 261
sludge, xi, 3, 4, 6, 31, 33, 34, 35, 36, 39, 40, 41, 42, 44, 45, 46, 49, 51, 52, 53, 54, 55, 56, 58, 59, 60, 61, 62, 63, 64, 65, 66, 67, 68, 69, 70, 71, 72, 73, 75, 181, 187, 191, 192, 193, 197, 198, 199, 200, 201, 202, 203, 204, 205, 206, 207, 211, 212, 213, 214, 215, 216, 217
smoke, 176
sodium, 174, 188, 202, 222, 227, 273, 274, 280, 281, 284
sodium hydroxide, 222
soil, vii, ix, 1, 2, 3, 7, 9, 12, 14, 15, 16, 17, 20, 25, 31, 32, 67, 72, 77, 78, 79, 80, 81, 82, 83, 84, 85, 86, 87, 88, 89, 91, 111, 135, 146, 167, 173, 179, 192, 193, 195, 198, 199, 209, 210, 211, 213, 214, 215, 217, 223, 224, 246, 247, 253, 262, 268, 281, 286
solid matrix, 247, 249, 251
solid phase, 49
solid state, 57, 66, 130, 139, 146, 150, 151, 152, 169, 179, 181, 182, 254
solid waste, xi, 2, 3, 8, 16, 31, 59, 60, 61, 62, 63, 64, 65, 68, 69, 70, 71, 155, 156, 157, 158, 167, 179, 183, 184, 191, 192, 193, 197, 198, 199, 204, 206, 207, 208, 211, 215, 268, 283
solubility, 40, 41, 44, 47, 87, 233, 280
solvents, 277, 278
sorption, 172, 178, 248, 251
South Dakota, 288
Soviet Union, 289
soybean, 15, 132, 169, 174, 177, 272, 287
species, 3, 4, 7, 39, 59, 80, 82, 118, 119, 132, 133, 134, 135, 136, 144, 148, 149, 152, 160, 194, 195, 198, 199, 203, 233, 246, 248, 249, 251, 253
specific heat, 242, 243, 251
specific surface, 111, 115, 234
spectrum, 245
spore, 34, 39, 46
stability, 35, 45, 50, 51, 75, 202, 204, 248, 280
stabilization, ix, 29, 32, 42, 52, 53, 54, 64, 66, 71, 74, 198, 213
stabilization efficiency, 42

stages, xi, 4, 7, 10, 31, 33, 40, 144, 155, 222, 226, 227
stakeholders, 193
standard deviation, 114, 123, 124, 227, 229, 238
standard error, 232, 233
standards, 42, 192, 211, 229, 246, 247, 255, 256, 257, 273
starch, xiii, 34, 65, 139, 140, 141, 161, 186, 265, 270, 272, 275, 276, 278
statutes, 26
steel, xi, 41, 111, 141, 191, 225
steel industry, 225
stock, 142, 227
stockpiling, 192
stomach, 165
storage, 6, 21, 38, 66, 156, 158, 159, 161, 168, 175, 268, 270, 272, 279
strain, 70, 138, 169, 277, 278
strain improvement, 138
strategies, 136, 150, 180, 184, 189, 225
strength, 35, 45, 54, 73, 74, 110, 221, 225, 227, 229, 230, 239, 240, 242, 257
stress, 224, 254
structure formation, 210
subsidy, 192
subsistence, 31
substitution, 50, 162, 287
substrates, xii, 34, 35, 71, 133, 134, 139, 141, 142, 145, 146, 169, 171, 179, 189, 191, 197, 199, 201, 203, 271, 282, 286
success rate, 208
sucrose, 173
sugar, viii, 17, 29, 37, 66, 148, 160, 162, 167, 169, 170, 202, 218, 266, 268, 270, 272, 275, 276, 284, 285
sugar beet, 37, 66, 148
sugar industry, viii, 29
sugarcane, 83, 126, 134, 143, 147, 170, 172, 186, 270, 272, 275, 280, 285
sulfur, 123, 141, 149, 272, 281
sulfuric acid, 278
sulphur, 85, 133, 134, 139
summer, 244
Sun, 275, 288
Superfund, viii, 19, 20, 21, 22, 23, 24, 25, 26, 27
supervision, 212
supply, viii, xiii, 9, 15, 29, 30, 31, 47, 48, 133, 134, 142, 193, 207, 265, 266, 270, 279, 283
suppression, 138, 216
surface area, ix, 77, 111, 126, 178, 197
surface chemistry, 178
surface structure, 234
surface treatment, 231

survival, 31, 55, 58, 66, 73, 161, 202, 203, 216
susceptibility, 47, 67, 133, 282
sustainability, 156, 174
sustainable development, 156, 170
Sweden, 258
Switzerland, 261
symbiosis, 36
symptoms, 87, 144, 165
synthesis, 50, 132, 136, 164, 188, 189, 277, 278
synthetic polymers, 69
systems, x, 3, 4, 6, 7, 9, 12, 41, 43, 67, 76, 91, 149, 168, 189, 198, 204, 229, 266, 283, 284

T

Taiwan, 133
tanks, 41, 163
tannins, 139, 160, 161, 165, 166, 171, 179, 183
tar, 176
technological developments, xiii, 265
technology, ix, 9, 14, 29, 31, 35, 39, 59, 62, 65, 70, 72, 75, 125, 126, 138, 142, 176, 197, 203, 217, 260, 270, 272, 275, 278, 281, 282, 283, 288
temperature, ix, 29, 33, 34, 37, 39, 40, 41, 42, 43, 44, 45, 46, 47, 48, 49, 50, 51, 52, 53, 55, 56, 57, 58, 59, 60, 61, 62, 63, 65, 69, 70, 74, 111, 113, 115, 116, 117, 118, 119, 126, 141, 167, 168, 175, 177, 178, 206, 216, 226, 227, 237, 243, 244, 246, 249, 250, 252, 255, 279, 280, 284
tensile strength, 14
tension, 33
territory, 223
testosterone, 4
Texas, 22
textbooks, 158
thermal activation, xii, 219
thermal destruction, 55
thermal properties, 243
thermal resistance, 16
thermal treatment, 175, 178, 186
thermodynamic equilibrium, 249
thermodynamic parameters, 177
thermodynamics, 172
thermoplastics, 67
thinking, 130
thorium, 246
threats, 246
threonine, 139
threshold, 23, 224
thresholds, 22, 26
time, ix, xii, 6, 23, 24, 31, 36, 37, 45, 51, 52, 53, 54, 55, 61, 63, 82, 85, 87, 89, 112, 113, 116, 120, 121, 132, 140, 143, 166, 168, 175, 176, 191, 193, 198, 204, 205, 207, 209, 211, 220, 245, 246, 247, 248, 252, 253, 255, 256, 265, 273, 279, 280, 283
time frame, 23
tissue, 86, 130
tobacco, 273, 282, 289
toluene, 218
total energy, 270
total product, 274
toxicity, xi, 86, 87, 129, 132, 135, 136, 143, 146, 148, 191, 205, 275, 278, 282
toxicology, 149, 246
toxin, 135, 199
trace elements, ix, 26, 77, 86, 87
trachea, 163
trade, xi, 155, 174
traits, 49, 165
transesterification, 174, 271, 272, 273, 274, 282, 285, 287
transformation, xiii, 197, 214, 265
transition, 44, 49, 50
transition temperature, 44, 49, 50
transport, 17, 34, 207, 209, 211, 212, 223, 246, 247, 248, 249, 252, 253, 258, 262, 267
transport costs, 223
transportation, 156, 168, 170, 223, 229, 266, 268, 270, 271, 275, 276, 281, 282, 286
trees, 2
trial, 44, 53, 82, 164, 181, 203
triggers, 22
triglycerides, 174
trimmings, 3
tryptophan, 138, 139
tuberculosis, 8
turbulence, 37
Turkey, 62
turnover, 50, 57, 66

U

ultrasound, 282
UNESCO, 17
uniform, 168, 177, 247, 248
United Kingdom (UK), 8, 10, 40, 53, 55, 56, 60, 69, 152, 153, 211, 214, 215, 258, 259, 260, 261
United Nations (UN), x, 11, 147, 210, 259
United Nations Industrial Development Organization, x
United States, 2, 8, 15, 26, 27, 31, 62, 89, 90, 209, 272, 275, 277, 286, 287
uranium, 223, 246
urban areas, 220
urbanization, 130
urea, 38, 132, 162, 165, 182

uric acid, 137, 164
urine, 3, 4, 7, 164
USDA, 20
USSR, 278

V

vacuum, 111, 115, 173, 222, 225, 227, 239, 276, 280
values, xi, 7, 10, 12, 15, 132, 142, 155, 159, 162, 167, 168, 172, 175, 178, 203, 227, 228, 229, 231, 232, 233, 238, 239, 242, 243, 244, 246, 247, 248, 250, 257, 268, 280
vapor, 177, 275, 278, 282
variability, 6, 48, 269
variables, 34, 239, 252, 254
variation, 133, 220, 244, 246, 247, 268
vector, 15, 63, 136
vegetable oil, 58, 174, 207, 266, 270, 271, 282, 284, 285
vegetables, viii, 1, 29, 81
vegetation, 193
vehicles, 270, 272
velocity, 249, 250, 252, 254
versatility, 39, 43, 54, 56
vessels, 113
vibration, xii, 219, 238, 239, 244
Vietnam, 156
viruses, 55, 56, 75
viscera, 74
viscosity, 37, 41, 52, 251, 279, 280
vitamins, 47, 130, 132, 139, 142, 197
volatility, 271
volatilization, x, 109, 125
vomiting, 143

W

walking, 136
waste disposal, xiii, 26, 31, 64, 207, 265, 266
waste management, viii, 9, 29, 30, 31, 33, 56, 64, 214, 217
waste treatment, 30, 35, 37, 39, 42, 43, 49, 52, 53, 55, 56, 59, 62, 64, 65, 70, 75, 76
wastewater, 6, 31, 34, 35, 36, 46, 58, 59, 60, 61, 63, 67, 68, 71, 72, 75, 76, 125, 126, 141, 149, 172, 175, 181, 187, 192, 203, 215
water absorption, 227, 229, 231, 232, 233, 238, 239, 243, 257
water quality, 20, 85, 161
water resources, 20
waterways, 6, 12
weakness, 36
wealth, 193, 211
web, 194
weight changes, 118
weight gain, 162, 164, 165, 202
weight loss, 202
weight ratio, 177
welfare, 20, 21
wetting, 120
wheat, 38, 62, 72, 75, 79, 81, 83, 134, 152, 169, 171, 172, 182, 186, 187, 278, 279, 281, 288
wild type, 50, 139
wind, 249, 270
windows, 224
withdrawal, 180
witnesses, 25
wood, x, 9, 12, 14, 16, 109, 110, 111, 112, 116, 118, 119, 120, 125, 127, 134, 135, 143, 168, 177, 184, 206, 222, 267, 275, 280, 281, 287
wood species, 222
wood waste, x, 12, 109, 110, 111, 112, 116, 118, 120, 125, 143
wool, 60, 63
workers, 161, 162, 164, 166, 171, 176, 177, 178, 179
World Bank, x
World Wide Web, 193
worms, 195, 199, 201, 203, 204, 206, 207, 208, 209
writing, 285

X

X-ray diffraction, 125, 240, 241

Y

yeast, 130, 133, 137, 138, 139, 148, 150, 152, 170, 188, 213, 275
Yeasts, 133
yield, 9, 11, 14, 15, 35, 40, 45, 47, 48, 62, 67, 70, 79, 82, 83, 85, 115, 138, 164, 168, 176, 196, 209, 226, 258, 273, 274, 275, 277, 278, 279, 280, 286

Z

zinc, 137, 172, 177, 178, 202, 208